Letters of

Sigmund Freud

Letters
of
Sigmund
Freud

Selected and edited by ERNST L. FREUD

Translated by TANIA & JAMES STERN

Introduction by STEVEN MARCUS

BASIC BOOKS, Inc., *Publishers, New York*

Translators' Note

WE WISH TO EXPRESS OUR THANKS to Miss Anna Freud, Mr. James Strachey, Mr. W. D. Robson-Scott, and Mr. Louis Wilkinson for reading through this translation and making many helpful suggestions.

T. AND J. S.

Library of Congress Cataloging in Publication Data

Freud, Sigmund, 1856-1939.
  Letters of Sigmund Freud.

  Translation of Briefe, 1873-1939.
  Includes bibliographical references.
  1. Freud, Sigmund, 1856-1939. 2. Psychoanalysts—
Correspondence, reminiscences, etc. I. Freud, Ernst L.,
1892-1970, ed. II. Title.
BF173.F85A43 1975   150'.19'52 [B]   74-28069
ISBN 0-465-09704-9

Introduction

Steven Marcus

THE nineteenth century was the Homeric epoch of letter-writing, and among the epistolary titans of his time Freud easily held his own. Throughout his life Freud dealt with all his correspondence without any assistance. His practice was to answer every letter he received, "no matter from whom." He invariably wrote his letters in longhand, and he adhered to the rule that any letter he received was to be answered within twenty-four hours. It is impossible to determine how many thousands of letters he wrote in a long and arduous life. A large proportion of them have, in the usual way of things, been lost, probably for good. Another considerable segment of Freud's correspondence—his letters to his patients—are still guarded as private property, with justice of course.

Taking these and other omissions into account, we can still be impressed when we learn that his son, without any effort of comprehensiveness, examined some 4,000 letters in order to make this selection of little more than 300. Among other things, such a circumstance reminds us again that the nineteenth century had an altogether different conception of time, and of human possibilities within time, from what prevail today.

The Letters of Sigmund Freud is one of the major printed sources of Freud's correspondence. Among the others are Ernest Jones's three-volume biography, *The Life and Works of Sigmund Freud; The Origins of Psychoanalysis,* edited by Marie Bonaparte, Anna Freud, and Ernst Kris, which consists of selections from Freud's letters, drafts, and notes to his friend and colleague Wilhelm Fliess written during the years 1887–1902; and another biography, *Freud Living and Dying* by Max Schur. Separate collections of Freud's correspondence with other colleagues such as Carl Jung, Karl Abraham, and Oskar Pfister have also been published.

The present collection comprises 315 letters, the earliest of them dating from 1873, when Freud was seventeen years old, the last of them written on September 19, 1939, just four days before he died.

The editor was guided by a general principle of selection, and this volume, he tells us, is confined to "letters of a more personal kind in the hope that those who know Sigmund Freud only from his work shall receive here a portrait of the man, of the percipient, thinking, battling human being." As a result, this collection of letters is of necessity a kind of miscellany. It pretends neither to definitiveness nor even to scholarly thoroughness. And since the editor declares that he wanted his father's "every letter to stand by itself, unencumbered," so that "his voice, his spirit, alone, should . . . permeate this book," the collection contains almost no commentary and almost no notes. In the instance of Freud, such editorial reticence is not an unmixed blessing, and the present collection should be regarded as a preliminary installment on a much larger project—the collection and editing of all Freud's correspondence. Nevertheless, these letters are representative enough to allow the reader to gain from them an accurate sense of the remarkable man who wrote them.

Freud was not merely a tireless correspondent—there are still plenty of those. He was a very distinguished one, and as far as I know there is nothing like these letters anywhere else. The only relevant standard of comparison that can be applied to them is that which arises from the supreme correspondents, the great men of letters, the poets and novelists of the nineteenth century—for example, Keats, Byron, Flaubert, or Dickens—and in our own time, D. H. Lawrence. Even this comparison is slightly misleading.

What we respond to and admire in the correspondence of the great men of letters is not merely the experience they offer us of getting to know and understand them as private persons, or the opportunity they afford of connecting a writer's life with his work so that we can enlarge our understandings of the work. We respond to such letters as literature itself, and we are right to do so. As we read the collected letters of poets and novelists, we make out certain characteristic themes and patterns of recurrence; we trace out a whole superstructure very similar to the superstructure that we find in works of art or literature. But such is not quite the case with the letters of Freud. What is most striking and impressive about them is precisely the absence of superstructure, of embellishment, of most of those qualities which we think of as constitutive of literature (it would be too easy a temptation to say that these absences are exactly what make Freud's letters great literature; the matter is not so simple as to be turned by a neat paradox).

Freud's letters have a unique kind of nakedness which, even at this late date, remains moving and sometimes even frightening. His constant effort in life was to be himself, and for Freud this meant that his life was to be dedicated to as naked and unadorned a confrontation of reality as possible. His effort was to see through, expose, and banish illusion, and by the time he has reached his advanced old age he is as disillusioned a man—in the simple, radical sense of the term—as it is possible to imagine. It is no accident, one reflects while reading these letters, that one of his central notions was the "reality principle." That principle informs these epistolary utterances with such unwavering consistency that one would be inclined to suspect it, were it not so deeply native to Freud's temperament and so consistent with his achievement.

In one sense this attitude was made possible because from the very beginning Freud had the ability to regard himself in a peculiarly impersonal way. He was able to speak of himself, as it were, in the third person and could contemplate his destiny as if it were an object external to himself. In an early letter he typically writes, "The trouble is I have so little ambition. I know I am someone, without having to be told so." Or, recounting the elaborate preparations of dress and toilet he made before paying his first visit in 1886 to the home of Charcot, he concludes by saying that "altogether I spent fourteen francs on the evening," but then goes on to add, "As a result I looked very fine and made a favorable impression on myself."

This capacity to contemplate himself warmly but from a distance was the same capacity that Freud exercised in his transactions with the world about him. He himself connected this gift with his earliest experiences. "A man who has been the indisputable favorite of his mother," he wrote, "keeps for life the feeling of a conqueror, that confidence of success that often induces real success." The workings of that confidence, and the success which it induced, are one of the stories that these letters record.

INTRODUCTION

The career through which that success was achieved and expressed is exemplary in character, and the arc of development which Freud's life describes is altogether representative of the nineteenth-century's conception of the heroic proportions to which humanity can attain. More than a third of the letters in this volume were written by Freud to Martha Bernays, his future wife, during the four long years of their engagement. They were separated for most of this time and Freud was then an impoverished and struggling young doctor with a very uncertain future ahead of him. Freud pours himself out in these early letters, and from them we can see how very much a proper young Victorian he was.

He believed in the Victorian doctrines of work and self-discipline, regarding both as a kind of secularized personal salvation and as a means for winning through in "the struggle for existence." He was thoroughly committed to the Victorian ideals of marriage and the relations of the sexes; he believed in chastity and fidelity, in the family as the center of life; he wanted a large family to live and grow about him, and he wanted himself to be its *pater familias*.

And yet, as we read these letters, we see this Victorian physician steadily and inevitably advancing in another direction, and giving birth out of those very Victorian ideals which guided his career to so much that we cherish today as modern and liberated. Later on, in 1915, he has this to say: "Sexual morality as defined by society, in its most extreme form, that of America, strikes me as very contemptible. I stand for an infinitely freer sexual life, although I myself have made very little use of such freedom. Only so far as I considered myself entitled to." I can think of no more striking demonstration of the triumph of the nineteenth-century ideals at their best, or of the unbreakable continuity between that world and ours.

Freud's achievement was also Victorian in the sense that it was an achievement of character as much as it was of intellect. He never tired of insisting on this. "I am not even very gifted," he would characteristically write, "my whole capacity for work probably springs from my character and from the absence of outstanding intellectual weaknesses." By character Freud was referring to the courage, defiance, and tenacity on which he had to rely as, alone, he made his first explorations into the dark continents of the human soul. This collection of letters is an imperishable record of the character which made those explorations.

Preface

As a letter writer, my father was unusually prolific and conscientious. He dealt with his voluminous correspondence unassisted and in longhand. He answered every letter he received, no matter from whom, and as a rule this answer was in the post within twenty-four hours. His evenings he devoted to scientific writing, but every spare minute between analyses was dedicated to his correspondence. In the course of his long life strict observance of this routine resulted in the composition of many thousands of letters.

A considerable proportion of this great body of manuscript has been traced and inspected during the past few years. A far greater proportion is lost, some of it irretrievably. Letters to my father's many patients are jealously guarded by them—and rightly so—as their private property. Complete collections of letters, such as those to my father's half brother, Emanuel, and his nephew John, have obstinately refused to come to light; others, including those to my brother Oliver, have vanished in the upheavals of emigration. Through the kindness of friends and occasionally by some happy coincidence, approximately four thousand of the letters still extant have been made available to me for the purpose of this selection. Some long, others short, some sketching a mere outline

ix

of an idea, others full of detail, these letters are addressed to almost everyone who played a significant part in my father's life. They bear dates, moreover, from every phase of that life.

The material in this volume has not been selected from a biographical point of view, nor does it include any of those letters dealing in their entirety with the theory and practice of psychoanalysis. Ernest Jones, in his *Life and Work of Sigmund Freud*, has already written the definitive biography on the basis of the letters available to him. The correspondence with Wilhelm Fliess has enabled Ernst Kris to publish his *Origins of Psychoanalysis*. And finally, the complete collection of letters that passed between Freud and his early collaborators (Abraham, Eitingon, Ferenczi, Jones, Jung, Pfister, Rank, and Sachs) will no doubt be used at some later date as the raw material for a comprehensive work on the development of the psychoanalytic movement.

In this collection I have confined myself to letters of a more personal kind in the hope that those who know Sigmund Freud only from his work shall receive here a portrait of the man, of the percipient, thinking, battling human being. These letters are intended to show the breadth of his nature, the passion and single-mindedness not only of the young man in love but of the mature man in his lifelong search for scientific truth. They reveal the lightness of his touch, his humor and irony, the nature of his relationship to the men and women of his acquaintance, his friends, his family, as well as his reactions to emotional, artistic, and philosophical experiences.

The arrangement of the material is purely chronological; the gap between the dates of two consecutive letters is sometimes wide, sometimes narrow. I have refrained, nevertheless, from inserting a running commentary, for it has not been my intention to write another biography. Even the explanatory notes have been kept as brief as possible. I want my father's every letter to stand by itself, unencumbered. His voice, his spirit, alone, should be allowed to permeate this book.

ERNST L. FREUD

1873

1882

1 *To* EMIL FLUSS

Vienna, at night
June 16, 1873

Dear Friend

If I didn't shrink from making the silliest joke of our facetious century, I would just say: "The *Matura*[1] is dead, long live the *Matura*." But I like this joke so little that I wish the second *Matura* were over, too. Despite secret qualms of conscience and feelings of remorse I wasted the week following the written exam, and ever since yesterday I have been trying to compensate for the loss by filling up thousands of old gaps. You never would listen when I accused myself of laziness, but I know better and feel there is something in it.

Your curiosity over the *Matura* will have to be content with cold dishes as it comes too late and after the meal is finished, and unfortunately I can no longer furnish an impressive description of all the hopes and doubts, the perplexity and hilarity, the light that suddenly dawned upon one, and the inexplicable windfalls that were discussed "among colleagues"—to do this the written exam is already of far too little interest to me. I shall refrain from telling you the results. That I sometimes had good and sometimes bad luck goes without saying; on such important occasions kind Providence and wicked Fate are invariably involved. Events of this kind differ from the ordinary run of things. Briefly, since I don't wish to keep you on tenterhooks about something so unattractive: for the five papers I got *exc., good, good, good, fair.* Very annoying. In Latin we were given a passage from Virgil which I had read by chance on my own account some time ago; this induced me to do the

[1] The examination on leaving school; it consists of two parts, one written, the other oral.

3

paper in half the allotted time and thus to forfeit an *exc.* So some-one else got the *exc.,* I myself coming second with *good.* The Ger-man-Latin translation seemed very simple; in this simplicity lay the difficulty; we took only a third of the time, so it failed miserably. Result: *fair.* Two others managed to get *good.* The Greek paper, consisting of a thirty-three-verse passage from *Oedipus Rex,* came off better: [I was] the only *good.* This passage I had also read on my own account, and made no secret of it. The math paper, which we approached in fear and trembling, turned out to be a great suc-cess: I have written *good* because I am not yet positive what I was given. Finally, my German paper was stamped with an *exc.* It was a most ethical subject on "Considerations involved in the Choice of a Profession," and I repeated more or less what I wrote to you a couple of weeks ago, although you failed to confirm it with an *exc.* Incidentally, my professor told me—and he is the first person who has dared to tell me this—that I possess what Herder[2] so nicely calls an *idiotic* style—i.e., a style at once correct and charac-teristic. I was suitably impressed by this amazing fact and do not hesitate to disseminate the happy event, the first of its kind, as widely as possible—to you, for instance, who until now have prob-ably remained unaware that you have been exchanging letters with a German stylist. And now I advise you as a friend, not as an interested party, to preserve them—have them bound—take good care of them—one never knows.

This, dear friend, was my written *Matura.* Please wish me loftier aims and less adulterated success and more powerful rivals and greater zeal. Oh, to think of all the good wishes that could be heaped upon me without anything improving by one iota! Whether the *Matura* was easy or difficult I cannot judge; just assume that it was jolly.

I have been to the exhibition[3] twice already. Interesting, but it didn't bowl me over. Many things that seemed to please other people didn't appeal to me, because I am neither this nor that, not really anything completely. Actually, the only things that fasci-nated me were the works of art and general effects. A great compre-hensive panorama of human activity, as the newspapers profess

[2] Johann Gottfried von Herder (1744-1803), German poet and philosopher.
[3] International Exhibition in Vienna, 1873.

to see, I don't find in it, any more than I can find the features of a landscape in a herbarium. On the whole it is a display for the aesthetic, precious, and superficial world, which also for the most part visits it. When my "martyr" (this is what we call the *Matura* among ourselves) is over, I intend to go there every day. It is entertaining and distracting. One can also be gloriously alone there in all that crowd.

Needless to say, I am telling you all this with the purely malicious intention of reminding you how uncertain it is when you will be able to see all this splendor, and yet how painful the parting is bound to be when the moment comes. I can well understand your feelings. To leave one's beloved home, fond relatives—the most beautiful surroundings—ruins in the immediate neighborhood —I must stop or I shall become as sad as you—after all, you know best what you have got to part from! I bet you wouldn't mind if your future employer didn't tear you away from the joys of home for another month. Oh, Emil, why are you a prosaic Jew? Journeymen of Christian-Germanic disposition have composed the most beautiful poems under similar circumstances.

You take my "worries about the future" too lightly. People who fear nothing but mediocrity, you say, are safe. Safe from what? I ask. Surely not safe and secure from being mediocre? What does it matter whether we fear something or not? Isn't the main question whether what we fear exists? Admittedly more powerful intellects are also seized with doubts about themselves; does it therefore follow that anyone who doubts his virtues is of powerful intellect? In intellect he may be a weakling yet at the same time an honest man by education, habit, or even self-torment. I don't mean to suggest that if you find yourself in a doubtful situation, you should mercilessly dissect your feelings, but if you do you will see how little about yourself you are sure of. The magnificence of the world rests after all on this wealth of possibilities, except that it is unfortunately not a firm basis for self-knowledge.

In case you cannot understand me (for my thoughts are following a certain drowsy philosophy), just ignore what I have said. Unfortunately I wasn't able to write during the day, in twenty-three days that day will come, the longest of days on which, etc. Since in the brief interval I am supposed to quaff knowledge in great gulps, I am left with no opportunity of writing commonly intelligible letters. I take comfort in the thought that I am not writing to a common intelligence, and beg to remain with every possible expectation

Your
Sigmund Freud

For several days Herr Bretholz, his elder daughter and his nephew, a sage from Czernowitz,[4] have been our daily visitors. Do you know the last? He really is a sage; I have enjoyed him very much.

Sig. Freud

[4] Capital of the then Austrian province of Bucovina.

2 *To* WILHELM KNÖPFMACHER

Vienna
August 6, 1878

Dear Friend

It is unfair of you to jeer at my haste. The filthy lucre runs through my fingers at such frightening speed that, with the honest man's fear of failing in my duties, I was trying to place my riches in your safekeeping. By my doing this today the proverbial weight is taken off my mind.

I thank you for the friendly way in which you helped me out of my embarrassment. If there is a God he will take note of your deed and repay you a thousandfold, and if there is not, then there exists at least one man who will remember it and consider it one more reason for remaining fond of you. I am also sending you herewith my collected works,[1] not my complete ones as I have reason to suspect, for I am awaiting the correction of a third,[2] and a fourth and fifth keep appearing in my prescient mind, which is startled by them like Macbeth by the ghosts of the English kings: "What! Will the line stretch out to the crack of doom?"

[1] Freud's first published papers (1877): "Über den Ursprung der hinteren Nerven- wurzeln im Rückenmark von Amnocoetes (*Petromyzon Planeri*)" ("On the Origin of the Posterior Nerve Roots in the Spinal Cord of Amnocoetes") and "Beobachtungen über Gestaltung und feineren Bau der als Hoden beschriebenen Lappenorgane des Aals" ("Observations on the Formation and More Delicate Structure of Lobe-shaped Organs of the Eel Described as Testicles").

[2] "Über Spinalganglien und Rückenmark des Petromyzon" ("On the Spinal Ganglia and Spinal Cord of the Petromyzon"), 1878.

6

During these holidays I have moved into another laboratory,[3] where I am preparing myself for my real profession:[4] "flaying of animals or torturing of human beings,"[5] and I find myself more and more in favor of the former.

With cordial greetings
Your
Freud

[3] Laboratory of the Physiological Institute at the University of Vienna.

[4] Freud was still wavering between zoology and medicine.

[5] Quotation from the preface to *Max und Moritz*, by the German humorist Wilhelm Busch (1832-1908).

3 *To* MARTHA BERNAYS

Vienna
June 19, 1882

My precious, most beloved girl

I knew it was only after you had gone[1] that I would realize the full extent of my happiness and, alas! the degree of my loss as well. I still cannot grasp it, and if that elegant little box and that sweet picture were not lying in front of me, I would think it was all a beguiling dream and be afraid to wake up. Yet friends tell me it's true, and I myself can remember details more charming, more

[1] On June 17, 1882, Freud became engaged to Martha Bernays, who left Vienna the following day for a previously planned visit to relatives. At the time of their engagement Freud (born 1856) was twenty-six, Martha Bernays twenty-one.

The following timetable shows the periods of separation and their meetings during the years of engagement:

June 17, 1882: Engagement.

June 18, 1882: Martha leaves for Wandsbek.

July 16-27, 1882: Freud visits Martha at Wandsbek.

Sept. 2, 1882: Martha returns to Vienna.

June 14, 1883: Martha's mother moves with her daughters to Wandsbek.

Sept. 2-28, 1885: Freud visits Martha at Wandsbek on his way to Paris.

Dec. 22-30, 1885: Freud visits Martha at Wandsbek, interrupting his stay in Paris.

Feb. 28-Mar. 2, 1886: Freud visits Martha in Wandsbek on his way from Paris to Berlin.

Mar. 27-28, 1886: Freud visits Martha at Wandsbek from Berlin.

Letters written by Freud to Martha during the final four months of their engagement have unfortunately not been preserved.

Sept. 14, 1886: Due to a gift from Martha's aunts, marriage was made possible at this date.

7

mysteriously enchanting than any dream phantasy could create. It must be true. Martha is mine, the sweet girl of whom everyone speaks with admiration, who despite all my resistance captivated my heart at our first meeting, the girl I feared to court and who came toward me with high-minded confidence, who strengthened the faith in my own value and gave me new hope and energy to work when I needed it most.

When you return, darling girl, I shall have conquered the shyness and awkwardness which have hitherto inhibited me in your presence. We will sit alone in that nice little room again, my girl will settle down in the brown armchair (out of which we were so suddenly startled yesterday), I at her feet on the round stool, and we will talk of the time when there will be no difference between night and day, when neither intrusions from without nor farewells nor worries shall keep us apart.

Your lovely photograph. At first, when I had the original in front of me I did not think so much of it; but now, the more I stare at it the more it resembles the beloved object; I expect the pale cheeks to flush the color our roses were, the delicate arms to detach themselves from the surface and seize my hand; but the precious picture does not move, it just seems to say: "Patience! Patience! I am but a symbol, a shadow cast on paper; the real person is going to return, and then you may neglect me again."

I would so much like to give the picture a place among my household gods that hang above my desk, but while I can display the severe faces of the men I revere, the delicate face of the girl I have to hide and lock away. It lies in your little box and I hardly dare confess how often during the past twenty-four hours I have locked my door and taken it out to refresh my memory.

And all the while I kept thinking that somewhere I had read about a man who carried his sweetheart about with him in a little box, and having racked my brain for a long time I realized that it must be "The New Melusina," the fairy tale in Goethe's *Wilhelm Meister's Wanderings*, which I remembered only vaguely. For the first time in years I took down the book and found my suspicion confirmed. But I found more than I was looking for. The most tantalizing superficial allusions kept appearing here and there, behind the story's every feature lurked a reference to ourselves, and when I remembered what store my girl sets by my being taller than she is I had to throw the book away, half amused, half annoyed, and comfort myself with the thought that my Martha is not a mermaid but a lovely human being. As yet we don't see humor in the same things, which is why you may possibly be disappointed when you read this little story. And I would prefer not to tell you all the crazy and serious thoughts that crossed my mind while reading it.

These pages, darling Marty, have not been written at one sitting. Both yesterday and this evening Eli[2] and Schönberg[3] were here, yesterday in fact several girls as well, and to avoid arousing any suspicion I managed to be quite sociable, although I would far rather have been alone. Only the sight of Schönberg gives me comfort, for the sight of his honest, lively features evokes in me, with sound and color, a host of precious memories. What sorceresses you women are! I like him better by the hour. I got your last greetings from the station and today from Eli the longed-for news of your safe arrival. Your brother seems to like being with us; I have not got to know him much better, as I have not been alone with him since you left. Otherwise I drug myself with work and console myself with the certainty that Martha will remain mine as long as she remains Martha.

My beloved little bride. If at one time I hesitated to bind you to me for life, I would not let you go now even if the most ghastly misfortune befell me and I had to drag you along.

Do please try to steal from your fond relations all the photographs of you as a child; it occurs to me that I could have held onto that old photo owned by your mother, at least until your return.

Should you need something from here or want something done, please don't favor anyone but me with your commissions. That is how exclusive I am when I love. Let me know all about what you are doing at the moment; it will make it easier for me to put up with your absence. And make good use of your stay in Hamburg for your health. I would so much like to see you with those full round cheeks the childhood pictures show.

Now the day has come to an end, the page is covered, and I must check the desire to go on chatting with you.

[2] Eli Bernays (1860-1922), Martha's brother.
[3] Ignaz Schönberg, a Sanskrit scholar and friend of Freud's, was already engaged to Martha's younger sister Minna (1865-1941).

Farewell and don't forget the poor man you have made so bliss-
fully happy.

Your
Sigmund

Minna sent me greetings via Schönberg.

4 *To* MARTHA BERNAYS

Tuesday morning in laboratory
June 27, 1882

My sweet girl

I have torn a few pages out of my copy book to write to
you while my experiment is taking place. The pen has been stolen
from the professor's[1] desk, the people around me think I am com-
puting my analysis; just now someone came over and made me lose
ten minutes. Beside me a silly panel doctor is testing an even
sillier ointment to see if it contains something harmful; in front of
me in my apparatus sizzle the gas bubbles which I have to filter.
The whole thing once more spells resignation, waiting; two-thirds
of chemistry consists of waiting; it is probably the same with life
and the nicest thing about it is what one grants oneself in secret,
as I am doing now. Your sweet letter came quite unexpectedly and
was therefore doubly welcome, and I enjoyed the tall trees and
the lovely garden as well as the charming confusion in your dear
sentences. Look out, girl, the drawers[2] are being put back in order,
in a new order, I hope, but—I was about to say something else when
an utterly idiotic neighbor involved me in a conversation about
quicksilver salt. May God punish him for it.

Well, your letter makes up for today's bad weather; within me
the sun is shining from a blue sky, outside there is fog and drizzle.
Why do you think the address you used this time is conspicuous?
Here it is the most convenient, or do you mean it is conspicuous in
Wandsbek? Your letter (I am no longer going to say "sweet," I am
going to apply to the Berlin Academy for an increase of affectionate adjectives—I am so in need of them) bore the postmark Ham-

---

[1] Ernst Wilhelm von Brücke (1819-1892), Professor of Physiology at the University of Vienna and Director of the Physiological Institute.

[2] Refers to a dream of Martha's on which she had reported in her previous letter.

10

burg. Is Wandsbek that near? Have you seen the sea yet? Please give it my best regards—we will meet yet. May land and sea cooperate to keep my girl flourishing and make her stay in foreign parts a pleasant one. I am so conceited I no longer want to recognize them as her home. How bold one gets when one is sure of being loved!

Poor Minna had to invent a five-page letter on the spur of the moment. What are those dangerous things Marty wrote to her? Let me know what Eli writes about me. It must be rather funny.

You are now even making me lazy, Marty. I do work all day, but in the evenings I am quite incapable of looking at a book. Fiction I do not care for; I know a beautiful fairy tale which I have experienced myself, and as for lofty science I bow low and say: "Your Highness, I remain your humble, most devoted servant, but please don't hold it against me; you have never looked kindly upon me, never said a comforting word to me; you don't answer when I write to you, listen when I speak, but I know another lady to whom I mean more than I do to you, who repays my every service a hundredfold, and who moreover has but one servant and not, like you, thousands. You will understand if I now devote myself to the other so undemanding and gracious lady. Keep me in pleasant memory until I return. I have to write to Martha."

I expect this will change when I can see and speak to Marty every day. The two ladies will get along well together and the proud, unapproachable one will have to be willing to pay the taxes for the lovely, modest one.

Yesterday I went to see my friend, Ernst v. Fleischl,[3] whom hitherto, so long as I did not know Marty, I envied in every respect. But now I have an advantage over him. I believe he has been engaged for ten or twelve years to a girl of his own age, who was prepared to wait for him indefinitely, but with whom he has now fallen out, for reasons unknown to me. He is a thoroughly excellent person in whom nature and education have combined to do their best. Wealthy, skilled in all games and sports, with the stamp of genius in his manly features, good-looking, refined, endowed with many talents and capable of forming an original judgment about most things, he has always been my ideal, and I was not satisfied until we became friends and I could properly enjoy his value and

---

[3] Ernst von Fleischl-Marxow (1847-1891), Assistant at the Physiological Institute.

abilities. On this occasion I brought him a criticism of a pamphlet by him, he taught me the Japanese game "Go"[4] and astounded me with the news that he was learning Sanskrit. I had to promise him to keep this a secret, but I knew at once that for Martha this had as little validity as other and more important secrets. Then I looked around his room, fell to thinking about my superior friend and it occurred to me how much he could do for a girl like Martha, what a setting he could provide for this jewel, how Martha, who was enchanted even by our humble Kahlenberg,[5] would admire the Alps, the waterways of Venice, the splendor of St. Peter's in Rome; how she would enjoy sharing the importance and influence of this lover, how the nine years which this man has over me could mean as many unparalleled happy years of her life compared to the nine miserable years spent in hiding and near-helplessness that await her with me. I was compelled painfully to visualize how easy it could be for him—who spends two months of each year in Munich and frequents the most exclusive society—to meet Martha at her uncle's[6] house. And I began wondering what he would think of Martha. Then all of a sudden I broke off this daydream; it was perfectly clear to me that I could not relinquish the loved one, even if to be with me were not the right place for her. A part of the happiness Martha renounced in the hour of our engagement we will make up for later. My girl must promise to keep young and fresh as long as possible, and even after nine years to be so charmingly surprised by everything new and beautiful as she is now. Martha will not allow herself to be absorbed by household worries, Martha is not a Lisette.[7] Can't I too for once have something better than I deserve? Martha remains mine.

A fond greeting to the dear one
from
Sigmund

[4] Japanese game played on a board.

[5] Hill in the Wienerwald, close to Vienna.

[6] Michael Bernays (1834-1897), Professor of the History of Literature at the University of Munich.

[7] Character in a poem by the German author Christian Fürchtegott Gellert (1715-1769).

5 *To* MARTHA BERNAYS

Vienna, Friday
July 14, 1882

Fair mistress, sweet love,

I beg leave to inform you that your gracious letter wherein you allow me to take a pilgrimage to your fair eyes hath made me mighty happy and that I am packing my satchel in order to learn if it be merely a fond glance I can expect from you or a kiss from your lips as well. And in so far as a traveler and stranger enjoys all manner of privileges and rights, you must not take it amiss if I desire more than one. Remember the words of an Anglo-Saxon poet who invented many gay and sad plays and himself acted in them, one William Shakespeare:

> Journeys end in lovers meeting,
> Every wise man's son doth know

and how he goes on to ask:

> What is love? 'tis not hereafter;
> Present mirth hath present laughter;
> What's to come is still unsure:
> In delay there lies no plenty;
> Then come kiss me, sweet and twenty,
> Youth's a stuff will not endure.

But should you not understand these frolicsome lines, consult none other than A. W. Schlegel's[1] translation of *Twelfth Night; or, What You Will.*

So if it pleaseth you let us descend from the lofty art of poetry to common prose and allow your servant to tell when he hopes to be near you. Your brother Eli hath amicably stretched out a helping hand with a free ticket as far as the frontier of this empire. Thereafter beginneth the realm of poverty, as the man of your choice hath more claims to the kingdom of heaven than to the treasures of this earth. Cannot therefore carry on as I commenced

---

[1] August Wilhelm von Schlegel (1767-1845).

and if I turn my back on this town at the hour of 8 on Sunday morning, you must not expect me in your Hamburg before Tuesday at 5:46 in the afternoon. Can even be that I arrive later, for railway complexities are a hard little nut for my weak head to crack and none of our other allies knows how to find a way out of such an entanglement of trains. After I have refreshed and washed myself in the early morning so that you will not take me for a Moor, I shall hasten to Wandsbek where enemies are holding you in—I trust uncertain—safekeeping. Allow me to hope that you will still be in the grove, for I would so much like to greet you unwitnessed by other human eyes. You have omitted, alas, to inform me about the length of the road and means of conveyance as well as your presence in the grove, but perhaps you will do so in your letter which I expect tomorrow.

Once we have seen each other the future will take care of itself and I will write no more about it.

If your cousin Max[2] will prove himself a friend and conduct you into town, I will be eternally grateful to him, although by so doing he would only be fulfilling an ordinary obligation toward mankind. Hope, however, he will not consider that three is company; he certainly will not find any support for his love of company from your uncompanionable lover and will be asked in a friendly manner to leave us alone. Do not care to kiss you beneath the gaze of a stranger, and would not know what to say to you in his presence. He will not be able to deny that he owes it to humanity to leave us alone.

So that you may be warned about your lover, you must not expect to make a great parade of him. He wears an unsightly shapeless gray jacket, light-colored breeches, will today acquire a gray felt hat like that of your brother but of less value. Your brother's traveling bag holds as little linen as is consistent with a man's property; as for the greatcoat, you have frequently sanctified it with your touch. You also know the ungainly walking stick, the wallet with your picture, the finger with the ring on it, to all of which has been added a little heap of money to support us in your inhospitable native town. May suffice to introduce ourselves as an engaged couple to the sun which brings everything to light, and to produce a likeness for our younger brothers and sisters. A

---

[2] Max Mayer, former admirer of Martha.

gem lies in wait for your birthday, keeps catching my eye as I go by, but I do not dare to acquire it now and bring it along; it will have to wait here until August 4. Thus your knight errant will bring nothing but his loving heart, he will come without weapons, having left poison and dagger at home for a rival. He cannot wait to see you and tell you how devoted he is and that if need be he is ready to protect and defend you against friend and foe. You know already that he was pleased to come off well in a skirmish, and he hopes that his enemy in Hamburg will spare him any hostilities by an honest renunciation.

Oh, this wretched medieval style, today but never again! I feel so much like a knight errant on a pilgrimage to his beloved princess who is being kept prisoner by her wicked uncle.[3] You must have been rather bored by it, sweet Marty; be tolerant. If you only knew how mad things look within me at the moment. But I will arrive in a quite sensible condition. To my joy, darling, Schönberg has returned.

Once again a kiss on credit, my angel, once again; perhaps tomorrow I can write from Mödling,[4] then cash payment.

To our happy reunion.

Your

Sigmund

[3] Elias Philipp (1824-1894), brother of Martha's mother.
[4] Small town, one hour by train from Vienna.

6 *To* MARTHA BERNAYS

Tetschen,[1] Sunday, 8 o'clock
July 16, 1882

My sweet little bride

If you only knew how lovely it is here and how incomparably more lovely it would be with you! The Elbe flows by, still a modest little river, showing me the way to you, to you. High mountains, some overgrown, some bare in strange formation, nice little houses which do not look as though they were meant to be lived in but to be put up and knocked down as in a game of bricks,

[1] Tetschen-Bodenbach, a town on the Austrian-German frontier.

all in a row along the river, a few proud buildings gazing down from the mountain slopes as though they had nothing to do with the rest. One of them all by itself on a hill, a castle or a monastery or something like that—it's really all the same to me. On the left lies Bodenbach, on the right Tetschen and between them two bridges, one for the railway and the other for "vagrant scholars" on their way to their sweethearts. On the second bridge I had to pay a toll of two kreutzer, but this I did willingly; I was glad I had not broken a leg.[2] I have been telling such an awful lot of lies lately. I crossed the bridge to Tetschen because in Bodenbach there was no coffeehouse from which I could write to you. It turns out that I have to stay here till 2 A.M. and won't reach Hamburg until 2:30 P.M. on Tuesday and don't know if I can see you even on Tuesday, and I am completely roasted or grilled—no, not completely done, only half, like an English roast beef. But to return to Bodenbach. There is a kind of holy Sunday stillness over everything and the bells are ringing, I don't quite know why, and the streets are so clean, the people so polite, the old ones looking as I have always expected Christian Fürchtegott[*] Gellert to look and the young ones so modest, as if nowadays they themselves lived in fear of God. In the middle of the market place there is a square stone, could be the tomb of an old Saxon king, but it probably isn't and I really don't care what it is. It is enough that I can walk around here without being asked, "Who gave you that ring you're wearing on your finger?" No, this I am not going to take off till I am once more under restraint in Vienna. I was about to tell you that I was bent on finding a coffeehouse. And then I saw in the street a round, rosy-cheeked girl and I asked: "Fair lady"—not: "Let it not offend you,"[3] etc., but: "Could you tell me where I can find a coffeehouse?" And imagine, I was standing right in front of it, and the girl was the waitress or the proprietor's daughter. And here I am, the only guest, in a little room with several chairs and tables; it takes a quarter of a century to get a coffee and there is very little sugar with it; my Marty will have to give me more sugar in my coffee. But the cake is good—I am eating two slices, spendthrift that I am—one for Marty, and now I must hurry up and

---

[2] Allusion to the poem "The Farmer and his Son," by Gellert.

[*] Literal translation: "fear-God."—*Tr.*

[3] Quotation from Goethe's *Faust*, I.

stop or I will have to leave all my money behind in the coffeehouse: for light and ink and use of the furniture, and all the beautiful things I still have to say will have to be left unsaid. But we are going to compete as to who will set eyes on Martha first: I or this scribble. We will be traveling by the same train, and then the happy time begins, the great unique happy time of being with the beloved, which is so near at hand, and I am already getting used to the idea of having experienced it, because I just couldn't believe it and all along have felt the fear of which the poet sings: "Earth, sink not down,"[4] etc.

But farewell now, sweet Marty.

Goodbye
Your blissful lover
Sigmund

[4] Quotation from the poem "Homecoming," by the German poet Ludwig Uhland (1787-1862).

7 *To* MARTHA BERNAYS

[Hamburg[1]] Sunday
July 23, 1882

"The Jew is called Nathan. (A strange Jew—h'm!) Continue, worthy Nathan. . . ."[2] (Or something like that; I can't go to the public library just now to verify the quotation. The man in the *Gänsemarkt*[3] will forgive me.) This was the beginning of our acquaintance. I had suddenly grown very fond of a little girl and suddenly found myself in Hamburg. She had sent me a ring which her mother had once received from her father; I had a smaller copy of this ring made to fit her tiny finger, but it appeared that the true ring[4] had stayed with her after all, for everyone who saw and spoke to her loved her, and this is the sign of the true ring. I didn't like this much and I wondered for a long time how I could make

[1] This letter, without address or signature, was written in Hamburg and probably delivered personally.

[2] Quotation from the drama *Nathan the Wise*, by the German playwright and critic Gotthold Ephraim Lessing (1729-1781).

[3] The monument to Lessing stands in the *Gänsemarkt*.

[4] Allusion to the fable of the rings in *Nathan the Wise*.

her less attractive, so that no one would fall in love with her any more, until one day it occurred to me that what mattered was whether she loved several people, not whether several or everyone loved her. Once this idea had occurred to me I was very happy in Hamburg. The mornings were always warm and beautiful, the evenings seemed close to the mornings, and I was grateful to the day for filling the gap between the beautiful morning and beautiful evening. True, the tyrannical temperament that makes little girls afraid of me could not be subdued. I wanted exclusiveness, and since I had attained it in great and important matters, I strove to achieve it in small and symbolic ones. My girl came from a family of scholars, and wrote—for the time being only letters—with untiring hand, thus spending the little money she had on notepaper. So I decided to acquire some notepaper for the dear industrious child and chose some on which she could write to me only. An M and an S intimately entwined, as the generosity of engravers grants us, renders every page useless for intercourse save between Marty and me. The man from whom I ordered this despotic paper on Friday could supply it only on Sunday; "for on Saturday," said he, "we are not here. It is one of our ancient customs." (Oh, I know that ancient custom!) He was a jovial old gentleman whom I took to be fifty-four; with this error I won his heart, as not long before I had won another heart with another error. He was seventy-four years old and boasted of his capacity for enjoyment and work and declared that he had no intention of departing from this life so soon. I liked the man. I happened to be in a similar mood. On Sunday I saw him again. He was very proud of the elegance of the monogram, but he did not wish to treat me as a mere customer. He showed me the building of the Deutsche Bank opposite his shop. "That's where the merchants of Hamburg keep their money which they don't want to leave at home; these cellars are full of gold—and silver." I expressed the hope that one vein of this rich metal deposit reached as far as his shop. But merchants invariably dissimulate. He then explained why so many people swarm into this building. If you were to owe me some money, he said, instead of paying me in cash, it would be transferred in the bank from your account to mine. I felt uneasy; apart from being in debt, I know absolutely nothing about banking. But he would not let me go, I had to take a chair beside him while he questioned me as to where I had already been, and recommended to me this

and that excursion: "I'd like to come along with you myself, but I
am an old Jew, and just look at me." I looked. His beard was
shaggy. Yesterday they were not allowed to be shaved. "You know,
of course, which Fast Day is upon us?" I knew all right. Just because years ago at this season (owing to a miscalculation) Jerusalem had been destroyed I was to be prevented from speaking to my
girl on the last day of my stay. "But what's Hecuba to me?"[5]
Jerusalem is destroyed and Marty and I are alive and happy. And
the historians say that if Jerusalem had not been destroyed, we
Jews would have perished like so many races before and after us.
According to them, the invisible edifice of Judaism became possible
only after the collapse of the visible Temple. So, said my old Jew,
nine days before Tisha B'av[6] we deny ourselves every pleasure.
We here are a number of men of the old school all of whom adhere
to our religion without cutting ourselves off from life. We owe our
education to one single man. Years ago Hamburg and Altona
formed one Jewish community, later they separated; until the Reform movement came to Germany, instruction was carried out by
inferior teachers. Then it was realized that something had to be
done, and a certain Bernays was called and chosen to be *Chacham*.[7]
This man has educated us all.—The old Jew was about to embark on
his achievements, but I was more interested in Bernays the man.
Was he from Hamburg? No, he came from Würzburg, where he
had studied at Napoleon's expense. (Oh, the myth-forming power
of mankind!) He came here as a very young man, thirty years ago
he was still living here. Did you know his family? "Me? I grew up
with the sons."

I now remembered two names, Michael Bernays in Munich,
Jacob Bernays[8] in Bonn.

That's they, he confirmed, and there was also a third son,[9] who
lived in Vienna, and died there.

I also knew something about this third brother, whose name remained so much in the background.

---

[5] Quotation from *Hamlet*.

[6] Jewish Fast Day in memory of the destruction of the Temple in Jerusalem
(586 B.C.).

[7] Title of the head of the Sephardic community, Martha's grandfather Isaac
Bernays (1792-1849).

[8] Jacob Bernays (1824-1881), Martha's uncle, Professor of Classical Philology in
Breslau, later in Bonn.

[9] Berman Bernays (1826-1879), Martha's father.

The father's rich talents were divided among the sons, the old man continued. The father had been a linguist, an interpreter of the Scriptures, and had left behind him some distinguished children. Thus one son chose languages, the material of which became the scientific work of his life, the second one is still teaching the appreciation of the subtleties and the wisdom which our great poets and teachers have put into their writings. The third son, a serious, reserved man, dealt with life on a level even more profound than is possible for science and art: he was above all a human being and created new treasures instead of interpreting old ones. Glory to the memory of him who presented me with my Marty!

Imagine if my old Jew, who was now talking with such enthusiasm about the teachings of his master, could have guessed that his customer, allegedly a Dr. Wahle[10] from Prague, had this very morning kissed the granddaughter of his idol! He went on to recall the memories of his own youth, and traits of Nathan the Wise now began to appear in what he said. Bernays had been a quite extraordinary person and had taught religion with great imagination and humaneness. If someone just refused to believe anything—well, then there was nothing to be done about him; but if someone demanded a reason for this or that which was looked upon as absurd, then he would step outside of the law and justify it for the unbeliever from there. Take the law about food, for example: what could be less important than what one eats? To which he would say: Let us go back to the story of Creation. A fable it may be, but what the whole of mankind has believed for centuries surely cannot be nonsense; it must have some meaning. When God had created the first human beings and put them in the Garden of Eden, was not the first commandment he gave them a commandment about eating? From this tree mayest thou eat, but not from that one. Why was it not a command about morals? And if the first commandment God made was one about food, can it be of no importance what one eats?

My old Jew provided several more ingenious attempts of this kind to explain and support the Scriptures. I knew the method: the Holy Scriptures' claim to truth and obedience could not be supported in this way, there was no place for reform, only for revolu-

10 Former admirer of Martha and friend of Freud.

tion; but in this method of teaching lay enormous progress, a kind
of education of mankind in Lessing's sense. Religion was no longer
treated as a rigid dogma, it became an object of reflection for the
satisfaction of cultivated artistic taste and of intensified logical
efforts, and the teacher of Hamburg recommended it finally not because it happened to exist and had been declared holy, but because
he was pleased by the deeper meaning which he found in it or
which he projected into it. It was criticism, even though willfully
manipulated and directed toward definite aims, but well suited to
give his disciples the decisive direction which my old Jew was
still following while I was fetching our monogram for the granddaughter of his master.

His teacher, he continued, had been no ascetic. The Jew, he said,
is the finest flower of mankind, and is made for enjoyment. Jews
despise anyone who lacks the ability to enjoy. (I couldn't help
thinking of what Eli, to his credit, once disclosed about his philosophy when in his cups: *Homo sum.*) The law commands the Jew
to appreciate every pleasure, however small, to say grace over every
fruit which makes him aware of the beautiful world in which it is
grown. The Jew is made for joy and joy for the Jew. The teacher
illustrated this with the gradual importance of joy in the Holy
Days.*

At New Year the Christian says: Let us hope we have a better
time in the new year than in the old. For the Jews first comes Rosh
Hashana,[11] when the lots are drawn for the whole year. It is then
that we have reason to fear the divine decision: this is the feast of
the fear of God. At Yom Kippur[12] we fast all day to show our love
of God; only love can bring such a sacrifice. This is the day dedicated to the love of God. But then comes Succoth,[13] of which it is
written: "The Jew must be only joyful during these days, and one
day is called the Joy of the Law. This is the Day dedicated to the
Joy in God."

A customer arrived and Nathan became a merchant again.
When I took my leave I was more deeply moved than the Old
Jew could possibly guess. If he ever came to Prague, he said, he
would give himself the pleasure of looking me up. He won't find

---

* *Steigerung der Feste.*
[11] Jewish New Year.
[12] Day of Atonement.
[13] Feast of the Tabernacles.

me in Prague, but as a substitute I will offer him another pleasure. If my Marty wishes to take with her to Vienna some gifts in the form of notepaper, she must go to the Adolphsplatz, to our old Jew, disciple of her grandfather, and mention her name. Let him see that the stock of his master has not deteriorated since he sat at his feet. And as for us, this is what I believe: even if the form wherein the old Jews were happy no longer offers us any shelter, something of the core, of the essence of this meaningful and life-affirming Judaism will not be absent from our home.

8   *To* MARTHA BERNAYS

Vienna, Monday
August 14, 1882

My sweet Marty

I found no leisure to write to you all day today, so this has to be once again a nocturne, anyhow it is quite a while since there has been one. As you know, the poor human being is always more affectionate in the evening than in the morning, because, because— well, there are so many reasons that I don't need to mention even one.

My precious darling, for the first time in ages we have been to the Prater[1] again—we, the family, not the *Bund*.[2] We were invited by our old man[3] in order to make up for some less good days. When he isn't exactly grouchy, which alas is very often the case, he is the greatest optimist of all us young people. The day aroused memories pleasant in themselves, but melancholy in their recurrence. Here it was that we had been close to each other every day and every day had grown fonder of each other; here we had all eaten and drunk beer together and finally we had even pressed each other's hands and I had not been able to wait for the moment to get up, which would give me my girl again, and on the whole I had been so shy and had kissed my Marty so rarely because as

[1] Park on the outskirts of Vienna.

[2] Circle consisting of Freud's friends Schönberg, Eli Bernays, the Brothers Fluss, and the brothers Wahle.

[3] Freud's father, Jacob Freud (1815-1896).

yet I could not quite grasp what has now become the first and most natural condition of my life: that I had suddenly won for myself a unique, incomparable girl. Oh, the Prater is a paradise indeed; only the Wandsbek grove is more beautiful because there we were alone like Adam and Eve, except for a number of animals (which were harmless enough): gentle, venerable clergymen, inquisitive yet discreet old women and some useful animals, too— cows which gave milk, and waitresses who produced butter and cakes, etc. Eve wore a brown dress as befitted the changed conditions and a great big hat that never wanted to stay on and the Almighty had placed seats under the lovely tall trees, all of which were ours, and nowhere to be seen an angel with a flaming sword, only one little delicate angel with emerald eyes and two sweet lips which, refusing to remain closed, had to be closed with kisses and yet were kissed so rarely just because it was morning—and yet all in all so perfectly beautiful, but even more beautiful things are bound to come. Are you already thinking of the day you are to leave, it is no more than a fortnight now, must not be more or else, yes, or else my egoism will rise up against Mama and Eli-Fritz and I will make such a din that everyone will hear and understand. And when you do return you are coming back to me, you understand, no matter how your filial feelings may rebel against it. From now on you are but a guest in your family, like a jewel that I have pawned and that I am going to redeem as soon as I am rich. For has it not been laid down since time immemorial that the woman shall leave father and mother and follow the man she has chosen? You must not take it too hard, Marty, you cannot fight against it; no matter how much they love you, I will not leave you to anyone, and no one deserves you; no one else's love compares with mine.

What is it like in Wandsbek? Does anyone still remember your "admirer"? Are people turning up claiming to have seen you with him? You were so delightfully daring, my adorable girl. Will you be willing to take some risks here, too? You simply *mustn't* be as rash here as you were there, for that I could never ask, but now and again, when we haven't seen one another for a long time, you will be able to think something up, won't you? "Oh, don't let's talk now about what it's going to be like in Vienna!" But, intolerant you, what if I am compelled to think about it all the time? Are you enjoying the choral contest[4] and are you turning the ring around a little less often? Well, there has to be ebb and flow in Hamburg. Here no longer.

Today I gave myself a clean bill of health and tomorrow am going to start work. Short steps and a long way, but we will get there all right, and then we will be able to wander on, arm in arm. How lovely that will be!

If only I knew what you are doing now. Standing in the garden and gazing out into the deserted street? Ah, I am no longer passing by to press your hand, the magic carpet that carried me to you is torn, the winged horses which gracious fairies used to send, even the fairies themselves, no longer arrive, magic hoods are no longer obtainable, the whole world is so prosaic, all it asks is: "What is it you want, my child? You shall have it in time." "Patience" is its only magic word. And in saying so forgets how things get lost when we cannot have them then and there, when we have to pay for them with our own youth.

> Goodnight, my beloved Marty.
> Forever your
> Sigmund

[4] German Choral Festival in Hamburg, 1882.

9 *To* MARTHA BERNAYS

Vienna, Thursday
August 17, 1882

My beloved girl

A month ago today my delighted eyes spied you sitting on the veranda of the Philipp's house and you didn't recognize me, and two months ago you had just become my fiancée. Since then little, very little has happened to make the union for which we are striving a reality. And yet we have made some use of this time. We were strangers to one another, had to get to know each other, experience things together—this we have achieved; and if we both keep healthy and some demon does not disrupt our feelings for one another, the ensuing monthly memorials should find us well on the

way toward our longed-for goal. For you, poor darling, hope for
the future must compensate you for the many sacrifices you are
imposing on yourself at the moment; for me, the courage to court
you has already found reward in the awareness of my sweetest good
fortune. If I may repeat a request today, please don't be taciturn
or reticent with me, rather share with me any minor or even major
discontent which we can straighten out and bear together as honest
friends and good pals. I have always acted like this, sometimes at
the expense of your delicate nature, and you have told me that you
agree. If in doing so I must have often hurt your feelings, I know
you have not misunderstood my efforts to make you my own as
intimately as possible, and if this be egotistical, love after all cannot be anything but egotistical.

It is only the reflex of my usual wretched mood that makes me
talk of such things, for just now there aren't any disagreements between us and I am not afraid of any, nor for that matter of any
event that could separate us. It only pains me to think I should
be so powerless to prove my love for you; but so long as you believe in me and love me—and both I know to be true—there is no
doubt that we will both remain fit and capable of enjoying better
times. Don't scold me for being serious, Marty, you know I can be
gay when you are with me.

With fond greetings and in restless anticipation of this beastly
month soon vanishing into the past.

Your
Sigmund

10 *To* MARTHA BERNAYS

Vienna, at night
August 18, 1882

Why do I o'er my paper once more bend?
Ask not too closely, dearest one, I pray:
For, to speak truth, I've nothing now to say;
Yet to thy hands at length 'twill come, dear friend.

Since I can come not with it, what I send
My undivided heart shall now convey,
With all its joys, hopes, pleasures, pains, today:
All this hath no beginning, hath no end.[1]

My beloved girl

A friend, normally a hardened sinner with whom I am fond of commiserating on the absurdity of this world, suddenly turned soft today; and, striding into the next room, took from the bookcase Master Goethe's incomparable poems and read out to me some lines of such ardent emotion (which had more meaning for me than for him) that I had to run away in order not to betray myself and to be alone with my thoughts. This afternoon I could not return to work and soon ran into another friend whom I used to know at the University and who since then has been diverted by a sad misfortune from his original aims. Contact with friends holds for me nowadays a special charm—the seriousness of life seems to have disclosed itself to us almost simultaneously; what in the beginning seemed to us dear and desirable, but easily accessible, has now withdrawn into the far distance although still remaining dear, and perhaps some of them carry, as I do, a new cherished aim locked up in their hearts. Dejected as I am, weary and looking as I do with so little hope into the future, I nevertheless cannot think of a soul with whom I would like to change places; I still haven't lost faith in myself, and as for Marty, my Marty, what could anyone have to offer in her stead?

We are all poor and promise to help one another whenever we can. They are all decent fellows, or I would not have them as friends; we can do so little for one another, and yet I rarely leave any one of them without feeling that he has helped me, that the interest he takes in me, the hope he places in me, have lifted me out of my despondency, have offset some part of the injustice done to me, and that perhaps I have been able to do the same for him. Although not so blissful as the knowledge that one is loved by an exceptional girl, I would not renounce the feeling that so many men stand quietly by me and help me to live! It also helps me to accept the fact that we are so poor. Just suppose, darling girl, that success corresponded precisely to the merit of the individual, would not love tend to lose some of its purity? I would not be sure

---

[1] Quotation from Goethe's *Sonnets,* IX, "The Loving One Once More."

whether it were me you loved or the recognition given to me, and
if misfortune befell me the girl could say: I do not love you any
more, your unworthiness is proven. It would be as horrible as in
the world of uniforms where everyone wears his merit written on
his collar and chest. But since fortune's manner of rewarding or
ignoring merit is so capricious—and unjust—love may remain
faithful to the poor man without becoming false, and if I seem
insignificant and unimportant to other people, with you I am allowed to feel rich and to enjoy unlimited praise and recognition.

Oh, my darling Marty, how poor we are! Suppose we were to tell
the world we are planning to share life and they were to ask us:
What is your dowry?—Nothing but our love for each other.—Nothing else?—Now it occurs to me that we would need two or three
little rooms to live and eat in and to receive a guest, and a stove in
which the fire for our meals never goes out. And just think of all the
things that have got to go into the rooms! Tables and chairs, beds,
mirrors, a clock to remind the happy couple of the passage of time,
an armchair for an hour's pleasant daydreaming, carpets to help
the housewife keep the floors clean, linen tied with pretty ribbons in
the cupboard and dresses of the latest fashion and hats with artificial flowers, pictures on the wall, glasses for everyday and others
for wine and festive occasions, plates and dishes, a small larder
in case we are suddenly attacked by hunger or a guest, and an
enormous bunch of keys—which must make a rattling noise. And
there will be so much to enjoy, the books and the sewing table
and the cosy lamp, and everything must be kept in good order or
else the housewife, who has divided her heart into little bits, one
for each piece of furniture, will begin to fret. And this object must
bear witness to the serious work that holds the household together,
that object to a feeling for beauty, to dear friends one likes to
remember, to cities one has visited, to hours one wants to recall.
And all this, a small world of happiness, of silent friends and proofs
of lofty human values, is as yet only in the future; not even the
foundation of the house has been laid, there is nothing but two poor
human creatures who love one another to distraction.

Are we to hang our hearts on such little things? Yes, and without
hesitation, so long as some event beyond our control does not knock
on the silent door. And of course we will have to go on telling each
other every day that we still love each other. There is something
terrible about two human beings who love each other and can find

neither the means nor the time to let the other know, who wait until some misfortune or disagreement extorts an affirmation of affection. One must not be mean with affection; what is spent of the funds is renewed by the spending itself. If they are left untouched for too long, they diminish imperceptibly or the lock gets rusty; they are there all right but one cannot make use of them. Oh, at the moment there are not even two poor human beings who love each other. Only one is here, the other is far away, continually curbing her feelings in the goodness of her heart. The poor sweet child, who has already suffered so many sad things which she doesn't mention, and hardly was she able to breathe again when she gave herself to the poor luckless man, renouncing so willingly her own little share of life's pleasures. But you have got to bring me luck, for me you are luck itself, without you I would let my arms droop for sheer lack of desire to live; with you, for you, I will make use of them to gain our share in this world so as to enjoy it with you.

You have my most affectionate greetings; perhaps at this moment you are thinking of me; it is the hour when you used to wait for me in the garden.

Your
Sigmund

11 *To* MARTHA BERNAYS

Vienna
September 25, 1882

For my beloved Marty

I am beginning these notes without waiting for your answer, my girl, in order to tell you more about myself and my activities than our personal contact would allow. I am going to be very frank and confidential with you, as is right for two people who have joined hands for life in love and friendship. But as I don't want to keep on writing without receiving an answer I will stop as soon as you fail to respond. Continuous inner monologues about a beloved person that are not corrected or refreshed by that person lead to false opinions about the mutual relationship, and even to estrangement when one meets again and finds things to be different

from what one had thought. Nor shall I always be very affectionate, sometimes I will be serious and outspoken, as is only right
between friends and as friendship demands. But in so doing I hope
you will not feel deprived of anything and will find it easy to choose
between the one who values you according to your worth and
merit and the many who try to spoil you by treating you as a
charming toy.

Please don't think, my sweet darling, that I have any wish to
find fault with you. On the contrary, all I want is that there should
be no touchiness and no secrets between us. You know that from
the moment we entered into this alliance we both had to change
to some extent in order to become for each other what we wish to
be; and perhaps I may be allowed to say and to explain when the
old Marty does not seem to have given way completely to my beloved girl.

"So he is not satisfied with me" you will think and—shed a tear?
No, you won't. Because we must face all this as equals. Would I
shed a tear if you remonstrated with me? We have taken upon
ourselves a difficult task and in carrying it out we must support
and correct one another. Words of love alone cannot do this; living together does not mean hiding unpleasant things from one
another or glossing over them; helping one another means sharing
everything that comes along. It seems to me that up to now all of
you have demanded and expected only pleasant things from friendship. You have all been quite content if you could end up by
saying: He or she was so nice and pleasant today. In August when
I wasn't well and Eli came to see me, he asked me reproachfully
why I—being so seriously ill!—didn't go to the hospital instead of
being a burden to my family. This won't be our way of thinking,
my beloved. I don't want to spend only pleasant hours with you,
all I want is to keep feeling and making you feel that we love each
other and that we are trying to adjust to each other as far as is
possible for two human beings.

I hope for my part I shall succeed. There was one instance in
which you were not quite fair to me and offended me deeply;
it was when you refused to drop your "friendship" with Fritz
Wahle for my sake. I was patient and finally you did do me justice.
And even at the time I was fully conscious that you nobly kept
your independence and truthfully reported everything to me. You
will eventually agree with me more fully on this point. You were

not quite sure enough in your judgment; let us both hope that such things will never happen between us again. And you will understand me when I say that even for a beloved girl there is still one further step up: to that of friend, and that it would be a ghastly loss for us both if I were compelled to decide to love you as a dear girl, yet not as an equal, someone from whom I would have to hide my thoughts and opinions—in short, the truth. Please accept the hand which I hold out to you in fondest affection and confidence and do with me as I am doing with you.[1]

[1] No signature—probably delivered personally.

12 *To* MARTHA BERNAYS

Vienna, Thursday
October 5, 1882

To whom else but to my deeply beloved, most ardently worshiped Martha should I report on the outcome of my visit to Prof. Nothnagel?[1] Don't be cross, my lovely girl (whose charm at noon today is still confusing me), if I initiate you into the intricate byways and conditions to which my struggle for existence has brought me. It is after all not just my battle and interest, we are so intimately connected, I am so unspeakably happy that you are mine, so certain of your interest, that everything becomes important to me only when you share it. Even if the outcome was not exactly what I had desired, it was nevertheless quite honorable and I see no reason to abandon hope for a better future so long as you, my angelic girl, can put up with me.

Well, I went to see N. with my collected works and a recommendation from Meynert.[2] The house he lives in is new, hardly finished, the flat reeks of varnish, the waiting room simply magnificent. On the wall hangs a picture showing four children, a beautiful boy who in twenty years will be snatching the best jobs from the medical students, a little girl with hints of potential beauty for whom within ten years the young men will be fighting at students' balls: both

[1] *Hofrat* Professor Dr. Hermann Nothnagel (1841-1905), Director of the Second Medical Clinic, Vienna.
[2] Professor Dr. Theodor H. Meynert (1833-1892), brain anatomist, Director of the Psychiatric Clinic where Freud worked.

with brown hair from which I concluded, rightly as it turned out, that their mother is dark; then the elder girl, an unattractive blonde with her father's features, holding in her arms a baby of indeterminate sex. Soon I also found on the walls books written by the father of this promising brood, a large portrait of a serious, dark-haired woman on an easel-like contraption, and standing beside her the man who holds our fate in his hands. It gives one quite a turn to be in the presence of a man who has so much power over us, and over whom we have none. No, he is not one of our race. A Germanic cave man. Completely blonde hair, head, cheeks, neck, eyebrows, all covered with hair and hardly any difference in color between skin and hair. Two enormous warts, one on the cheek and one on the bridge of the nose; no great beauty, but certainly unusual. Outside, I had felt a bit shaky, but once inside, as usual in "battle," I felt calm.

"I have been asked to bring you greetings from Prof. Meynert, and to express his regrets at having missed you the other day. And on my own behalf I am taking the liberty of handing you this card."

While he was reading the card, I sat down. I knew what was on it: "Dear Professor, I am herewith warmly recommending to you Dr. Sigmund Freud on account of his valuable histological work and would be grateful if you would give him a hearing. In the hope of seeing you soon. Yours—Theodor Meynert."

"I set great store by a recommendation from my colleague, Meynert. What can I do for you, Herr Doktor?"

When speaking, he made a very pleasant impression; he talked like a man who means what he says and who weighs his words, reserved but trustworthy.

"You've probably guessed already," I said. "It is known that you're about to engage an assistant, and it is also said that before long you will have a new job to offer. I also understand that you set great store by scientific research. I have done a certain amount of scientific research, but at the moment I have no opportunity to continue, so I thought it advisable to present myself as an applicant."

"Have you some offprints of your papers with you, Herr Doktor?"

"Yes," said I, putting my hand in my pocket.

While he was glancing through the papers, I explained my position. "At first I studied zoology, then I changed to physiology, and have done some research in histology. When Prof. Brücke told me he couldn't give his assistant notice and advised me, a poor man, not to stay with him, I left."

Now N. began. "I won't conceal from you that several people have applied for this job, and as a result I can't raise any hopes. It wouldn't be fair. I will mention you as a candidate, however, and put your name down in case another job turns up. As I've said, I won't make any promises, but this you will hardly have expected. *Qui vivra verra.* I'll hold onto your papers, if I may."

All this was said in a friendlier manner than I can reproduce here; he was not so gruff, if anything rather reserved in a friendly way. One thing emerged clearly: the first job, to be occupied immediately, has been taken (by a son of a Prague professor, so rumor has it); as for the second, not yet vacant, he does not want to commit himself, but he did take me seriously.

"One more thing," I said. "At the moment I'm serving as an *Aspirant* in the General Hospital, and if you can't offer me any hopes or prospects of an assistant's job, I could serve as an *Aspirant* with you."

"What exactly is an *Aspirant?*" he asked. "I'm not yet familiar with the terms used here."

I now gave a brief explanation (something my girl must also bear with here) to the effect that a hospital consists of two things: clinics and departments—clinics, where the professor and his assistants teach the students; departments, where the *Primarius* and his *Sekundarii* (without students) treat the patients. The professor has the choice of his assistants, but the *Primarius* cannot choose his *Sekundarii.* Any doctor can become an *Aspirant* while waiting for the position of a *Sekundarius* to fall vacant, and during this time he is called, as I am, an *Aspirant.* This interim period, however, can be spent in a clinic as well as in a department. Understand, Marty? Professor N. did not appear to understand entirely, for he said: "If you have any prospects of a job as a surgeon's assistant (which I haven't), then don't hesitate to accept. But I advise you to go on working in the scientific field, and when it's time to hand in an application, I will consider your case."

"But I cannot afford to go on working in the scientific field in this way, I've got to branch out and go through the medical curriculum as fast as possible in order to set myself up in practice,

probably in England, where I have relations. I have worked long
enough for nothing. As it is, I've got to abandon a chemical paper
which I had started."

"I am not referring to publications," he replied. "Just go on working in the scientific field; after all, medicine can be practiced scientifically, too."

"I know that, and it differs little from the working methods of
the physiologist."

"It's the same," he interrupted.

"But I feel I must pursue what is most necessary for the medical
practitioner."

"Do that, it won't prejudice you in my eyes when the opportunity
turns up."

"If I understand you correctly then, I am to act as though there
won't be any hope of my working with you in the immediate
future?"

"Exactly," he said. "Take what you can get; I can't promise you
anything, it wouldn't be fair. Incidentally, are you thinking of
deciding on an academic or a practical career?"

"My inclinations and my past experience point toward the former,
but I've got to—"

"Of course, first you've got to live. Well, I'll keep you in mind.
Once more: *Qui vivra verra*." And with that he got to his feet.

"In any case, I thank you very much. And may I come and fetch
my papers after a while? They are my only copies."

"I'd like to read them. Could you come and pick them up in
three or four weeks? I'm very busy at the moment."

"I can quite believe it, Professor. As a matter of fact, the gist of
what I've written can be found in the annual report and in
Schwalbe's *Neurology*."

One more bow, and that was that. Well, my girl? For the moment
all this has led to nothing. The first job is gone, and for the second
my application will certainly be considered, for the man spoke
honestly. In a few days Meynert, for whom N. has great respect,
will intercede personally for me, and if he gets to know the other
friends I have among the professors, I will rise in his estimation.
For the time being, however, I shall go on working as though there
is no hope. What I am going to tackle next I am not quite sure. I
am considering dermatology, not a very appetizing field, but for
general practice very important and interesting in itself. I intend

calling at that department tomorrow; if there are no *Aspirant* jobs vacant, I shall go to Meynert.

I hope from now on to be on better terms with your poor mother, whom I like despite our conflicting interests, and you I hope to see at 10 A.M. on Saturday in the Prater.

Your faithful
Sigmund

$$\frac{1883}{1884}$$

13 *To* MINNA BERNAYS

Vienna
February 21, 1883

Dear Minna

Like so many things in this world, my motives for answering you so promptly are mixed; half because I was so pleased by your unexpected letter, and half because, feeling numb and weary after the extraction of a tooth, I cannot work and yet feel I must be alone. (Martha has gone off with my sisters to visit Dr. Herzig's[1] fiancée and niece.) No, I am doing you an injustice; it is seven eighths because I am pleased and one eighth because I am tired. It was very sensible of you to write as you did. As a happy man at the moment I feel a strong desire to do something for the temporarily separated couple,[2] and considering the abundance of power which the good Lord has so trustfully laid in my hands, I have little choice but to talk to the one and write to the other. Since Elise[3] left and Anna[4] and Eli are reconciled, conditions here during these past weeks have been rather less interesting and rather more satisfactory. There are still clouds in the domestic sky; in the one corner growls your Ignaz, in the other sulks your Mama; there are, however, definite signs that, "weary of their long strife,"[5] this couple too will eventually make peace.

But these are matters which deserve to be treated differently. You

[1] Dr. Wilhelm Herzig, an early friend of Freud's, later Professor of Chemistry at the University of Vienna.

[2] At this time Martha was in Vienna, Minna in Hamburg, thus separated from Schönberg, her fiancé.

[3] Intimate friend of Martha's.

[4] Freud's eldest sister (1858-1955), later married to Eli Bernays.

[5] Quotation from the ballad "Lenore," by the German poet Gottfried August Bürger (1747-1794).

probably know what it's all about. Your Mama has decided to go to Hamburg with you, first to reconnoiter, then to settle there. Whereupon Schönberg told her she is selfish and that he has failed to find in her the mother he had hoped for. Relations were broken off—the bitterness was great. I will tell you at once that I have adopted a definite attitude about this, but I would like to ask you urgently, provided you concede that I have some influence in the matter, not to side with Mama in your letters nor to believe all the complaints that you hear about us from her. Now, I don't want you to think that I feel hostile to her or that I have altered my high opinion of her or that I am on less affectionate terms with her. I do not think I am being unfair to her; I see her as a person of great mental and moral power standing in our midst, capable of high accomplishments, without a trace of the absurd weaknesses of old women,[6] but there is no denying that she is taking a line against us all, like an old man. Because her charm and vitality have lasted so long, she still demands in return her full share of life—not the share of old age—and expects to be the center, the ruler, an end in herself. Every *man* who has grown old honorably wants the same, only in a woman one is not used to it. As a mother she ought to be content to know that her three children are fairly happy, and she ought to sacrifice her wishes to their needs. This she doesn't do, she complains that she is superfluous and neglected, which we certainly give her no reason to feel; she wants to move to Hamburg at the behest of some extraordinary whim, oblivious of the fact that by so doing she would be separating you and Schönberg, Martha and myself for years to come. This certainly isn't very noble-minded, nor is it downright wicked; it is simply the claim of age, the lack of consideration of energetic old age, an expression of the eternal conflict between age and youth which exists in every family, in which no member wants to make any sacrifice and each one wants a free rein to go his own way. The very physical proximity of our living conditions makes conflict inevitable. This is how I feel and I appreciate her no less on this account; the feeling that we belong to each other is not interfered with, and I don't want to offend her by making it clear that any act of hers could break the personal tie between us. If she is being unfair, to me it

[6] At this time Martha's mother, born in 1830, was fifty-three.

means that my mother[7] is being unfair; but it has not made her a stranger to me. On the other hand I demand less from her; after all, she is not the woman of my choice; in short, you are probably already thinking that with my usual modesty I am setting up my behavior as a model for Schönberg to emulate. He, however, makes demands on your Mama as though she were a young and pliant bride not yet exhausted by sacrifices and sorrows, a woman still expecting great happiness from life and still possessing the élan, the alertness and self-criticism of youth. A little of this I think was your fault; you have allowed them to come too close to each other. The harmony Schönberg seeks can be achieved, with great self-denial, between two people, but between three, never; besides, for this number it is quite superfluous. I think you should be a little more jealous of him; it should never come to the point where your relationship with him can be clouded by that to any third person.

If at times he would show her less admiration, he would at other times offend her less deeply. Recently she also played a mean little trick on me in connection with the family's settling in Hamburg; I didn't have it out with her because I was too little affected by it. As a result I am sure she is not too pleased with me; I think she is like our dear, by now extinct friend Fritz Wahle, a trifle jealous of other people's emotions in which she has no share; but with our system of closing ourselves off we are in a better position than Schönberg and certainly than you. We have no intention of sharing the painful pleasure of stirred-up passions, we have a practical aim: to live and work together in a close union *à deux* and in a somewhat looser one with other couples, and we feel entitled to dismiss everything we consider useless for this purpose.

In all justice, I can only agree with my friend in this case. The move to Hamburg is a loss to us all and—don't let's be ashamed of human weakness—a danger. Only dire necessity could justify it. We are going to oppose this project with all the means at our disposal, but I do think it's a luxury to fall out over it a year in advance. I still hope that for Mama, who is our guest in Kaiser Josef Strasse 33[8] at the moment, the summer months in Hamburg will be sufficient to discourage her.

[7] Freud means his mother-in-law.
[8] Street where Freud's family lived.

The arguments to which we have been compelled to listen, in favor of this visit, did not impress us much, nor were they convincing enough even partly to conceal the dominant desire behind them.

[*The letter is continued by Martha.*]

14 *To* MARTHA BERNAYS

Vienna, 2 A.M.
July 13, 1883

Gardener Bünsow,[1] lucky man, to be allowed to shelter my darling sweetheart! Why didn't I become a gardener instead of a doctor or writer? Perhaps you need a young chap to work for you in the garden, and I could offer myself so as to bid good morning to the little princess and perhaps even demand a kiss in return for a bunch of flowers.

But this letter is not to Gardener Bünsow, rather to you, my Marty, my Cordelia-Marty. Why Cordelia? This will be explained later. Are you longing to know, darling? Your sore throat will be better, is bound to be better, by the time you get this letter—it was nice of you to write to me about it, but not at all nice of you to get it. If it's nothing worse, don't let yourself be excessively coddled, wrapped in shawls and all that kind of thing, my child; I believe that a little toughening, which can surely be risked in such harmless circumstances, is better in the long run. And I am looking forward to your news and hope you will eat well, if necessary in secret,[2] and if you need money for this, my sweet, just let me know, for I have some again.

Today was the hottest, most excruciating day of the whole season, I was really almost crazy with exhaustion. Realizing that I was badly in need of refreshment, I went to see Breuer,[3] from whom I have just returned, rather late. He had a headache, the poor man, and was taking salicyl. The first thing he did was to chase me into

---

[1] Martha had gone for a rest to Düsternbrook, a village in the vicinity of Kiel, and was living in the house of a gardener called Bünsow.

[2] From her mother, who was a strict adherent of the Jewish food laws.

[3] Dr. Josef Breuer—see *List of Addressees*.

the bathtub, which I left rejuvenated. My first thought on accepting this wet hospitality was: If Marty were here, she would say, "This is just what we must have, too." Of course, my girl, and no matter how many years it will take, we shall have it, but the only miracle I am counting on is that you will love me as long as that.—Then we had supper upstairs in our shirtsleeves (at the moment I am writing in a somewhat more advanced négligé), and then came a lengthy medical conversation on moral insanity and nervous diseases and strange case histories—your friend Bertha Pappenheim[4] also cropped up—and then we became rather personal and very intimate and he told me a number of things about his wife and children and asked me to repeat what he had said only "after you are married to Martha." And then I opened up and said: "This same Martha who at the moment has a sore throat in Düsternbrook, is in reality a sweet Cordelia, and we are already on terms of the closest intimacy and can say anything to each other." Whereupon he said he too always calls his wife by that name because she is incapable of displaying affection to others, even including her own father. And the ears of both Cordelias, the one of thirty-seven and the other of twenty-two, must have been ringing while we were thinking of them with serious tenderness.

But now fond greetings, for I have gone to sleep, Marty.

Your
Sigmund

[4] Breuer's patient, mentioned in Breuer and Freud, *Studien über Hysterie* (*Studies on Hysteria*), Deuticke, Vienna, 1895.

15 *To* MARTHA BERNAYS

Vienna, at night
August 22, 1883

My beloved Marty

I have little more to say in answer to your letters than that I am delighted with all the good news they contain, that I readily agree with you in every respect, that I would prefer to tell you all I think with a kiss, and on the whole—glancing back—can't we claim to have surmounted all our difficulties and say that our love has now become lasting and deep?

I will use this late hour to tell you some factual news, otherwise these little items accumulate and I will never be able to drop back into a light and intimate chat.

Well, today I heard from Simon[1] that he has instructed his bank to remit to Anna 100 florins to buy herself something for the wedding.[2] It is none of my business, but this really isn't much from a rich uncle, nor is it very delicately given. If we ever get rich, sweetheart, we will do things differently. I know that you for one won't need any encouragement.

Paneth[3] wrote today out of his dream of bliss; he sends greetings to you and his bride to me, the whole thing a panegyric to this best of all possible worlds.

Since Holländer[4] left I see only two friends—Herzig, who occasionally comes in from the country where the girls[5] also are, and Robert Franceschini, a friend as yet unknown to you, who began studying medicine with me—there were only three in the whole of our class to take up medicine, and we scribbled our names on a skeleton which we bought together—he then had to abandon medicine and zoology on account of a protracted illness, then became tutor to some rich people whom he also had to leave because of his illness, and finally after some improvement he found a job as a railway official. At the moment he is living very modestly with his old mother, writes feuilletons under the initials R. F. for all the papers on all kinds of subjects, in which he is helped by his numerous talents and his erudition. Lately he has embarked on the study of philosophy and is writing a thesis to take his degree. But the only person one can be with and talk to intimately is Herzig, whose value you know or at least can guess. On Tuesday for the first time we went to the electricity exhibition, just to find our bearings, for we intend to revisit it frequently. Everything is as yet unfinished, noise and uproar from the machines which are partly still in the process of being set up, no light as yet, telephone booths inaccessible. Most of the things exhibited one cannot understand anyhow without explanation or previous study. However, we have acquired the available books on electricity and are busy studying

[1] Simon Nathansohn, brother of Freud's mother.
[2] Freud's sister Anna married Martha's brother Eli.
[3] Dr. Josef Paneth (1857-1890), fellow student of Freud's.
[4] Dr. Bernard Holländer, Assistant in Meynert's Clinic.
[5] Herzig's fiancée and niece.

them. So far I haven't seen anything especially interesting except
a row of small rooms containing, under the pretext of electric light-
ing, some charming furniture, as well as a very nice exhibition by
Jaray.[6] The sight of these rooms made me lose all my philosophy.
Herzig remained cold and said he had already finished that chapter, knowing he would never own such things. I was in ecstasy,
imagining your delight at sight of these lovely furnishings. Looking at the things were several pretty girls whose thoughts I think
I guessed: Oh, if only I could marry a man who would give me
that! I envied the young ladies this beautiful dream and was sad
at the thought that my girl, on seeing all this, would not be able to
look so hopefully into the future. I was quite glad you weren't
there. This mood soon passed, however, and I fell to thinking
more sensibly: how unhappy we could be on this splendid sofa
and how happy in an old leather armchair, and that the wife should
always be the most beautiful ornament in the house, and that all
these rooms were empty and lifeless because their mistress wasn't
there.

What we see on future visits to the exhibition will be faithfully
reported to you.

During these past few days I have been having some serious differences of opinion with Pfungen,[7] and I have treated him too
harshly, which was very unfair of me. I am afraid I do have a
tendency toward tyranny, as someone recently told me, and added
to this is the fact that I am all too gay nowadays; I let myself go
in a kind of youthful high spirits of immaturity, which used to be
quite alien to me. I also have the capacity, in other respects praiseworthy, of hating someone on intellectual grounds, just because
he is a fool, and this is what the otherwise excellent man unfortunately is. He is quite *meschugge*[8] and all his thoughts are crazy.
But I must alter my attitude toward him, for he is really a very
decent man.

This same kind of gay moodiness, as I would like to call it, also
leads to my not making good use of my time: I read a lot, fritter
away much of the day. For instance, I now possess *Don Quixote*,
with Doré's[9] large illustrations, and concentrate more on it than

6 Well-known furniture store in Vienna.
7 Assistant in Meynert's Clinic.
8 Yiddish—slightly mad.
9 Gustave Doré (1832-1883).

on brain anatomy. You are quite right, little princess, it is no reading matter for girls, I had quite forgotten the many coarse and in
themselves nauseating passages when I sent it to you. No doubt it
achieves its aim in a remarkable manner, yet even this is somewhat remote from my princess. But the incidental stories are
charming, all these you really must read. While in the midst of the
book today I nearly split my sides; I haven't laughed so much for
ages. It is so beautifully done.

Now farewell, my lovely princess. In my silence about our love
please see once more the sympton of my unworried and healthy
certainty of possession, and go on loving me as I will always love
you, and then we will compete as to which of us can be the more
loving.

With affectionate greetings till we write again
Your
Sigmund

16 *To* MARTHA BERNAYS

Vienna
August 23, 1883

Treasured princess

Just back from my country practice to find your sweet letter
with the good news that you are feeling well and all the pleasant
things you don't tire of telling me every day. I had a talk today with
a colleague in the hospital, Dr. Widder, who said he considers it a
great mistake to marry as long as one has no money and that it
will take me eight years to get anywhere. All this he was saying
not from worldly wisdom, etc. but out of the innocence of his heart,
as he sees it. Defending my case valiantly, I told him he just doesn't
know my girl, who is willing to wait for me indefinitely, that I
would marry her even if she had turned thirty—a matron, he interrupted—that I would bring it off by starting work elsewhere,
that a man has to take some risks and that what I stand to gain
is worth any risk. He admitted that within two years I could be
earning two thousand gulden, and showed me a letter from a Dr.
K. in Brünn, who hopes to earn five to six thousand florins in the
course of a year, etc. He was not entirely serious about the gloomy

picture he painted. The most beautiful part of it all I of course did
not tell him—that it makes one so unspeakably happy to feel oneself loved, even if we don't yet belong to each other formally and
completely, and above all if one is lucky enough to have abducted
a little princess! Courage, my treasure! You will become my wife
much earlier and you won't have to feel ashamed of having had
to wait so long. One quite small piece of good news I will let you
know today: unless I am very, very much mistaken I think my
"latest method"[1] is going to work; I wrote to you before that I am
putting my hope in the light of the sun—it really seems to be effective. But don't be disappointed if I write again that it doesn't
work; discovery requires patience and time and luck; if something
is to succeed it always has to start like this. So courage, my little
princess.

My patient is no worse; I am busy dispelling all kinds of minor
complaints; so far nothing has happened that I cannot cope with,
and when I give an order I often hear that Breuer has contemplated
doing the same. His wife, of whom you wished to hear, has much
the same serious charm that you have, my angel, but is not quite
so sweet; I admire her because she has excellent observation, nurses
him with such patience and is so good at cheering him up. I really
do hope he will improve; Breuer doesn't think he will, and fears the
next six years of slow deterioration.

So you have run out of notepaper, Marty? In September you shall
have a slice of my salary and with it order some more notepaper
of the same kind. No, no, rather spend it on yourself, it is such
ages since you have had any money, and at the moment I cannot
send you more than a few little marks, but I can't exchange them
today, in fact not before Saturday, as tomorrow I am on duty.

Now I must break off for the evening, I will continue this letter
to my beloved at night.

Forgive me, dearest, if I so often fail to write in a way you deserve, especially in answer to your affectionate letters, but I think
of you in such calm happiness that it is easier for me to talk about
outside things than about ourselves. And then it seems to be a kind
of hypocrisy not to write to you what is uppermost in my mind: I
have just spent two hours—it's now midnight—reading *Don Quixote*,
and have really reveled in it. The stories of the indecent curiosity

---

[1] To harden and dye slices of the brain for microscopic examination.

of Cerdenio and Dorothea, whose fate is interwoven with Don Quixote's adventures, of the prisoner whose story contains a piece of Cervantes' life history—all this is written with such finesse, color, and intelligence, the whole group in the enchanted tavern is so attractive, that I cannot remember ever having read anything so satisfactory which at the same time avoids exaggeration. All the happy couples, the ladies who promptly love one another like sisters and receive the poor Moorish girl so affectionately, the knight lashed to the window and ordered to prevent wicked giants from breaking in: none of this is very profound, but it is pervaded by the most serene charm imaginable. Here Don Quixote is placed in the proper light through being no longer ridiculed by such crude means as beatings and physical maltreatment but by the superiority of people standing in the midst of actual life. At the same time he is tragic in his helplessness while the plot is being unraveled. Sancho, with his sly motives and in the way he keeps tumbling from the dream world into reality, is wonderful. And then Doré's illustrations; they are superb only when the artist approaches his subject from the fantastic angle: when for instance he picks out a few words of the tavern-keeper's wife to show how a wretched little knight has cut in half six giants with one blow of his sword, the lower halves of the bodies still standing while the upper halves roll in the dust. This picture is really of a marvelous absurdity and a splendid contribution toward dispelling all the romantic nonsense about chivalry. He succeeds too with the Oriental scenes, the strange and grandiose architecture, also with the harshness of nature in the dark mountains; and he is good wherever the text lends itself to caricature, for instance when the ghosts bewitch the knight and lock him up in a cage. It's enough to make one die with laughter. But in other scenes, those in which the true character of the knight is revealed, the subtle irony is missing. Here the caricature is mostly exaggerated and the illustrations fall far short of the text. But I can well imagine how magnificent his illustrations for *Orlando Furioso*[2] must be, material that would seem to be made for Doré, and even several things out of the Bible, especially the legendary and heroic scenes.

Now, my dearest most beautiful sweetheart, please take these comments in your stride, don't consider me ungrateful or reproach

---

[2] *Orlando Furioso,* by the Italian poet Lodovico Ariosto (1474-1533).

me for thinking too little of you or seeming too cool. The more
intimate your letters become, the more silent I get; as I read them
something like a continual tacit assent goes on within me; yes, that
is how I want my Marty to be, as she is now. Long may she remain
this way and healthy to boot.

Well, I wonder what you got for your birthday? And what does
Minna mean by saying that you had three this year?[3] I am afraid
you have been treated rather poorly by me this time. Just wait,
though, till things are going well with me, and I will celebrate your
birthday properly. We have after all so many dates to celebrate, I
have seen you on so many days—and often wasn't grateful enough—
and the memory of having seen you is quite enough to make a memorial of the occasion.

> Goodnight, my princess, keep well and remain fond of
> > Your
> > Sigmund

Please thank Minna for her sweet, intelligent letter which can
only receive a less brilliant answer, for which she won't have to
wait very long, however. Ask her to stop writing to Schönberg for
once, so that he can answer me, too.

Am I so sleepy or is it just that my handwriting is so bad today?
I can hardly read it. I even omit words, too, don't I? One more affectionate greeting, Marty.

[3] Martha's birthday was celebrated according to both the secular and the Jewish
calendars.

17 *To* MARTHA BERNAYS

Vienna, Tuesday night
August 28, 1883

My precious girl
      I came to my patient today completely at a loss how to find
the necessary sympathy and attentiveness for him; I felt so limp
and apathetic. But this feeling vanished when he began to complain and I to realize that I have a function and an influence here.
I don't think I have ever attended him with greater care, nor made
such an impression on him; work really is a blessing. And now I

feel well and calm; I have decided to be severe with myself so as not to fall back into such a state of weakness; the awareness of calm preparedness is surely the finest thing a man can find in himself. It is what the poet described in the lines:

> New strength and heart to meet the world incite me,
> The woe of earth, the bliss of earth, invite me. . . .[1]

The mood for which an even greater poet found the loftiest expression with the words:

> Let us consult
> What reinforcement we may gain from hope;
> If not, what resolution from despair.[2]

But I have no use for this mood, it must not be spent on one decisive battle, rather be saved for a long, tenacious struggle with small, isolated tasks.

I am well again now and able to enjoy things and am glad that even in the bad days I did not think of you with any less tenderness than I do now. There may easily be a more accommodating love than mine for you, but hardly a more serious one, in cold blood. When I am angry with you nowadays, as I was about the traveling project,[3] it is gone as soon as I have spoken my mind, and I don't like to leave it unspoken, for it all burrows its way into me and cannot be cauterized away, proofs of which you have had. But—then no more about me—introspection and presumption are also part of this mood.

It wasn't very easy to find any peace today; the moment I got home I was told that my mother had waited two hours for me, left a small parcel and a message for me to go to the Prater as Father is leaving tomorrow. . . . He is not leaving till tomorrow evening. I cannot stand anyone's company for long, least of all that of the family; I am really only half a person in the sense of the old Platonic fable which you are sure to know, and the moment I am not active my cut hurts me. After all, we already belong to each other and if we are going to have a tussle—this too is part of love—let it be at close quarters.

---

[1] Quotation from Goethe's *Faust,* I.
[2] Quotation from Milton's *Paradise Lost.*
[3] A visit to Elise.

What else happened today? Oh yes, my bookseller[4] came to see me to ask my advice whether he should accept a book which the author himself wishes to translate from the original English. Since my bill with him is pretty big, I am glad to establish a personal relationship with him. The book is beautifully illustrated and I am going to recommend it to him. I hope he will present me with a copy of the translation. Unfortunately it is not something I can send to you, it is a pathological histology! Oh, my precious sweetheart, what stupid, uninteresting things I write to you! I am going to tell you a funny little story, but you mustn't be sorry for me. When I got home I found a letter from a friend who frequently comes to see me (privately), asking me to lend him *another* gulden till the first of the month, to leave it with the janitor and if I don't have a whole gulden, then half a gulden, but at once; on the first everything would be paid up. Well, my entire fortune happened to consist of four—kreutzer, which I couldn't very well offer him. So I decided, since my ordinary bankers were not at home, to waylay a colleague who owes me some money, in fact quite a considerable sum for this time of the month. But he couldn't be found, I was getting hungry and had to go to the Prater, so what was I to do? Then fortunately another colleague appeared from whom I borrowed a gulden in no time. But by then it was too late to send part of it to the other friend, I just had to go to the Prater, so today he got nothing, but if my debtor pays tomorrow he shall have something. One day he and I will probably be rich, but don't you think this is a funny kind of gypsy life, Marty? Or does this sort of humor not appeal to you and make you weep over my poverty? Don't take it to heart; before you have a chance to sell your jewels to save me I shall be an affluent man again.

And now goodnight, sweet princess; if I have written more impersonally and less affectionately, I have a little purpose—and you are to guess what it is.

Your faithful servant
Sigmund

[4] Presumably Deuticke, bookseller and publisher in Vienna.

THE LETTERS OF SIGMUND FREUD

18 *To* MARTHA BERNAYS

Vienna, Wednesday evening
August 29, 1883

My beloved Martha

Your charming, intelligent letter and your excellent description of the Wandsbek Fair gave me great pleasure and suited my continuous improvement—if there weren't still some catarrh I could say my well-being. You think almost like Wagner in *Faust* during that beautiful walk and I ought to answer with gentle indulgence in the manner of Dr. Faust: "Here I am Man—dare man to be!" But no, beloved, you are quite right, it is neither pleasant nor edifying to watch the masses amusing themselves; we at least don't have much taste for it any more and our anticipated or already enjoyed pleasures, an hour's chat nestling close to one's love, the reading of a book that lays before us in tangible clarity what we think and feel, the knowledge of having achieved something during the day, the relief of having solved a problem—all these gratifications are so different that it would be affectation to pretend that one really enjoys the kind of spectacle you describe.

But now please forgive me if I quote myself; I remember something that occurred to me while watching a performance of *Carmen:* the mob gives vent to its appetites, and we deprive ourselves. We deprive ourselves in order to maintain our integrity, we economize in our health, our capacity for enjoyment, our emotions; we save ourselves for something, not knowing for what. And this habit of constant suppression of natural instincts gives us the quality of refinement. We also feel more deeply and so dare not demand much of ourselves. Why don't we get drunk? Because the discomfort and disgrace of the after-effects gives us more "unpleasure" than the pleasure we derived from getting drunk. Why don't we fall in love with a different person every month? Because at each separation a part of our heart would be torn away. Why don't we make a friend of everyone? Because the loss of him or any misfortune befalling him would affect us deeply. Thus we strive more toward avoiding pain than seeking pleasure. And the extreme cases are people like ourselves who chain themselves together for life and death, who deprive themselves and pine for years so as to remain faithful, and who probably wouldn't survive a catastrophe that robbed them of their beloved. In short, people like the Asra[1] who could love only once. Our whole conduct of life presupposes that we are protected from the direst poverty and that the possibility exists of being able to free ourselves increasingly from social ills. The poor people, the masses, could not survive without their thick skins and their easygoing ways. Why should they take their relationships seriously when all the misfortune nature and society have in store threatens those they love? Why should they scorn the pleasure of the moment when no other awaits them? The poor are too helpless, too exposed, to behave like us. When I see the people indulging themselves, disregarding all sense of moderation, I invariably think that this is their compensation for being a helpless target for all the taxes, epidemics, sicknesses, and evils of social institutions. I am not going to pursue this thought any further, but it would be easy to demonstrate how "the people" judge, think, hope, and work in a manner utterly different from ourselves. There is a psychology of the common man which differs considerably from ours. They also have more community spirit than we have; only for them is it natural that one man continues the life of the other, whereas for each of us the world comes to an end with our death.

My dearest girl, if you dislike this kind of talk, just tell me to stop. You don't realize the extent of your influence over me and you must not conclude from the harsh way I deal with certain things connected with the basic conditions and experiences of our relationship that I am generally intolerant. I am quite prepared to be completely ruled by my princess. One willingly lets oneself be dominated by the person one loves; if only we had got as far as that, Marty!

The girl in whose fate I took such an interest lost the moving effect she had on me after a few days. There were too many complications involved which did not correspond to our own relationship, and too many faults on her side. Being a physician certainly doesn't make one immune to human suffering, nor should it, but one does become less vulnerable if there is happiness in one's own life. . . .

I find myself in continuous professional friction with Pfungen,

[1] From a poem by Heinrich Heine (1796-1856).

and have now got to the point where I contradict him in front of Meynert; he of course backs me up, because Pfungen is full of delusions and eccentric ideas. But I must admit to myself that I do have a tyrannical streak in my nature and that I find it terribly difficult to subordinate myself. I am sure you know this already, but if you love me in spite of it I shall manage to be happy all the same.

I am spending every free hour of the day on my paper, the beginning of which is not unsatisfactory. I don't think, Marty, that I react to success and failure quite as excessively as you make out. I am not yet quite clear about my method; it works, but I am not always in control of it, it does not always produce the same results.

Goodnight, my sweet darling, my precious princess, you. Your letters cheer me up tremendously.

Go on being fond of
Your
Sigmund

19 *To* MARTHA BERNAYS

Vienna, Tuesday, at night
September 4, 1883

My dearest girl

I can well imagine why I have had no letter from you today. Two days ago you got the news that I was quite ill again, and being worried, you decided to wait for the next letter; I wrote that letter four days ago, but today I am quite well and want to discuss traveling plans with you. Not in your direction alas, my sweet child, for I am not rich enough for that. Actually, since I cannot come to you I have a good mind to drop the whole idea; but listen to this and tell me what you think. I have two invitations on hand: the one to go and visit Brust[1] in Baden;[2] I met him today in town, he had come in with his brother; also intends to call at the Kaiser Josef Strasse. I could stay eight to ten days—traveling expenses one gulden, living expenses there not more than here; Brust even

[1] Freud's fellow student, colleague, and friend of Freud's sister Rosa (see letter of Jan. 10, 1884).

[2] Health resort, about an hour by train from Vienna.

wants to try to introduce me at the inn as his brother as he did two years ago so that the waiters won't accept any tips from me, but that I will not tolerate this year.

The other suggestion is rather more original: Dr. Widder insists that on the fifteenth I accompany him for a fortnight to his native village, near Kaschau,[4] where the plan is to do nothing but eat grapes. I would be the guest of his family; the journey costs fifty to sixty florins, but as I should be traveling as an *Oberarzt*[5] I would pay only half, and very likely I could even get a free ticket via Zuckerkandl[6] as far as Pest,[7] so I could save even more. So all in all, it would cost me ten florins more than I spend here in a fortnight, but I would also have seen Pest, the Carpathians, and lead a regular gypsy life oblivious of medicine for ten days. The second suggestion has a lot to recommend it, but involves an expense of time and money, which the first does not. Now which one shall I decide on? I grant you a decisive vote and will enumerate the circumstances which I must take into account. First, Breuer must be back and take K. off my hands, then the latter must have paid up, and everything depends on whether the fee is nearer thirty or fifty. Second, Breuer may have some suggestion which could be either useful to me or which I wouldn't be able to refuse. At the moment, then, it is difficult to know what to do. Marty, I have just had a crazy dream: suppose K. were to reward my efforts for the month of August by paying me one hundred gulden, then I could leave Baden here and Kaschau there and travel to Wandsbek. Alas, my good girl, there is no hope of this. He would have to pay me seven florins (!!!) a visit. Oh, if only I had cured him: why on earth didn't I do that? And by the way, we still haven't agreed on the conditions for our reunion.

But to be serious again, my dearest treasure, I think I will accept one of the two invitations. Although endowed with a very strong constitution, I have not been in a continuously good condition these last two years; there have been so many hardships that it really required the joy and happiness of our relationship to remain healthy. I am like a clock that hasn't been repaired for a long time, dusty in every joint. As my miserable person has taken

---

[4] Small town in Hungarian Slovakia.

[5] Military rank as reserve officer.

[6] Dr. (later Professor) Emil Zuckerkandl (1840-1910).

[7] Budapest.

on a greater importance, also for myself, since I acquired you, I am more concerned with my health and do not intend to wear myself out. I would rather renounce my ambition, attract less attention, have less success, than endanger my nervous system. In the future, for the remainder of my apprenticeship in the hospital, I think I shall try and live more like the Gentiles—modestly, learning and practicing the usual things and not striving after discoveries and delving too deep. My happiness lies above all in my relations with you, later in making you mine. We must stick close together and make life beautiful for ourselves, and what we need for our independence we will achieve by decent, steady work without any gigantic efforts.

Once we are together and have strengthened and assured each other, then will be the time to aim at greater things again. Why aren't you here, darling, to give me your answer? I am afraid I do waver a lot in my plans, don't I? Please tell me, and what you are thinking and how you want things to be.

Schönberg was supposed to come today, but he hasn't, perhaps Brust's visit prevented him. And what are you doing, my silent sweetheart—but you are not silent, you are only far away and won't be able to answer this until September 8. I have been wanting to send you a paper, a special number of the *Illustrated,* but I have no money yet, still haven't been relieved in Pötzleinsdorf.[8] The poor man isn't at all well, but I did finally succeed in getting him to sleep; to comfort a wife with lies is so hard.

Today I met Frau Emma Pappenheim[9] with a child, on a bench in the Gersthofer Allee; this time I recognized her earlier. I thought to myself how strange it is that Hamburg children are wafted over here and the Viennese over there.

Farewell my dear sweetheart.

Your faithful

Sigmund

Love to Minna.

[8] Suburb of Vienna, where Breuer's patient lived.
[9] Friends of the Bernays family from Hamburg.

20 *To* MARTHA BERNAYS

Vienna, Saturday
September 8, 1883

My precious Marty

What can it be that you want and do not dare to mention?
I am so filled with curiosity, especially since Schönberg tells me
there is something similar on his horizon. What on earth can it be,
then? A tooth out of the Caliph's jaw, a jewel from Queen Victoria's crown, a giant's autograph, or something equally fantastic
which would mean putting on my armor at once and setting out
for the Orient?

Or is my sweetheart's desire nearer home; could it possibly be
a deed of self-conquest? Am I to fast at Yom Kippur or reconcile
myself to someone I don't like? Surely not. My Marty would not
abuse her power and persuade me to actions that lack sense as
well as honesty. I hope she wants something for herself and I hope
I can catch it and give it her. . . .

My greatest concern at the moment is to get Schönberg away
for the winter;[1] one of his brothers is making difficulties; I am
waiting for the arrival of the other to see what I can do; there's
even a vague possibility of his going to the Riviera as a tutor independent of the brothers. But I think I might be able to do something with the brothers. Your confession about your reading, princess, amused me greatly. It seems as though you don't quite want
to bite—like the peasant in the old proverb: "What he doesn't
know he won't eat."

But do finish *Don Quixote*; the second part contains many fewer
of the shocking qualities than the first and is far more fantastic.
I also quite agree that in winter, or when the weather begins to
get bad, you should write only every other day, so as to get some
exercise; after all, I am sure of you and think of you with gay and
unclouded happiness. But Marty, you will then have to write a
little more every other day, otherwise I might develop an uncon-
trollable hunger for news of you.

[1] Schönberg had been taken ill with tuberculosis.

I now no longer have to choose and you don't have to decide about that journey; Dr. Widder can't get away on the fifteenth because his chief is on vacation and he is in charge of the department. What actually makes me hesitate about Baden is whether, considering B's ambiguous attitude toward Rosa, it is a good idea to get on such intimate terms which could be interpreted as speculating on future relations. Or do you think this is too far-fetched? I am very well, by the way, and very lazy and, since I spend every evening with Schönberg, also very pleasantly stimulated. Breuer is not back yet; I am expecting him with impatience, for personal and professional reasons; the weather is now so bad that he can't very well stay away much longer. I have an idea he will turn up on Monday.

I can see quite clearly from your letters that you are well, but please tell me also what you look like, whether you have put on weight, whether you are feeling better and if your complexion is clearer than when you left—or else, or else I will ask Minna, or, having sold my library, I will come and waylay you in the grove, just to convince myself, and go off again the same evening. Would you like me to do that, my darling? And what about the cold baths; is it again too cold for them?

Affectionate greetings to my precious sweetheart from
your devoted
Sigmund

Best regards to Minna; I am going to compose a long letter to her in the near future.

21 *To* MARTHA BERNAYS

Vienna, Sunday, 3 P.M.
Sept. 9, 1883

My sweetheart

Don't you ever again say that you are cold and cannot find the right words; you write such unspeakably sweet, such movingly tender letters that I could answer them only with a long kiss, holding you lovingly in my arms. I hope one day it will be nothing but a pleasant memory when I tell you how I have yearned for you and I will never quite believe it when I really have you with me. I daren't think much about it for fear that my patience to bear it till then would melt away.

Now, in answer to all your dear questions, let me point out that I am not such a sick man as you think. Since those bad days that were more of an interruption of health than an actual illness, I feel very well; the inactivity suits me for a short time, and if nothing comes of the Baden project I am not to be pitied; as for Kaschau, it is off, as you will know today. I really have no intention of being lazy from now on, I only mean to renounce the exhausting pursuit of distinction and, as you say, keep myself productive and capable of enjoying our life together. I have always thought that there is a short and a long way of achieving something; if the short one is barred to me I will confidently take the long one, and this is precisely what is happening to me at the moment. I was enchanted to learn that you are so ambitious for me, my sweet child; in the beginning, I wasn't; I was seeking in science the satisfaction which the effort of searching and the moment of discovery offer; I never was one of those people who can't bear the thought of being washed away by death before they have scratched their names on the rock amidst the waves. But when I think what I would be like now if I hadn't found you—lacking ambition, lacking the joy in the lighter pleasures of the world, lacking any fascination with the magic of gold, and endowed at the same time with very moderate intellectual and no material means whatever—I would just have strayed miserably about and gone into a decline. You give me not only aim and direction, but so much happiness as well that I cannot be dissatisfied with the otherwise rather wretched present; you give me hope and certainty of success. I knew it before you loved me and I know now that you do love me and it is your doing that I have become a self-confident, courageous man.

Marty, my sweet treasure, our happiness rests finally in our love, I don't want any more than you want for us both, not out of cowardice but because I am aware of the insignificance of other desires compared with the fact that you are mine. And you are so sweet and so good. The only news I have, my darling, is that Schönberg comes to see me every day, that I am fairly satisfied with his condition, but that as soon as his brother Alois is back I am going to pursue the project of his Italian journey. Mother was in bed with a temperature yesterday, a minor recurrence of her old protracted lung trouble; today her temperature is normal again, but it means that I have to go home every day. Pauli[1] is not well, either. Yesterday I went on an excursion with Dolfi[2] to Pötzleinsdorf; she waited for me while I was seeing my patient, and then we walked back via Dornbach.[3] She is the sweetest and best of my sisters, has such a great capacity for deep feeling and alas an all-too-fine sensitiveness. Needless to say, we talked mostly about you and she will always be a most affectionate friend to you; her instinct allows her to guess what her judgment cannot provide.

Is it possible, Marty, that your wish about which I have been racking my brains for so long, is nothing more than a loofah? Surely you must want something else and will let me know, for by the time you read this I will have some money again. It is always worth sending you something simply for the pleasure you take in it, even if the object isn't worth it. . . . But you always see the intention behind it, don't you?

My sweet darling, I meant to write more today, but Schönberg and Franceschini have been here all the afternoon, then we had supper together and now I am sleepy and so sad that I have to write to you instead of being allowed to kiss your sweet lips.

So let us part for today with a fond goodnight.

In devoted love
Your
Sigmund

[1] Pauli (Pauline) Freud (1864-1942), Freud's sister.
[2] Dolfi (Adolfine) Freud (1862-1942), Freud's sister.
[3] Suburb of Vienna.

22 *To* MARTHA BERNAYS

Vienna, Sunday
September 16, 1883

My sweet little woman

I also want to ask for something: that you accept from me, as an atonement for some mean thoughts and some criticism which did not do you justice, the two things you have wished for. I will get the little dictionary, and you must let me know what the jacket is going to cost. If I cannot afford it now, I want to be allowed to do it later, next month. But don't deny yourself, my precious, any

little luxury; I don't. And you are so young and can be so pleased
with things, and I know that everyone who sees you wants to give
you some pleasure, so why shouldn't I be allowed to do the same?
Your letter moves me like the voice of an angel, and helps me to
rise above all my silly worries about you and above my deeply shat-
tered emotional condition. I did not want to talk about it on our
monthly memorial day, but I cannot conceal it from you any
longer: I have just returned from the funeral of my friend Nathan
Weiss.[1]

On the thirteenth, at 2 P.M., he hanged himself in a public bath
in the Landstrasse. He had been married hardly a month and had
returned ten days ago from his honeymoon. He left two letters,
one to the police asking them to inform his parents tactfully and
to suppress any word of it in the newspapers, the other to his wife.
By Thursday evening the news was already known in the hospital,
a colleague rushed to his apartment to take him to the hospital in
order to scotch the rumor; but the door was locked. His brother,
Sekundararzt in the hospital, confirmed the news. Early on Friday
morning Lustgarten[2] came to see me while I was still in bed, soon
afterwards two other colleagues, all with the same news, but we
just could not believe it, it was too difficult to conceive of a man
who combined in himself more restlessness and zest for life than we
had seen in anyone else, as dead and silent. Even now, though I
have just heard the thud of the earth on his coffin, I cannot get
used to the thought.

And why did he do it? He was well on the way to achieving
everything he had been striving for, he had become a *Dozent,* enjoyed a considerable reputation in his field; since directing a department in the hospital he was assured of a large practice, he had
just succeeded in getting married—but that was the trouble; the
details that drove him to his death are unknown to us, but that
they are linked up with his marriage is beyond doubt. I have forgotten how much I have told you of what preceded this marriage,
but I think I had better repeat here all I know about him, for his
death was by no means an accident, rather it was a logical outcome
of his temperament; his good and bad qualities had combined to

[1] Dr. Nathan Weiss, Assistant at the Neurological Clinic.
[2] Dr. Sigmund Lustgarten (1857-1911), Assistant at the Chemical Institute.

bring about his downfall; his life was as though composed by a writer of fiction, and this catastrophe the inevitable end.

His father is lecturer at the religious college of Vienna, a very brilliant scholar who, had he chosen to study Chinese instead of rabbinical law, would certainly have become a university professor; despite all this, however, he is a very hard, bad, brutal man. "My father's a monster," Nathan used to say. His mother is a decent, simple, good-natured woman, who bore many children and shared all the vicissitudes of life with her husband, although their life lacked any deeper relationship. In that home there was no love and bitter poverty, no education and endless demands. To satisfy the father's colossal vanity, all the sons had to study, most of them did not get very far, went to the dogs; six months ago one of them shot himself because he could see no other way out. Only Nathan and one brother, now working in the hospital, got anywhere. Nathan was the most gifted, he inherited all his father's talent, but he was a good fellow. He was not often looked upon as such; it was always said that he had a bad character and many of his actions confirmed it. This was due, I think, to his main driving force— self-love; I might almost say self-adulation. No doubt he was superbly equipped to get on in the world, and so long as things were difficult for him he never bothered about what means to choose. He was incapable of any self-criticism, overlooked, forgot, and forgave himself anything he had done badly or which showed him up in a bad light; on the other hand, anything that raised his self-importance he cultivated and exhibited in front of others. Breuer said rightly that he reminded him of the story of the old Jew who asks his son: "My son, what do you want to be?" And the son answers: "Vitriol, the stuff that eats its way through everything." Weiss really was vitriol and he really did eat his way through everything. His gigantic self-importance was matched by an energy of an unusual kind, an ability to burrow his way into things and never let go. But I don't think he owed his success to this ability; I always saw him in another light. I considered his extraordinary appetite for life to be the outstanding quality in his character.

He took pleasure in his own speech, in his own thoughts, yes, even in insignificant, indifferent actions of everyday life, and was convinced that no one could perform them as well as he. Everything he said and thought possessed a plasticity, a warmth, a quality of importance, which was meant to conceal the lack of deeper

substance. For his gifts were not remarkable, he knew little, never
delved very deep and he lacked completely the basic conditions
for scientific work: criticism and thoroughness. As a result his
achievements are of moderate value, and lack any original content. It was his temperament, his personality, the liveliness and
clarity of his presentation, which brought about his success. It
was rather like the story of the two wanderers in the well-known
poem by Anton Auersperg:[3] both talk about meadows, forests, sunshine, but how differently they express it! When Weiss talked of a
well-known phenomenon he gave the impression of a great discovery freshly made by him; when he addressed one in his strange,
witty gibberish as "a compromised central European," one really did
feel compromised; one could as little help believing in his assertions
as one can help laughing when someone else laughs, or yawning
when someone else yawns. Much of the high opinion in which
people held his ability was inspired by himself, for he was always on
the spot, buttonholed everyone, talked only about himself and of
himself as the most able expert on the subject with which he just
happened to be occupied. Another positive element in his talent
was the quickness with which he thought and his brilliance at putting two and two together. One could almost say that his self-confidence was the direct physiological result of the vivacity, quickness, and clarity of his thought processes. He behaved invariably as
we would after a lot of champagne: light, capable, and happy.
With his incessant restlessness, he gave the impression of a raving
maniac. For this reason it is so difficult for any of us to imagine
him dead; not for a minute had anyone ever seen him quiet.

He was always concentrated, always preoccupied with the same
subject, and as a result was so one-sided that he lacked not only
interest for any science outside a certain field of medicine but also
the ability to enjoy human and natural things. For fourteen years
he hardly ever left the hospital, whirled like a fast-moving auto-
maton out of the building and into the restaurant, into the coffeehouse, and back. His recreation consisted of playing cards and
chess, at which he was a master, and in spite of the agitation it
produced in him and which sometimes caused him to be exceedingly ruthless, it was a pleasure comparable to a theater perform-

[3] The Austrian poet Anton Alexander Graf Auersperg (1806-1876), who wrote
under the pseudonym Anastasius Grün.

ance to watch him at play and to listen to his original, biting wit. Even when he was well able to afford it, he could not be induced to travel and see something of the world. On his return from his honeymoon he said to me: "I'm not one of those people who can stare for hours into a lake and enthuse about it."—He avoided any kind of social life that might require him to make an effort, he didn't look at anything and didn't know what was going on in the world. As a result he was completely lacking in manners, and cynical, and when you and Minna saw him and found him odd enough, it was when he happened to be on his most civilized and decent behavior. Once, while still a student, he fell in love with a girl who didn't like him and who took another man who had everything he lacked. Since then no affection has softened his nature.

He paid for his success partly at the cost of his reputation, and had few friends, although for quite some time people had given up passing judgment on him and had accepted him as a phenomenon not subject to the usual laws. He was incapable of friendship and could talk to a man for years without once asking what he did, but he was very communicative and whoever he happened to see most often was the person in whom he confided most. His life seemed to be an open book; only after his death did we discover that he concealed a lot. For me he had more friendship based on respect than for many others, and he had grown fond of me. He talked of being permanently at my disposal, and in the event of his death of making me his heir. All this took place at a time when his ambition, mitigated by his inborn good nature, was directed at fairly noble aims. As soon as he didn't need them any more he abandoned his mean ways, his real achievements lessened the appearance of his presumptuousness, the recognition given him for his abilities made the buying of it superfluous. Then came the moment when he wanted to appear as a noble, unselfish human being, to achieve for his character what he had achieved for his ability. Hence his generosity towards me, hence the long list of good intentions which drove him to his death. Influenced perhaps by the happiness of lovers around him, he tried to create the same thing for himself, tried and tried and allowed himself no time to let it come his way. Whenever a colleague produced a fiancée, he would inquire about the latter's sister, but always came too late. He got himself introduced to the houses of a few wealthy people, but whether he didn't play there the role he aspired to, or whether by chance he just didn't find anyone, he declared in any case that he was going to marry a poor girl. He wanted to make a girl happy, and impress the world. On his list he had three possibilities, Helene F., the young Hammerschlag girl and—our Rosa, whom he may have seen once. (I learned this only yesterday.) He started on the conquest of the first probably because he did not look askance at taking in a little wealth on the side. I well remember the day, three years ago, when he said to me: "A woman came for a consultation today with her two daughters. Charming people; if I had the money and weren't ill [at that time he thought he was], I'd marry the elder girl at once." It was she who later became his wife, although he did not see her for some time. Finally he introduced himself to the family and began to court the girl. The family accepted him from the beginning, but for a long time the girl resisted him. He seems to have encountered a real Brünnhilde, a reserved, not very yielding, extremely demanding creature. She was considered intelligent and sensible. I saw two letters from her which gave me the impression of sound, sober respectability, but little feminine refinement in handwriting or expression. She was twenty-six, had turned down many good offers of marriage and didn't seem to feel any need for love. He now courted her ardently, and met with nothing but criticism and rebuff. She told him he was arrogant, mannerless, had a thousand faults which he would have to get rid of; he listened to every word, promised to improve, became gentle, refrained from using abusive language, so that one could actually introduce him to girls. At last she began to give in, thought she loved him, perhaps she had actually begun to feel some affection for him; she couldn't be certain. After all, no girl in love for the first time knows whether or not it is the real thing. He filled the world with the news of his happiness; when asked about the dowry he always answered that he was not worrying about that. Gradually, however, his mood changed, he grew depressed, and finally confidential. There had been differences of opinion between them—he omitted to mention the reasons—and now the girl was melancholic, she wept, wouldn't speak, took no pleasure in his company. He also divulged that all the sisters were hysterical. I tried to comfort him, told him the girl was evidently sensitive and conscientious, and realized that her affection for him was not strong enough in view of the imminence of the wedding. It couldn't be otherwise, I said, after so short an acquaintance; he ought to give her time and not press her too much. But now he was determined to win her, he wooed her more and more ardently, spent about a thousand gulden on presents for her, contributed another huge sum toward her trousseau, converted all his savings into cash so as to furnish their apartment magnificently and made it well-nigh impossible for her to refuse him—and things went from bad to worse.

When he told me that she had asked him to marry her sister instead, and that she was temporarily relieved when the wedding was postponed, I was convinced she did not like him. And I told Breuer about it. He said that if a girl entered marriage in this spirit the greatest misfortune could occur, and that such conditions usually end by someone in the family turning up and declaring: "I will not allow you to marry."—This person, however, did not turn up, the entire family urged the poor girl on. She was sent off on a short trip, but returned unchanged. Now I implored him to accept the fact that she did not love him and to take a trip himself, that on his return he would feel more detached, that the girl would have clarified her mind, and that then he could come to some definite decision. But he just could not bear the thought that a girl could refuse him, and he sacrificed everything recklessly with the single object of not having to face the world as a failure. Her family then brought so much stupid pressure upon the girl that she, who could not find the courage for a definite refusal, renounced the postponement. Five days after he had promised me to go on a trip, the wedding took place. She is supposed to have said: "It's a question of now or never." It is not difficult to guess why she had hesitated. I think he dropped his self-restraint too early, and physical aversion and moral disapproval quickly stifled all affection in the still cool and prudish girl. He on the other hand had believed that he could force love as he had forced all his other successes, and a false shame prevented him from letting the world know that he had been rejected.

I saw him once after the honeymoon; he was not alone and couldn't speak freely. Paneth saw him as late as the twelfth, and when asked about his marriage, he said he had known better things, accused himself of being a wretched failure,* but again someone was present who prevented any further confidences. He wasn't seen anywhere; on the other hand no one wanted to disturb the young

---

* *Pfründner.—Tr.*

couple; all one knew was that her family were continuously in and out of the house. On the thirteenth he hanged himself. What drove him to it?

As an explanation the world is ready to hurl the most ghastly accusations at the unfortunate widow. I don't believe in them. I believe that the realization of an enormous failure, the rage caused by rejected passion, the fury at having sacrificed his whole scientific career, his entire fortune, for a domestic disaster, perhaps also the annoyance at having been done out of the promised dowry, as well as the inability to face the world and confess it all—I believe that all this, following a number of scenes which opened his eyes to his situation, may have brought the madly vain man (who in any case was given to serious emotional upheavals) to the brink of despair. He died from the sum total of his qualities, his pathological self-love coupled with the claims he made for the higher things of life.

Over his corpse began the feud of the families and on his still-open grave there sounded a loud discordant scream for revenge, as unfair and as reckless as if he had uttered it himself. The lecturer Friedmann, a relation and colleague of his old father, began: "Thy name was Noah, and thy parents associate with it the dictum: 'Thou shalt be the comfort and the support of our old age.' And all this comfort now lies here. And it is written: 'If a corpse be found, and one does not know by whose hand he died, then one must turn to the next of kin; they are the murderers.' But we, his parents and brothers, have not shed his blood—"And now in clear words he began to accuse the other family of having dealt the fatal blow. And all this he spoke with the powerful voice of the fanatic, with the ardor of the savage, merciless Jew.

We were all petrified with horror and shame in the presence of the Christians who were among us. It seemed as though we had given them reason to believe that we worship the God of Revenge, not the God of Love. Pfungen's thin voice was lost in the reverberation of the wild accusation of the Jew.

Both his widow and his father have issued special mourning announcements. The papers print two interpretations, both false, one from her family, the other from his. I am afraid there are some ugly revelations still in store for us.

Thus his death was like his life, cut to a pattern: he all but screams for the novelist to preserve him for human memory.

Well, lucky the man who is tied to life by a sweet darling. I just cannot write any more today, Marty.

With fondest love
Your
Sigmund

23 *To* MARTHA BERNAYS

Vienna, Saturday
October 6, 1883

My dearest treasure

Now you are going to laugh at me: for the first time in ages I don't know what to write to you. I am so engrossed in the reading of papers, medical papers of course, and the trying out of methods, that I felt inclined to begin this letter: "Today's mail due from Wandsbek has not arrived." And besides, I love you so immensely since receiving your last letter that I cannot think of anything to say but: it is a shame that I am sitting here alone and that you are not with me.

Today was a quiet working day; I had to skip the morning rounds because I was at the *Journal*[1] till 9 o'clock. Then I feverishly tried out some new methods till I had discarded my last idea; now I have one left for tomorrow which I shall certainly dispose of, too. Tomorrow is Sunday; fortunately I will be on duty;[2] what else can one do on a Sunday when the only person in whose presence one would like to rest is inaccessible? Patience. If I remain as fit and healthy as I am at the moment, something good must come my way. But you are quite, quite right, never again will I interrupt work in your absence.

And you, what are you doing, what do you look like, how, what, I want to know everything about you, and best of all I would like to hear everything all over again every day. My den is getting quite cozy. It's a pity I cannot name a spot where you have sat. But even if there were such a place, it would be covered with periodicals. I am reading my way into medicine. My first contri-

---

[1] Administrative duty in the hospital reception office.
[2] Medical duty in the hospital.

butions were printed today, unsigned of course. The deeper I penetrate into medicine, the more difficult writing for publication becomes. Not because I have to fulfill demands that are greater than they used to be. No, because most publications require so much self-denial. If authors had more self-criticism, nine-tenths of them would not be authors. I have to read a great deal of indifferent and even more inaccurate stuff, and cannot of course write like this myself. In medicine a greater part of one's intelligence is spent discarding useless things; however, this is an inconspicuous way of being intelligent. Anyhow I hope that a deeper absorption in the subject matter will give me the desire and ability to produce something useful.

Marty, does it annoy you to hear me talk about such things? Oh, you won't be annoyed, you are so good and—between ourselves— you write so intelligently and to the point that I am just a little afraid of you. I think it all goes to show once more how quickly women outdistance men. Well, I am not going to lose anything by it.

> Farewell, my girl.
> With many affectionate greetings from
> Your
> Sigmund

24 *To* MARTHA BERNAYS

Vienna, Tuesday
October 9, 1883

My beloved Marty

What I am doing now? I am more industrious than ever and feel better than ever. Most of the time I work my way through a mountain of papers, reading partly for myself, partly for the *Medical Weekly;* I sit in the laboratory, where my Method is actually working and looks very fine, although several things still need correcting, and from early in the morning till eleven o'clock (I had almost forgotten to tell you) I function in the wards as a *Sekundararzt,* busy learning, writing, and occasionally acting as surgeon. The whole situation, my darling, has something heavy about it, akin to a dream or delirium; these are the right conditions to help one survive a long separation; whether they are pleasant it is hard to say; personal feelings don't get much chance of a hearing. Continually having so much to do acts as a kind of narcotic, but as you know I have lately been looking for something to rescue me from my great emotional and excitable state. Now I have it. It seems as though the waves of the great world do not lap against my door; at other times I have to fight against the sensation of being a monk in his cell, as described by Scheffel.[1] Strange creatures are billeted in my brain. Cases, theories, diagnostics, formulas have moved into brain accommodations most of which have been standing empty, the whole of medicine is becoming familiar and fluid to me, here bacteria live, sometimes turning green, sometimes blue, there come the remedies for cholera, all of which make good reading but are probably useless. Loudest of all is the cry: tuberculosis! Is it contagious? Is it acquired? Where does it come from? Is Master Koch[2] of Berlin right in saying that he has discovered the bacillus responsible for it?

When a letter from you arrives the whole dream fades, life enters my cell. Then all the strange problems creep away, the mysterious pictures of diseases fade, and gone are the empty theories "according to the present status of science," as they are invariably called.

Then the world turns so warm, so gay, so easy to understand. My sweet darling is no illusion, she does not have to be proved by chemical tests; in fact she can, although no giant, be seen by the naked eye. Fortunately she has nothing to do with diseases—I hope she is very well—except that she was incautious enough to take a doctor for a lover. Oh Marty, it is so much more lovely to be a human being than a warehouse for certain monotonous experiences. But one is not allowed to be a human being for an hour unless one has been a machine or a warehouse for eleven hours. And there we are, back where we began.

I hope to hear from you tomorrow, my precious girl. Farewell, try to be a little gay.

Your devoted
Sigmund

[1] Allusion to the historical novel *Ekkehard*, by the German author Josef Victor von Scheffel (1826-1886).

[2] Professor Dr. Robert Koch (1843-1910), who first isolated the tubercle bacillus in 1881.

25 *To* MARTHA BERNAYS

Vienna, Tuesday, 7 P.M.
October 15, 1883

My beloved Marty

Your sweet letter of congratulation found me at the *Journal* speculating on how to pin down and improve the Method. Probably the whole of next week will be spent on further experiments after careful preparation. Today I am going to show Breuer the specimens during consulting hours. I have no doubt that I have got the thing although just recently it has been behaving rather capriciously. However, behind such whims of nature there usually lurks a chance of learning more. I am thinking of changing my working hours by taking a course with Ultzmann[1] from eleven to twelve instead of the one with Urbantschitsch[2] from four to five, thus leaving the afternoon free for my own work. Also for the pupil, if only she would come, for I already see with horror the day approaching when etc., etc.

Today is the sixteenth monthly memorial of our engagement and an especially affectionate greeting is due to the sweet girl whose letters have continually grown better and cleverer and nobler, although she herself has always possessed these qualities. Darling Martha, until now we have been in the ascent, haven't we? and have more and more reason to be pleased with each other. Which is why we can look forward confidently to the following month; it was only a month ago that I had to complain, but today this seems almost forgotten.

What my work, for which you so sweetly wish me luck, consists of, I cannot tell you without a terribly long explanation. One thing, however, I can divulge: it has to do with a method for the chemical treatment of the brain. Clear? No.

Well, "as is well known," the brain must first of all be hardened (in spirits, for instance) and then finely sliced in order to show

---

[1] Dr. Robert Ultzmann (1842-1889), Professor of Urology at the University of Vienna.

[2] Dr. Viktor Urbantschitsch (1847-1921), Professor of Otology at the University of Vienna.

where the fibers and cells lie in relation to one another, where the fibers lead to, etc. The fibers are the leading ducts of the different parts of the body, the cells are in control of them, so respect is due to these creatures. Now, on the sliced segments of the hardened brain very little is visible, but more appears when they are colored with carmine, because then the cells and fibers grow redder than the other less important parts. Even so, it is still very difficult to see all the fine fibers or even to get very clear pictures. It is well known that silvering and gilding produce beautiful pictures on other specimens—that is, quite different coloring for the different elements; now this is also being tried out on the brain. I believe that so far I have succeeded best. These are technical tricks which exist in every craft, but which science cannot do without. Is my darling princess satisfied now?

Such dirty notepaper and such an enormous envelope, she will think, but this is life at the *Journal.* Fortunately it is just 9 A.M. About to be relieved.

Farewell, princess, and I hope I shall often be able to give you good news.

Your
Sigmund

26 *To* MARTHA BERNAYS

At the *Journal,* Tuesday
October 23, 1883

My beloved Marty

I dare to say "my beloved" although I do occasionally have bad thoughts and write so angrily. If I have offended you again, please put it down on the list with the others and think of my longing, my loneliness, my impatient struggle and the shackles that are imposed upon me. Now and again I have something like attacks of despondency and faintheartedness which you, my dear and kind one, must not share. At these times you must laugh at me and remember how quickly I regain my elasticity and my unclouded judgment. This afternoon, girl, I once more had good results, found a new gold method which promises to be more lasting than the previous one, but if this also turns out to be capricious, I can nevertheless foresee the final result: I shall discover completely or almost completely what I am looking for.

These difficult times will not discourage me so long as we remain healthy and are spared exceptional misfortune. Then we are certain to achieve what we are striving for—a little home into which sorrow may find its way, but never privation, a being-together throughout all the vicissitudes of life, a quiet contentment that will prevent us from ever having to ask what is the point of living. I know after all how sweet you are, how you can turn a house into a paradise, how you will share in my interests, how gay yet painstaking you will be. I will let you rule the house as much as you wish, and you will reward me with your sweet love and by rising above all those weaknesses for which women are so often despised. As far as my activities allow we shall read together what we want to learn, and I will initiate you into things which could not interest a girl so long as she is unfamiliar with her future companion and his occupation. All that has happened and is happening will, by the interest you take in it, become an added interest for me. You will not judge me according to the success I do or do not achieve, but according to my intentions and my honesty; you will not regret having sacrificed the beautiful years of your youth to fidelity, and I shall be proud of you. You will be able to read me like an open book, it will make us so happy to understand and support each other. You will prevent me from doing anything petty, from anger, envy, and the desire for unimportant things, and if you worry about having interfered with my scientific career I will laugh and tell you the story of Benedikt Stilling,[1] a doctor who died a few years ago in Cassel, practiced science in his youth and was then compelled to take a job as a doctor. But for thirteen years he worked every morning on the human spinal cord, the result of which was a great work, and every evening he continued to work on the brain, and he is known as the foremost among the scientists to whom we owe the knowledge of this noble organ. All this shows the industry, the tenacious enthusiasm of the Jew, not even coupled with the talent normally expected from Jews. This we can also do.

My beloved Martha, part of what you will be to me you are already. But I expect you to become more and more. Others keep

[1] Dr. Benedikt Stilling (1810-1879), German anatomist and surgeon, founder of the theory of the vasomotor nervous system.

71

going only under fortunate circumstances; we, Marty and I, will do the same, although separated and not at all fortunate.

Goodnight, my dearest little woman, keep pouring out your heart to me; I feel so sad when you haven't done so for some time.

Your
Sigmund

27 *To* MARTHA BERNAYS

Vienna, Tuesday evening
October 25, 1883

My darling girl

Yes, it is true, we have made a discovery which may not be insignificant, and you must forgive me if I talk a lot about it today. Yesterday in my joy I went to see Breuer as late as 9:30, and on my way there I thought up all kinds of compliments for his wife, so that she should not be bored by our conversation. For instance, "Not only women are beautiful, chemical preparations can be, too!" and a second one which I actually managed to make use of, as you will learn. Finding no one at home, I settled down in the consulting room and picked up the nearest book—I looked in vain for the cigar box. (I have been granted these rights once and for all.)

The book I picked up I liked so much I decided to send it to my Martha. Poor sweet princess, it is already ordered for you, but just now when I would so much like to give it to you as a celebration of our success, I am quite poor. As things never work out as one expects them to, Frau Mathilde[1] came home first and informed me that I wouldn't be able to see her husband until eleven o'clock, that he actually was downstairs on his way to fetch the children, who had gone to the Karl Theater, where the Meininger[2] are now giving a special performance. She hoped I wasn't annoyed. I was not in the least annoyed. I just mentioned that I had had such a pleasant day I would like to end it in the best possible company (my second compliment, which actually earned me a handshake), and then ran downstairs, where I met Breuer. "Let's go for a walk," he said. We went arm in arm toward the Karl Theater and when I

---

[1] Wife of Dr. Josef Breuer—see List of Addressees.
[2] Famous group of actors from the Court Theater at Meiningen.

broke my news and talked of it for a long time and finally asked his forgiveness for holding forth on a subject which might not interest him, he was good enough to say: "Few things interest me more."

This afternoon at three I went to see Fleischl, whom I found again in a miserable state; I showed him the preparations one after the other: first the silver and then the gold made with the unreliable method, then the new ones. While I was in the midst of the first gold preparation, Brücke arrived. "Anything to be seen?" "Yes, brain gildings." "Ah, that's very interesting, especially since gold has the reputation of not being much use for this." "But this is a new method, Herr Hofrat." "I see, your methods alone will make you famous yet." And with that he went off. Fleischl was quite beside himself with delight. Sanguine as he is, he congratulated me over and over again and advised me to concentrate for the next seven years on the exploitation of this discovery. I laughed out loud and told him I would starve long before then. "You won't starve," he said. Then he confided in me that he too had a discovery up his sleeve: to build a new kind of accumulator for electricity; if it came off he would make a lot of money and give me enough to allow me to concentrate on this work without any worries. Needless to say, this suggestion cannot be taken seriously, but it is rather significant of the warmth of our relationship nowadays. I thanked him accordingly and asked him, in the event of his discovery succeeding, to give me just enough to travel to Hamburg in the summer. This was granted. Then I asked him if he would use the same method for an examination, say of the retina, that fine sensitive little skin at the back of the eye, which is actually part of the brain, and to my great joy he promised to start on it as soon as the exhibition had closed. To my joy because to teach an old teacher something is a pure, unmitigated satisfaction.

Then I went to Breuer, whom I found rather cantankerous after his luncheon: his microscope was not quite in order. As a result I wasn't able to show him everything, but what he did see drew from him quite a number of admiring comments. Then he said: "Now that you have the weapon, I wish you a happy war." No doubt it will mean a great deal of work before the first paper can appear, of which my little woman will receive an offprint. The great question is: Will this method also be suitable for tracing the fine nerve fibers in the tissues, in the skin, in the glands, etc.? If this is the case, then indeed a new prospect will be opened. My material situation would no doubt also profit from it; the years of waiting for my darling would be shortened. Should it not come off so well, it will still be of great use for the central nervous system. I am sure it won't fail; I gave it another test today. What I am afraid of is that I may succeed with new methods, which would entail so much work that my head would spin with the commotion.

Apart from its practical importance, this discovery has an emotional significance for me as well. I have succeeded in doing something I have been trying to do over and over again for many years. When I survey the time since I first began to tackle this problem, I realize that my life has progressed. I have longed so often for a sweet girl who might be everything to me, and now I have her. The same men whom I have admired from afar as inaccessible, I now meet on equal terms and they show me their friendship. I have remained in good health and done nothing dishonorable; even though I have remained poor, those things which mean something to me have become available, and I feel safe from the worst fate, that of loneliness. Thus if I work I may hope to acquire some of the things that are still missing and to have my Marty, now so far away and lonely as her letter shows, close by me, have her all to myself, and in her tender embrace look forward to the further development of our life.

You have shared my sadness; now today share with me my joy, beloved, and don't get the idea that there is ever anything in the center of my thoughts but you.

With affectionate greetings and kisses
Your
Sigmund

[*On the back of this letter's envelope, in English*]: HOPE AND JOY.

28 *To* MARTHA BERNAYS

Vienna, Thursday, 5 P.M.
November 15, 1883

My sweet princess
        This is to be your name from now on. I have been thinking of you more than usual during these past days, and just want to

remind you—by wishing myself luck and success for the return of
the date which gave you me—of the special coincidence that this is
the seventeenth monthly memorial and that the seventeenth is again
a Saturday. But I won't have to renew my courtship, will I? Today
is a holiday[1] and I have done no work whatsoever, in order to
refresh myself. The weather is quite horrible; this evening I think
I will go and see Hammerschlag.[2] I am so weary that it will do me
good if someone is friendly to me. What's more, they will ask after
you and I will have a chance to talk about you.

What you said in your last letter about Mill[3] and his wife should
have inspired me on the spot to tell you something about them
both. The essay by Brandes[4] gives only a personal impression of
the man, it is far from being an evaluation of his whole position in
our contemporary history. I got the idea of reading him when
Gomperz[5] entrusted to me the translation of his last work. At the
time I found fault with his lifeless style and the fact that in his work
one could never find a sentence or a phrase that would remain in
one's memory. But later on I read a philosophical work of his which
was witty, epigrammatically apt, and lively. Very possibly he was
the man of the century most capable of freeing himself from the
domination of the usual prejudices. As a result—and this always
goes hand in hand—he lacked the sense of the absurd, on several
points, for instance in the emancipation of women and the question
of women altogether. I remember that a main argument in the
pamphlet I translated was that the married woman can earn as
much as the husband. I dare say we agree that housekeeping and
the care and education of children claim the whole person and
practically rule out any profession; even if simplified conditions
relieve the woman of housekeeping, dusting, cleaning, cooking, etc.
All this he simply forgot, just as he omitted all relations connected
with sex. This is altogether a topic on which one does not find
Mill quite human. His autobiography is so prudish or so un-
earthy that one would never learn from it that humanity is divided

[1] Name day of Leopold, Patron Saint of Lower Austria.

[2] Samuel Hammerschlag (?-1904), Freud's teacher of religion at school as well
as his fatherly friend.

[3] John Stuart Mill (1806-1873), English philosopher and economist.

[4] Georg Brandes (Georg M. Cohen, 1842-1927), Danish literary historian and
critic.

[5] *Hofrat* Theodor Gomperz (1832-1912), Professor of Classics at the University
of Vienna (see letter of Nov. 12, 1913).

between men and women, and that this difference is the most important one. His relationship to his own wife strikes one as inhuman, too. He marries her late in life, has no children from her, the question of love as we know it is never mentioned. Whether she was the wonderful person he revered is generally doubted. In all his writings it never appears that the woman is different from the man, which is not to say she is something less, if anything the opposite. For example he finds an analogy for the oppression of women in that of the Negro. Any girl, even without a vote and legal rights, whose hand is kissed by a man willing to risk his all for her love, could have put him right on this.

It seems a completely unrealistic notion to send women into the struggle for existence in the same way as men. Am I to think of my delicate, sweet girl as a competitor? After all, the encounter could only end by my telling her, as I did seventeen months ago, that I love her, and that I will make every effort to get her out of the competitive role into the quiet, undisturbed activity of my home. It is possible that a different education could suppress all women's delicate qualities—which are so much in need of protection and yet so powerful—with the result that they could earn their living like men. It is also possible that in this case it would not be justifiable to deplore the disappearance of the most lovely thing the world has to offer us: our ideal of womanhood. But I believe that all reforming activity, legislation and education, will founder on the fact that long before the age at which a profession can be established in our society, nature will have appointed woman by her beauty, charm, and goodness, to do something else.

No, in this respect I adhere to the old ways, to my longing for my Martha as she is, and she herself will not want it different; legislation and custom have to grant to women many rights kept from them, but the position of woman cannot be other than what it is: to be an adored sweetheart in youth, and a beloved wife in maturity.

There is so much more to say on this subject, but I think we see eye to eye anyway.

Farewell, my sweet girl. Your letter won't come today, so I will go out.

Affectionate greetings and kisses
from your
Sigmund

29 *To* MARTHA BERNAYS

Hotel Stadt Freiberg, Leipzig, 7:30 P.M.
December 16, 1883

My sweet princess

Isn't this absurd? I have just been laughing out loud. Isn't it amazing that I am suddenly writing to you under a printed letterhead and smoking a ten-pfennig cigar tasting of straw? And that I have only six kreutzer in my pocket! Instead of the usual Austrian money I have silver and gold marks whose value impresses me so little that I would be tempted to spend the whole lot today if it weren't so dark that I daren't go out. I am going to write to you now with as much leisure and peace as I wrote last in a hurry, for it is only 7:45 and I have four hours to myself before meeting my brother;[1] by then I should normally feel sleepy, but who could sleep under such strange circumstances? Marty, you will already have observed that at heart I am still a child; I can be so happy simply because I am in another place, have different money in my pocket, because bread here is free, and because my windows look out on Halle'sche Strasse instead of on Courtyard III.[2] Laugh if you like, I intend to remain like this as long as possible and then one day we shall laugh together under similar conditions. On the other hand, there is one thing I must try not to think about, and that is our almost certainly thwarted reunion, otherwise—but you were right in your letter, my darling. I will miss your letters of tomorrow and the day after, very much.

I must say I do find traveling third class at night pretty unpleasant. The human being, whose astronomical geography is in his blood and nerves, begins to feel sleepy at a certain hour and is dissatisfied when he finds nothing prepared for it. The morning was cold and gray and drizzly, fit only for ducks. Not till we reached Saxony did the sky begin to clear and it turned quite nice in the afternoon. I did not pay much attention to Saxon Switzerland because I know it so well from my wanderings on foot; I preferred

[1] Freud was in Leipzig to meet his half brother Emanuel (1834-1915), who lived in England.
[2] In the General Hospital in Vienna where Freud resided.

to read the book by Dr. Luther. But I took a good look at the Elbe
and if you arrive in Hamburg in time its waters will whisper many
thousand greetings. Between Dresden and Riesa I had my first
great adventure, unpleasant at the time, pleasant in retrospect.
You know how I am always longing for fresh air and always anxious
to open windows, above all in trains. So I opened a window now
and stuck my head out to get a breath of air. Whereupon there
were shouts to shut it (it was the windy side), especially from one
particular man. I declared my willingness to close the window
provided another, opposite, were opened, it was the only open
window in the whole long carriage. While the discussion ensued
and the man said he was prepared to open the ventilation slit instead of the window, there came a shout from the background:
"He's a dirty Jew!"—And with this the whole situation took on a
different color. My first opponent also turned anti-Semitic and
declared: "We Christians consider other people, you'd better think
less of your precious self," etc.; and muttering abuses befitting his
education, my second opponent announced that he was going to
climb over the seats to show me, etc. Even a year ago I would have
been speechless with agitation, but now I am different; I was not
in the least frightened of that mob, asked the one to keep to himself
his empty phrases which inspired no respect in me, and the other
to step up and take what was coming to him. I was quite prepared
to kill him, but he did not step up; I was glad I refrained from
joining in the abuse, something one must always leave to the others.
With the compromise of ventilation-slit versus window, Act I
came to an end. The conductor summoned by me took neither side
but offered to escort me to another compartment, which I declined.
Later, when several people opened the window in order to get out,
and left it open, I settled down boldly beside it, for I felt very
ready for a fight. The anti-Semite, this time with ironic politeness,
renewed his request. No, said I, I'd do nothing of the kind, told him
to turn to the conductor, and I held my own as far as the next station. There the conductor again refused to say anything, but another
official, who happened to have heard of the issue but not of the
scene, decided that in winter all windows had to be closed. Whereupon I closed it. After this defeat I seemed to be lost—a storm of
jeers, abuses, and threats broke out, until I turned round and yelled
at the ringleader to come on over and make my acquaintance. I was
not at all sure of the outcome. The answer was that no one was

talking about me at all, they had no intention of having their conversation interrupted, but—and from then on everything was quiet.
I do think I held my own quite well, and used the means at my
disposal courageously; in any case I didn't fall to their level. After
all, I am no giant, haven't any hackles to show, no lion's teeth to
flash, no stentorian roar, my appearance is not even distinguished;
all this would have had a lightning effect on that mob, but they
must have noticed that I wasn't afraid and I didn't allow this experience to dampen my spirits. So much time and space has been
spent on this silly story. Now I must order another sheet of paper.

The journey here from Dresden is endless; it was 5:30 when I
arrived and already dark. I hired a porter as a guide first to the
post office to see whether Emanuel might possibly have changed his
traveling plans; when this was not the case I asked to be taken to
a hotel near the Magdeburg station where I had ordered a room
with two beds for him and myself and another room for Mr. Robin-
son.[3] By a strange coincidence it is called the Hotel Stadt Freiberg,
the town where Emanuel and I met for the first time, where I was
born. (Needless to say, it is not called after the same Freiberg.)
There I made myself look a little more human and hurried to the
mirror to see what I actually looked like. My self-confidence had
been somewhat increased by the battle with the infidels, but sank
again when I saw myself in the mirror. No, I don't look at all
noble; neither the blackest coat nor the whitest shirt could conceal
my obvious plebeianism. But an elegant princess loves me nevertheless, and when I have money, which is as good as certain (my
self-assurance tells me so) then I shall dress her in the most beautiful clothes and it will never occur to a soul that she could have
married anyone but a prince. Then I made my way slowly toward
the Magdeburg station and bought myself some cigars; by then,
however, it was time to satisfy a gigantic appetite with a meal
which didn't bear much resemblance to my Viennese suppers. There
hadn't been time for a proper luncheon. In the Stadt Freiberg
restaurant I sat among the Leipzig Philistines, listening to their
talk and watching their faces. They talk just as much rot as the
people at home, but they look more human; I don't see so many
grotesque and animal-like faces, so many deformed skulls and
potato noses here. On the contrary, if I were in Vienna I would

[3] Emanuel's nickname for his brother Philipp. See letter of December 20, 1883.
Note 2.

think I was in the company of men of letters, professors, and architects. However, not much seems to lie behind these finer, sharper features. But I heartily dislike the Saxons' way of talking about things, they are continually discussing topics which we would never seriously talk about except at rare moments. Those ethical truths with which we are all pretty familiar these fellows pronounce like so many maxims lifted out of an anthology; they spread ideals, as it were, like butter on bread, and yet they are certainly no more deeply influenced by them than our Philistines are.

The room from which I write, beloved, is quite nice, high up, but has only single windows, and judging by the way I feel, my poor Emanuel will be pretty cold. At eleven o'clock however, they are to come and light the stove; this I have arranged. But what if he does not turn up today? In that case I shall have to meet every train tomorrow, and if he still doesn't come, I shall have to pay the whole bill, travel to Dresden and there borrow something from Hammerschlag's brother-in-law. Charming prospect! But that he won't come is out of the question; it is only my habit of preparing myself for every eventuality that makes me think of it. I feel so well and adventurous. But now comes the bitter side. I won't be able to see you, my treasure. I have the little ring for you in my pocket, and long to slip it on your finger myself. It also needs some explanation; because a simple plain gold ring—it turns out to be a little wider than I expected—looks rather like something which one day, to my delight, you will wear, but cannot wear yet. We will have to have it altered; our monogram put on it, for example, then covered with enamel. Let me know soon, Marty, if you want me to have this done in Vienna. For you must be able to wear the ring.

Tomorrow I will hardly have time to write more than a card, I have also got to pay a visit to the Altschuls[3] in Dresden. So I shall be leaving here tomorrow evening or early Tuesday. Oh, how disgusting it is that I cannot travel farther.

Farewell, my dearest beloved girl, and keep well and be a little gay and take courage and love me.

Your most devoted
Sigmund

It's 9:45, I am wandering about the town; hope I won't lose my way.

[3] Relatives of Hammerschlag.

30 *To* MARTHA BERNAYS

Vienna, Wednesday evening
December 20, 1883

My precious darling

In today's quiet I can at last tell you more about Dresden;
just the pleasantest of my impressions there have not been mentioned. Right next to the castle we discovered a wonderful cathedral, then a theater, and finally a spacious building—square, with
a large courtyard and in each corner a tower, built in the style of
our Belvedere[1]—which Philipp[2] seriously insisted must be the castle
because it was so beautiful. However it was the so-called Zwinger,
which houses all of Dresden's museums and art treasures. At last
we found the picture gallery where we spent about an hour, the
old boys chiefly to rest themselves, I to bring home with me a
few fleeting impressions of these famous works of art. I believe I
acquired there a lasting benefit, for until now I have always suspected it to be a silent understanding among people who don't have
much to do, to rave about pictures painted by the famous masters.
Here I rid myself of this barbaric notion and myself began to admire. There are magnificent things in the Zwinger; some I knew
from photographs and reproductions and could point out to the two
Englishmen, for instance, the painting by Van Dyck showing the
children of the unfortunate Charles I, the later Charles II, James II,
and a young, plump little princess. Then I saw the Veronese with
the most beautiful heads and bodies, madonnas, martyrs, etc.; I
hardly had time to glance at them all. In a small room by itself I
discovered what, according to the way it is displayed, must be a
gem. Looking closer, I saw it was Holbein's "Madonna." Do you
know this picture? Kneeling in front of the Madonna are several
ugly women and an unattractive little girl, to the left a man with a
monk's face, holding a boy. The Madonna holds a boy in her arms
and gazes down on the worshipers with such a holy expression. I
was annoyed by the ordinary ugly human faces, but learned later

---

[1] Baroque palace in Vienna built by Johann Lukas Hildebrand (1668-1745).

[2] Philipp Freud (1838-1912), the younger of Freud's half brothers, who **lived**
in England.

they were the family portraits of the Mayor of X, who had commissioned the painting. Even the sick, misshapen child whom the Madonna holds in her arms is not meant to be the Christ child, rather the wretched son of the Mayor, whom the picture was supposed to cure. The Madonna herself is not exactly beautiful—the eyes protrude, the nose is long and narrow—but she is a true queen of heaven such as the pious German mind dreams of. I began to understand something about this Madonna. Now, I happened to know that there was also a Madonna by Raphael there and I found her at last in an equally chapel-like room and a crowd of people in silent devotion in front of her. You are sure to know her, the Sistina. My thoughts as I sat there were: Oh, if only you were with me! The Madonna stands there surrounded by clouds made up of innumerable little angel heads, a spirited-looking child on her arm, St. Sixtus (or is it the Pope Sixtus?) looking up on one side, St. Barbara on the other gazing down on the two lovely little angels who are sitting low down on the edge of the picture. The painting emanates a magic beauty that is inescapable, and yet I have a serious objection to raise against the Madonna herself. Holbein's Madonna is neither a woman nor a girl, her exaltation and sacred humility silence any question concerning her specific designation. Raphael's Madonna, on the other hand, is a girl, say sixteen years old; she gazes out on the world with such a fresh and innocent expression, half against my will she suggested to me a charming, sympathetic nursemaid, not from the celestial world but from ours. My Viennese friends reject this opinion of mine as heresy and refer to a superb feature round the eyes making her a Madonna; this I must have missed during my brief inspection. But the picture that really captivated me was the "Maundy Money," by Titian, which of course I knew already but to which I had never paid any special attention. This head of Christ, my darling, is the only one that enables even people like ourselves to imagine that such a person did exist. Indeed, it seemed that I was compelled to believe in the eminence of this man because the figure is so convincingly presented. And nothing divine about it, just a noble human countenance, far from beautiful yet full of seriousness, intensity, profound thought, and deep inner passion; if these qualities do not exist in this picture, then there is no such thing as physiognomy. I would love to have gone away with it, but there were too many people about, English ladies making copies, English ladies sitting about and whispering, English ladies wandering about and gazing. So I went away with a full heart.

At three o'clock I accompanied the brothers to the Reichenberg train, drank a farewell glass of Rhine wine with Emanuel, had my luggage taken by porter to the Altstädter station and walked through the whole town and the crowds of shoppers as far as the Bismarckplatz—right beside the station from which I leave—where the Altschuls live. It took a little while before someone appeared; the room in which I sat brought back to me the peculiarities of the family. Then there arrived a delicate, dark girl, not beautiful, but sympathetic and gentle; this was Fräulein Emmy, for whom I had special greetings. A pleasantly animated conversation was interrupted first by the appearance of little Marie, an even more delicate creature with large gray eyes. Then came the father, a powerfully built man with a gray beard and easygoing manners, then his wife, and then the others. The wife, the third sister whom I now know (Frau Hammerschlag, Frau Schwab, and her), not much older than Frau H., looks quite worn out, did not say much and made a dignified impression. My Viennese friends have assured me that she is a most remarkable woman, and I am quite prepared to believe it. The husband strikes one as a powerful personality; he has lived many years in America, then in England. Almost all the children were born in England; at the moment two sons are in America and the others are thinking of going there, too. The father received me very warmly, insisted on my prolonging my stay (with which my purse did not agree), involved me in a conversation on scientific questions for which he showed a lively and considerable understanding. He is an ardent freethinker and has brought up all his children outside of the faith. Apart from Emmy, who gives English lessons, there was the younger girl, Clare, who arrived only yesterday from Berlin where she teaches singing, and two boys with the most delicate Jewish faces. I stayed till 11:30, the conversation was easygoing, the tone set largely by the stories of the plain and animated girl from Berlin. Someone was sent to fetch the eldest son, who is a junior counselor-at-law, but he couldn't be found. As it was striking eleven and the old man was about to take me next door to tell me about his illness, the young man arrived, accompanied me to the station, and stayed with me till one o'clock. We found we were of the same age and talked freely together. Every member of the family sent greetings to you. Needless to say, they knew all about you, the Viennese had sent them my dossier, and my engagement to a noble lady from the far north was no doubt strongly emphasized as one of my peculiarities.

Now, my darling Marty, how about this for a good gossip! I hope I will hear from you again tomorrow. Farewell and keep well for

Your
Sigmund

Most beautiful Marty, don't be cross, the little ring won't leave till tomorrow.

31 *To* MARTHA BERNAYS

Vienna, Monday
January 7, 1884

My beloved princess

I cannot initiate your innocent mind into the secrets of the hospital's administration as late in the day as this (by the time I have finished this letter the date will no longer be true). Let us have a chat instead, Marty! The daily report says I have been hard at work till 7 P.M., that I then let myself be inveigled into a game of taroc,[1] and then again worked a bit, and that I am not even tired. Today I put my case histories in order at last and started on the study of a nervous case; thus begins a new era! In the evening I meant to examine something through the ophthalmoscope which I had carried off from Meynert's clinic, but realize that I am out of practice, which saddens me. I must practice again. In the evening the newspaper boy brought me for the first time a few lovely offprints and books, which I shall review and then keep. Even a little piece of material for the Method arrived today. Otherwise the Method is idle, for as yet nothing has hardened; I have managed to find only a very small improvement. This is all there is to say about me, and the Princess can see that in my godforsaken boredom I am doing the one thing that is good and wholesome for me.

Now over to you, my darling. I am so glad that for quite a while now there has been no mention in our letters of any "mutual" in-

---

[1] Freud's favorite card game.

disposition, also that this time we have skipped our little monthly squabble which used to appear with such impressive regularity at the end of every first week, so that by the seventeenth[2] we both had a chance to forgive each other. Let us hope we have now grown out of this peculiarity. Your suggestion, by the way, that I state things clearly instead of alluding to them, made me laugh. How would it be if we reversed the roles once more? Very amusing. But I am no longer so silly as to write you lengthy dissertations now (for which you don't even thank me), if we are not going to meet till July.

I must say I don't find the notion that every girl has a silent admirer who eventually marries her very convincing; I would be less surprised if some girls had from five to 36,000, and that the majority, about whom there is nothing much to admire, had none at all—but it is one of nature's more charitable institutions. Almost as wonderful is the institution that every man finds a girl who sees in him the perfection of manhood, whereas in reality he is a miserable devil living by the grace of God's patience. But what are we philosophizing about? Let us not make the world more complicated than it is already, and if one is in love oneself and has a lot to do, one ought to leave such clichés to unemployed novelists. That's what you would say, and you are quite right. As for me, I am glad you have only one suitor and not so many thousands, if only because out of so many there might be one who is better than I, and to slay 25,000 rivals (you note that I place you in the first category) would be difficult for me just now, as I am so busy.

Dearest little woman, today I celebrated my eighth successive day of duty[3]—i.e., I have been afflicted with a week's duty, but tomorrow I am going to walk ostentatiously out, and—exchange five marks. If you are absolutely broke, you must let me know. If I make a discovery during the next three months, you shall have the golden snake[4] which I promised you in Nothnagel's time. Farewell my sweet treasure.

Sigmund

[2] Freud and Martha had become engaged on a seventeenth.

[3] To make up for time lost while ill.

[4] The Aesculapian serpent, symbol of the medical profession. Freud had jokingly told Martha that girls engaged to Lecturers on Medicine had to wear golden bracelets to distinguish them from the fiancées of ordinary doctors, who wore silver ones.

32 *To* MARTHA BERNAYS

Vienna, at the *Journal*
Thursday, January 10, 1884

My precious darling

Comfort is the last thing I am entitled to, and it is not because you would have taken care of my comfort that I miss you. If you look at the little book we planned to keep as a chronicle of our engagement and which, as a result of insufficient participation by one of us, has not been kept up, you will find an entry to the effect that you promised to do all in your power not to leave me. I took this promise as a firm guarantee that you would stay with me. Whether you could have done something about it or not I cannot judge, whether you put up enough resistance when the project of moving arose, whether one person's[1] determination was more decisive for you than my wish—all this I shall not investigate, and I really would not know how to start the investigation. But you must not say that I shouldn't have let you go. How could I demand from you sacrifices that had advantages only for me? No, that wasn't possible, and perhaps the other solution wasn't possible, either. Now we are separated, my dear Marty, and we must not belittle my work, which alone can bring us together again. . . .

Yesterday I met Father in the street, still full of projects, still hoping. I took it upon myself to write to Emanuel and Phillipp urging them to help Father out of his present predicament. He doesn't want to do it himself since he considers himself badly treated. So I sat down last night and wrote Emanuel a very sharp letter. Sorry that I have to write about such sad things! Earlier I had been to see the Hammerschlags, where I was very warmly received. The old professor took me aside and charged me with a delicate mission concerning his young son Albert, who is a medical student; then he informed me that a rich man had given him a sum of money for a worthy person in need, that he mentioned my name and he was herewith handing it to me. I am describing the situation to you in all its crudity. The good professor himself, as he has often told me, has experienced great poverty in his own youth and sees nothing

[1] Martha's mother.

shameful in accepting support from the rich. Nor actually do I,
but I intend to compensate for it by being charitable myself when
I can afford it. It is not the first time the old man has helped me in
this way; during my university years he often, and unasked, helped
me out of a difficult situation. At first I felt very ashamed, but later,
when I saw that Breuer and he agreed in this respect, I accepted the
idea of being indebted to good men and those of our faith without
the feeling of personal obligation. Thus I was suddenly in the possession of fifty florins and did not conceal from Hammerschlag my
intention of spending it on my family. He was very much against
this idea, saying that I worked very hard myself and could not at
the moment afford to help other people, but I did make it clear to
him that I must spend at least half the money in this way. Then
the conversation turned on conditions at home, and I felt no com-
punctions about giving him some idea of the circumstances and
pointing out that the girls were in need of earning a little money.
I then asked his permission to bring Rosa[2] up to see them, and after
joining the others he started talking about my sisters in a way that
made me realize that he had arranged beforehand with his wife
to question me about them. I don't know any people kinder, more
humane, further removed from any ignoble motives than they. I
hope Rosa will make a friend of Anna Hammerschlag,[3] an admirable
girl; perhaps they will recommend her and she will find it
easier than I to discuss with Frau H. what can be done about
the two other girls. You mustn't forget that the Hammerschlags
themselves are very poor, have nothing but his pension and what
the eldest children earn, the son as a tutor and the girl as a school
teacher. The other one, Albert, the medical student, has a large
stipend and is an assistant of Ludwig,[4] the professor of chemistry.
I have always felt more at home with this family than with the
wealthy Schwabs, quite apart from the deep-seated sympathy which
has existed between myself and the dear old Jewish teacher ever
since my school days. Well, now you know everything and I am
wondering if you will thank me for my frankness. Knowing you,
my one and only girl, I think you will. . . .

[2] Freud's sister Rosa (1860-1942), later married to Dr. Heinrich Graf (see letter
of Sept. 4, 1883).

[3] Only daughter of the Hammerschlags, later married to Mr. Rudolf Lichtheim of
Breslau.

[4] Dr. Ernst Ludwig (1842-1915), Professor of Chemistry at the University of
Vienna.

I am not entirely dissatsified with the progress of my work. I work steadily, teach myself and jot down the interesting observations, and then read them up; in any case I am learning a lot, not least about myself. But in time I also hope to be able to find some material for publication.

You are right, this is the department where poor Nathan[5] was *Sekundarius;* I am even going to move into the room where he lived for eighteen months and which at night is probably haunted by his ghost. But I sleep well and it will not worry me.

I hope the reading club will be a great success; I would love to surprise the ladies at it. Will gentlemen be invited in time? Surely not all the Wolfingen girls[6] are engaged?

"The Chimes"[7] is charming, movingly beautiful, quite difficult in the beginning. "The Battle of Life"[7] would be easier and more suitable for all of you. But you know this already in part.

Today I am both at the *Journal* and on duty, which is not so much an accumulation as a saving, for one can leave the *Journal* now and again in order to drop in at the department, and this saves a day.

I will write more cheerfully tomorrow, my precious darling, but you must not keep from me the thoughts that cross your mind while reading this letter.

Goodnight.

Your
Sigmund

[5] Nathan Weiss (see letter of Sept. 16, 1883).
[6] Nickname for Martha's cousins called Wolff.
[7] Two short novels by Charles Dickens (1812-1870).

33 *To* MARTHA BERNAYS

Vienna, Wednesday
January 16, 1884

My precious darling

Yesterday was Breuer's birthday; I had hoped to see him at the Club,[1] but he did not turn up and I have just written him a few lines. I had a lazy day yesterday; my discovery evaporated in the

[1] The Physiological Club.

chemical laboratory, and I was rather annoyed about it. It is hard
to find material for publication, and it infuriates me to see how
everyone is making straight for the unexploited legacy of nervous
diseases.

At the *Journal*

My precious darling, I realize with horror that I didn't write to
you for the seventeenth; the reason is that I have been so caught
up in myself, and then I have days on end—they invariably follow
one another, it is like a recurring sickness—when my spirits decline
for no apparent reason and I tend to get exasperated at the slightest
provocation, even by the fact that my dear and good sweetheart
covers her pages of notepaper, which she sacrifices to me, with such
a sprawling hand. It is rather odd, and reminds me of your recent
remark that by the time we are sixty-eight, if we ever get that far,
we won't be able to boast of much resilience. Even when you use
endearments, I don't entirely like it. I think you are misjudging me
because we are so far apart and ascribe to me a measure of kindli-
ness, decency and I don't know what, which I have never possessed,
never will possess, and which you hardly could have found in me
when we were together. When we meet again you may be disappointed on finding that I look different from the lovely picture your
tender imagination has painted of me. I don't want you to love me
for qualities you assume in me, in fact not for any qualities; you
must love me as irrationally as other people love, just because I love
you, and you don't have to be ashamed of it.

I would rather you didn't make me out so good-natured; I can
hardly contain myself for silent savagery, and your latest letters are
so tame; if you weren't such a very sweet angel, I would love to have
a good squabble with you. It wasn't so bad, after all, to read something every month that came from the depths of passion. When
you are mine we must have a little quarrel at least once a week,
so that your love can always start fresh again. You are probably
laughing at all this nonsense, but I am having my bad days, am
working a lot but without that real enthusiasm which, according to
my calculation, will have returned by the time you read this letter.

I note the calming influence of your gentleness as I write; in fact
I already feel quite a bit calmer. But it annoys me that I always
seem to be talking about myself. I am not really as important to
myself as it must appear from my behavior.

As for your cousin, you really shouldn't find it difficult to understand her character. There is, after all, as you have told me yourself, quite a touch of hereditary mental weakness in the family, and what is evident in this girl is a plain simple-mindedness which in itself explains so many mysterious features in so many people.—By the time you are twenty-six, you too will have to consider all kinds of furnishings and acquisitions; forgive me for jumping from her to you, but this is always how my mind works; whenever I hear of a girl marrying I am first of all interested in her age, and then I work out how many years younger you are. Four with luck, four years of work and some success, that ought to do it. But how are you going to survive four years in the quiet of Wandsbek? I just can't stop worrying today. It is no good trying to change things by force. You will be so annoyed with me.

I have copied out the letter I wrote to Breuer; I think it comes off because I felt so strongly, and I am sending it to you because I seem to remember that we intended to write and congratulate him together:

"Dear and admired friend. A year ago I came to your house unaware that it was your birthday and felt then that I was a part of the center of the wide circle to whom you mean so much. One can't always be so lucky. This year I hoped to shake your hand in the Physiological Club, but on missing you there I was pleased to think that you were able to spend the evening alone with your wife and children. My wish for you is that you may keep everything you have and which you know so well how to appreciate. I have prevented my Martha from giving you a present only by pointing out your frequently expressed dislike for "female needlework." We have nothing else to offer. When I think of my relationship to you and your wife, I find myself most grateful to you for the esteem you have shown me, an esteem which has raised me far above my present station, and by which you either anticipate others or will remain isolated. Neither would be without analogy in your case. In the hope of seeing you again before the end of the month—I am working more intensely at the moment—Your, Dr. S.F." Rather incoherent, isn't it?

Goodnight, my darling, you may well be a little annoyed, but don't keep it to yourself, better to let go a bit

at your faithful
Sigmund

34 *To* MARTHA BERNAYS

Vienna, Friday evening
January 18, 1884

My sweet girl

I wish I could have any number of days like today, distinguished by small success in work and such affectionate love.

Your letter and your parcel have made me unspeakably happy. My precious Marty, you are so good and sweet, even when you don't give presents, but you know how to give so charmingly. The buttons I shall always treasure; I am going to put them on at once, although they don't show off to advantage under my high waistcoat. On the other hand, just last week I ordered an open waistcoat —but a moment ago I put them on after all, and they are magnificent. The tie produces for me the hitherto unheard-of luxury of a change of ties, for I still possess another decent one. The cigars are excellent; my friend Teych[1] is quite right, they are better than those I used to buy from him and I intend to buy from him in the summer. Schönberg, much as I like him, won't get one, my reason being that he shouldn't smoke. But the truth is: I give nothing away that comes from you.—Incidentally, there is a brief review of his *Hitopedescha*[2] in the evening edition of the *Neue Freie Presse*, well-meaning but polemic and therefore almost certain to have been written by Prof. Müller.[3] I expect he will send you and Minna a copy.—But the nicest thing of all is your letter, dearer and sweeter than any chocolate, and now, my darling Marty, I am once more your debtor. I should have led a less luxurious life, and saved something for you. I would never have believed that so much money can disappear so fast.

Now as to why I am in good spirits—because your letter made me happy, not just good-humored. Today at last I started working on nervous disorders; I hope I have found the material for my first small clinical publication.[4] For yesterday a poor tailor's ap-

---

[1] Tobacconist in Hamburg.

[2] Ancient Indian book of fables.

[3] Max Müller (1823-1900), Professor of Oriental Languages at the University of Oxford, editor of *The Sacred Book of the East.*

[4] "Ein Fall von Hirnblutung mit indirekten basalen Herdsymptomen bei Skorbut" ("A Case of Cerebral Hemorrhage . . ."), Vienna, 1884.

prentice arrived with scurvy, the well-known disease in which ecchymosis appears in all organs. Apart from some apathy, he didn't show any visible symptoms. Early this morning he was quite unconscious, which suggested a cerebral hemorrhage. So I went to see him again before luncheon and found a number of interesting symptoms from which could be deduced the locality of the hemorrhage (always our chief concern in brain disorders). So I sat beside him all the afternoon and observed the interesting and most variable development of the illness till seven o'clock, when symmetrical paralysis appeared, with the result that until his death at 8 P.M. nothing escaped my notice. The publication of this case is justified by several interesting and instructive phenomena, indeed it is imperative, especially if the autopsy tomorrow yields some satisfactory conclusions and confirmation of my diagnosis, which is based on localization. Now I need the *Primarius'* consent to publish, which I hope he won't refuse; I intend to keep at him. You see, it is not quite certain yet, nor is it very much, but it is at least a beginning which should make the others take notice of me. It will also bring in a few gulden, perhaps ten—appear possibly in the *Medical Weekly*—and thus by the end of the quarter I hope to be able to contribute a small sum to your spring outfit. Marty, if only I could give you everything I can think of and make you entirely, absolutely my own, how wonderful it would be!

I am not taking Rosa up to the Hammerschlags until Sunday because I understand they won't be home on Saturday. Tomorrow I may go and tell Breuer the good tidings.

With the most affectionate thanks and kisses

from your faithful
Sigmund
who is again eager to work and to live

35 *To* MARTHA BERNAYS

Vienna, Monday
January 28, 1884

A moment ago, my beloved treasure, I put the finishing touches to my first clinical publication. There it lies now, eighteen pages long, and it will spread itself in two to three issues. For

better or worse I have finished and there is a load off my heart.
Now I can tackle something else and my Method will once more
have a chance. Fleischl will arrange for the publication in *Brain*,[1]
and I still have one to two weeks to work on it, then I give my lecture, show my slides, send off my German and my English manuscripts, and then I am again like "Lucky Hans."[2] Life is hard, but I
am drugging myself in work.

Your last letter, my treasure, contained the news I have been
looking forward to for a long time. Of course I want your picture,
if possible life size, to consecrate my new room into which I still
haven't moved, and if I am the only one to receive a picture from
you, I shall be especially happy. Otherwise, I really haven't much
to say; my cold is still with me, my zest for work has returned, my
impatience is greater than ever and one day it will make me ex-
plode. I didn't go to Breuer till Saturday, didn't go home for lunch
because I was on duty. Schönberg, whom I had invited, waited for
me in the coffeehouse till 6:30, as I read on my blackboard. The
poor man, our rounds lasted till 7:15 P.M. I now have a compatriot
of yours in my department, one Rodewald, who owned an inn in
Hamburg, where he himself seems to have been the best customer.
He is, moreover, a nervous case and if he hangs on long enough
I may write a paper on him, too.

My patient was considerably improved yesterday—but he was
only paying me a friendly visit. I still don't have any others.

Look, Marty, do read something good and write to me about it,
I am getting such a barbarian with the pressure of my work.

I am not going to accept the job as Meynert's assistant; you wrote
very sensibly about it. I probably won't come to Fräulein Fanny
Philipp's[3] wedding; I will come only to see my girl, not to make
new relations, of whom I have quite enough already.

In my opinion the B on the notepaper is too ostentatious and the
M too modest. As you know, I am interested only in the M.

I have got to break off now, my darling, I must go back to the
lab. More this evening at the *Journal*.

My precious darling, the *Journal* is always a very pleasant break;
I can read a lot, make plans, look into the past and the future. On

---

[1] "A new Histological Method for the Study of Nerve-Tracts in the Brain and
Spinal Cord," published in the English periodical *Brain*, June, 1886.

[2] Figure from a German fairy tale by the Brothers Grimm.

[3] A cousin of Martha's.

the other hand, if I do this I usually get annoyed at not being able to carry out my intention of writing to you every day. It was so sweet of you to use that brief stop in Hamburg to send me a postcard. Everything is so sweet of you and about you. Do you think you can continue to love me if things go on like this for years: I buried in work and struggling for elusive success, and you lonely and far away? I think you will have to, Marty, and in return I will love you very much. Schönberg told me that Prof. Bühler's[4] wife waited fifteen years for her husband; and now she is older than you will be in fifteen years. Are you prepared to wait fifteen years for me, or would you rather not? In my thoughts I have let you become thirty, and have decided to find you then as young as you were when I met you. No, much younger, for at that time you actually gave quite a matronly impression. But will you really remain as young as that? Don't you feel rather proud of being able to make someone so far away so happy? Well, I can't; with me it all depends on luck, with you just on your temperament. How I am looking forward to your picture!

My work has taken me right into the middle of some quite interesting problems; I discovered by myself all the questions pertaining to the subject, and have arrived at the truth by the method of Herr Kannitverstan.[5] There is a Russian[6] working in the lab who wants to translate my Method. Oh dear, there I go again talking about my stupid work. I am badly in need of an hour's chat with you in order to come out of my shell again. Farewell, my darling. Goodnight.

Your
Sigmund
Journal-doctor pro tem

[4] Georg Bühler (1837-1898), Professor of Oriental Languages at the University of Vienna.

[5] Character from a story by the German poet J. P. Hebel (1760-1826).

[6] Dr. L. Darkschewitsch, whom Freud met again in Paris (see letter of Nov. 4, 1885).

36 *To* MARTHA BERNAYS

Vienna, Tuesday
January 29, 1884

Dear Fräulein Martha Bernays

At first I could not imagine what the solemn presentation of a red plush visiting card (it is plush, isn't it?) between old lovers

like ourselves could mean. I suspected it to contain some kind of
picture puzzle, preferably a photograph. Then I got the bright idea
that it could be a name card, an idea I found confirmed after reading your letter. Well then, a golden Martha Bernays on a background of red! I like looking at the name, but I know a better one:
Frau Martha Freud would strike my eye and my ear as far more
beautiful. Your letter, Marty, with its wisdom about love and life,
raised my spirits considerably; I have not been so merry and gay
for a long time and am so grateful to you. I was very amused to
realize how deeply involved you are in being engaged—so deep
that you consider all the knights at the Round Table to be engaged
without any further proof. If you were not "half married" yourself,
you would be just as ready to consider them all unattached. I feel
so gay today for no other reason than that produced by your letter,
and so much in the mood to hear you talk and to close your mouth
every now and again with a kiss to make you stop.

Why I never took you to see the Hammerschlags? I often meant
to, but sometimes you couldn't and in the end the hours were too
precious for me to have shared them with anyone but you. You
were not at all awkward at the Breuers'; on the contrary, you were
very talkative, more than you were with me at the time; you have
no reason to reproach yourself.

Now for my bit of news. I leave here tomorrow; unlike the first
room, this one is not associated with memories of sweet happiness.
On Tuesday and Thursday two weeks ahead I am going to give
my lectures in the Physiological and Psychiatric Clubs. My paper[1]
lies before me, finished to the last word. Tomorrow it will be sent
off or handed in.

By the end of this week I hope to have finished my paper on the
Method in two languages. Nothing new in the department at the
moment. I am soon going to choose a topic for a paper from among
the problems concerning nervous diseases. I am not worried about
failing to find a topic, and I can evidently continue working on
this subject on my own. Today the Club met; I sat behind Billroth[2]
and Nothnagel and was naughty enough to think: Just wait till
you welcome me as you are welcoming the others now! Billroth
doesn't know me; Nothnagel, by the way, was rather patronizing

---

[1] (See letter of Jan. 28, 1884.)

[2] Dr. Theodor Billroth (1829-1894), Professor of Surgery at the University of
Vienna.

last time. Meynert continues to treat me with great respect and advised me to give a lecture at the Medical Society as well, which I don't intend to do just now.

Goodnight, my sweet darling, you do still feel well and still love

Your Sigmund
don't you?

37 *To* MARTHA BERNAYS

Vienna, Thursday evening
February 7, 1884

Were I in the position, my Princess, to bestow decorations, you would receive as a reward for your last letter the most beautiful one, that of the White Carrier Pigeon, to be worn on a red ribbon. The letter arrived just as I had started to write my paper and when I had read it I felt so gay that my work went very fast. I had started at 3:30 and by nine I had finished; I leaped for joy—I never miss this exercise if there is the slightest reason for it—and then I meant to write to you. But I was interrupted by a visit—or rather, as a reward to myself I went to the Gasthaus, and so I am writing only today. I was not idle today, either. I copied out the excerpt for the Russian and gave it to him, then I finished the English manuscript and had it corrected by the American, and now I have still got to transcribe the latter and make a few corrections in the German manuscript, and then I am finished with it. Tomorrow I will take them both to Fleischl and then Amen.

I now have time to return once more to my patients and do some reading. I wonder how long it will be before I write something again. Not too long, I hope. A man must get himself talked about.

Silberstein was here again today; he is as devoted to me as ever. We became friends at a time when one doesn't look upon friendship as a sport or an asset, but when one needs a friend with whom to share things. We used to be together literally every hour of the day that was not spent on the school bench. We learned Spanish together, had our own mythology and secret names, which we took from some dialogue of the great Cervantes. Once in our Spanish primer we found a humorous-philosophical conversation between two dogs which lie peacefully at the door of a hospital, and appropriated their names; in writing as well as in conversation he was known as Berganza, I as Cipion. How often have I written: *Querido Berganza!* and signed myself *Tu fidel Cipio, pero en el Hospital de Sevilla!* Together we founded a strange scholarly society, the "Academia Castellana" (AC), compiled a great mass of humorous work which must still exist somewhere among my old papers; we shared our frugal suppers and were never bored in each other's company. Intellectually he did not like soaring very high; he remained in the human domain; his outlook, his reading, his humor, all were bourgeois and somewhat prosaic. Later when he was ill I became his doctor, and one day he invited all his old colleagues to a farewell party in Hernals,[1] during which he himself with his good-natured expression poured the beer from a barrel in order to conceal his emotion. Then while we were sitting together in a café and Rosanes[2] was telling odious jokes, also only, to prevent his sentimentality from overflowing, I was the first to break the ice and in the name of them all made a speech in which I said he was taking with him my own youth, little realizing how true this was. At first I wrote to him off and on; he was badly treated by his half-mad father, of which he complained; I on the other hand tried to arouse his romantic instinct and encouraged him to run away to Bucharest and look for a job more worthy of him; after all, in his youth he had been full of romantic dreams about Red Indians, Cooper's *Leatherstocking*, and sailor's stories. Even quite recently he kept a boat on the Danube, and invited all his friends on trips during which they had to serve as oarsmen and call him Captain. Then you appeared on the scene and everything that came with you; a new friend, new struggles, new aims. The drifting apart which had gradually developed between us became apparent again when I advised him from Wandsbek against marrying a stupid rich girl whom he had been sent to have a look at. And then we lost contact with each other. He obviously got used to the moneybags, although he is kept short enough as it is; he is prepared to marry this girl so as to establish his independence as a merchant. What has happened to me, you know. And now we have met again and no doubt both of us are

[1] District of Vienna.
[2] Dr. J. Rosanes (1857-1922), surgeon, Director of the Stephanie Hospital in Vienna.

thinking how strangely life has treated us, harnessing us both and sending us galloping off, the one in that direction, the other in this.

When he was still very young, Anna was his first love, then he had a liaison with Fanny, in between he was in love with every girl he met, and now he is with none. I was in love with none and am now with one. That is the story of my friend Silberstein, who has become a banker, because he didn't like jurisprudence. Today he is about to gather together again his old boon companions in Hernals, but I am on duty, and in any case my thoughts are not in the past, but elsewhere.

Farewell my beloved treasure, my mailbox was silent today; tomorrow I hope it will speak again to

> Your Sigmund

38 *To* MARTHA BERNAYS

Vienna, Thursday
February 14, 1884

My darling, my girl, my little woman

Do you realize that it is two whole days since I heard from you and that I am beginning to worry! Could you be ill or angry with me? I am only too willing to write to you more often again; best of all I would like to write to you all day long, but what I still prefer is to work all day so as to be able to hold you in my arms for years and years. Otherwise, why should I do so many things now that go against the grain: write papers, give lectures, examine patients, make up to people? But so far in my struggles for recognition I haven't done anything bad and I hope to avoid doing so in the future. Otherwise you wouldn't love me any more. You don't have to worry about this.

I really do wish you had been present to hear my lecture today, Marty. I haven't had such a triumph for a long time. Just imagine your timid lover, confronted by the severe Meynert and an assembly of psychiatrists and several colleagues, trying to draw attention to one of his earlier works, the very one which had been overlooked by Prof. Kupfer.[1] Imagine him beginning with allusions, unable to control his voice, then drawing on the blackboard, in

---

[1] Dr. Karl Wilhelm von Kupfer (1829-1902), Professor of Anatomy at the University of Vienna.

the middle of it all managing to make a joke at which the audience bursts out laughing! The moments in which he is afraid of getting stuck, each time fortunately concealed, become fewer, he slides into the waters of discussion where he sails about for a full hour, then Meynert with some words of praise expresses the assembly's vote of thanks, follows this up with a few appreciative observations, then dissolves the meeting and shakes him by the hand. Then the old gentlemen who hitherto had ignored him congratulate him and gather round him to make a few belated comments, and finally Meynert requests that he make an excerpt[2] for the Society's *Yearbook* and promises to correct the relevant passages with footnotes in his forthcoming book; at last he leaves in an elated mood, wondering whether his work won't after all succeed in making his girl his own. Oh, but now comes the worry about holding one's own, finding something new to make the world sit up and bring not only recognition from the few but also attract the many, the well-paying public.

But I am not going to do anything more today. If there were any justice I should have had a letter from you. Your last, it's true, was again so sweet that anyone less insatiable than I am could go on reading it for ages.

I seriously consider having breakfast in my room so as to save money and at the same time eat better food. Shall I decide on tea or coffee? There are such things as automatic coffee machines, as good as tea machines, and I really don't think there is much to recommend tea, least of all the tea one buys here. What's the opinion of your Highness, my delicate little princess?

> Fare thee very well and write again soon to
> Your
> Sigmund

2 "Die Struktur der Elemente des Nervensystems" ("The Structure of the Elements of the Nervous System"), *Jahrbücher für Psychiatrie*, 1884, 5 (3), 221.

39 *To* MARTHA BERNAYS

Vienna, Wednesday
March 19, 1884

Her Highness the Princess deigns to look out at me from her plush as if she had guessed again what I did today. Well, I bet she hasn't guessed right. Be prepared for the most unlikely thing you ever heard. In the morning I lay there in the vilest pain and looked at myself in the mirror till I shuddered at the sight of my wild beard. My rage rose and rose until finally it boiled over. I decided not to have sciatica any more, to become human again, and to abandon the luxury of being ill. In no time I was dressed and sitting at the barber's, literally breathed a sigh of relief as I looked once more like a well-trimmed garden hedge, and as the weather was so glorious I walked for a while in the courtyard. It grew more and more easy, after a warm bath I could walk quite well, then I dashed into the laboratory, made up my mind to start work again, in the afternoon played chess in the coffeehouse, and on receiving a brief visit from Prof. Hammerschlag I decided to return it in the evening. This I did; of course they were all rather concerned and soon threw me out again, but here I am in the saddle once more, have no pains despite the long day, only feelings of fatigue which is understandable, can work again and am immensely, immensely pleased that I have recovered by my own decision. I can't really explain it to myself, but it is a fact. Not that I expect the pains and the difficulties in walking to have completely disappeared by tomorrow, but if they are no worse than they were after my bold attempt today, I will be able to work and soon it will be gone for good.

Goodnight, my little princess, and not another word about my sciatica. I will add a few lines tomorrow.

Thursday, March 20, early

You vain little worm. You were ashamed of your photograph, and quite unreasonably! I don't have to mention how it is bound to please the one person who loves you; what I will mention is that it is more conducive than any other picture to inspiring respect in everyone else.

Any bad results of yesterday's venture ought to show themselves today. However, I am quite well, completely without pain, and have only some feelings of fatigue in the leg.

I think we can now put an end to this chapter in our correspondence. Today I am planning to go to the instrument-maker and start a new account. You can see how reckless I am, but I now have to take some risks. After that I have to do a little job that should bring me in fifteen gulden. Then I intend to refloat my four or five ships in the laboratory; apart from this I am going to stay at home and read. This is how I expect to spend the next ten days.

If Breuer doesn't turn up today, I am going to surprise him in the evening. Only yesterday he said I wasn't ill enough to be treated like a patient.

I trust the little princess is not indisposed because she announces herself as being tired? Please don't take advantage of my inability, caused by my own illness, to grumble about your state of health. On the other hand, if being ill gives a man the advantage of receiving more letters from his dear sweetheart, then I shall go to bed again.

How do you like that threat?

> With fondest greetings
> Your
> Sigmund

40 *To* MARTHA BERNAYS

> Vienna, Saturday
> March 29, 1884

Beloved sweetheart

If it were not against discipline I would say that there is no need to tell me when you aren't feeling well (but please note that I don't say it)—I can always recognize it so clearly in your letters. When writing your last letter you weren't well either, for your foreboding is exactly like those nightmares that torture one only when one is suffering from indigestion. Then on waking one is relieved that it was only a dream, and this is how we should also behave, quite apart from the probability that you would have found another man, whereas I wouldn't have found a professor's daughter.

Heavens above, little woman, how innocent and good-natured you are! Don't you realize that this very science could become our bitterest enemy, that the irresistible temptation to devote one's life without remuneration or recognition to the solving of problems unconnected with our personal situation, could postpone or even destroy our chances of sharing life—if I, yes, if I were to go and

lose my head over it? Now, this is out of the question; I feel in fine
fettle and intend to exploit science rather than allow myself to
be exploited in its favor. I have been anxious during these weeks
only because my experiments with brain anatomy comprise my
only work. By the time you read this I shall be back on duty, busy
with patients, and the new electrical appliances will help to keep
me in the clinic.

The experiments are going well, by the way; I find myself
in the position of making a few important assertions, partly confirming Meynert's disputed discoveries, partly new explanations
which I feel sure will increase, and I hope all this will lead to
some more good papers. My only embarrassment is what to do
about Holländer: we had agreed in the beginning to publish the
paper together, but now it would be far better for me to work
on my own, for not only is his usefulness in this work far less than
mine, he is no use whatever. He is quite incapable of adjusting
himself to things, he turns up every two weeks for a couple of days,
gets hold of a slide, lights a cigar (which one should never do in
this work), reads a book (which one should also never do), then
declares that the problems are very difficult—they are, thank God,
otherwise anyone could do them—or that the light is too bad, then
drops everything again and off he proudly strides. Good fellow
though he is, his dilettantism inspires little respect. He behaves
moreover, like a *grand seigneur,* takes no part in the technical
work, whereas I work almost every evening till eleven or midnight,
and it is not only that he is no use to me, I don't need him, either.

Since we are in the midst of scientific matters, one more word
about the *Dozentur.* There is no salary attached to it, but two kinds
of advantages. First the right (actually the only duty) to give
lectures on which, if they are well attended—whatever this depends
on!—I could manage to live, and this would enable me to relieve[1]
my poor harassed friend Breuer. Secondly, one rises to a higher
social level in the medical world and in the eyes of the public, has
more prospects not only of getting patients but of better-paying
ones—in short, it helps one to build up a certain reputation. Admittedly, there are also *Dozenten* without patients and in spite
of the fair success of my labors our whole future does indeed still

[1] Not to accept any further financial support from him.

look rather dark. At least we will do everything in our power, and
it will come out all right. . . .

On Monday the thirty-first it will be three years since I became an
M.D.; it hasn't brought me in anything so far, but it takes a lot to
finish off a doctor, especially one with a precious sweetheart who
protects him from idleness and silly escapades.

With the best wishes for my little princess's health

Your

Sigmund

41   *To* MARTHA BERNAYS

Vienna, Tuesday, at the *Journal*

April 15, 1884

My sweet beloved

What a strange turn things are taking! As you say, "all must
change, and live."[1] I am returning your optimism with news that
may signify this kind of a change—I don't like grandiloquence,
even though I am very deeply moved. In any case, let us say it
seems that we have started on the second volume of our highly
interesting family chronicle ("Riches").[2] Just listen—it really sounds
like a chapter out of Dickens: Paneth and his bride have invested in
my name a capital sum of fifteen hundred gulden, of which the
interest of eighty-four gulden a year is to be used for an annual
trip to Wandsbek, which sum however is to be at my disposal at
any time, especially if I were to take a decisive step toward our
union by starting, say, a medical practice here or in the country,
or by emigrating to America. Of this "foundation," the interest from
which has already started as of April 1, only you are supposed to
know. Even the Schwabs know nothing about it. Paneth broke
the news to me today and we exchanged some very friendly words
in the process. The idea behind it is to enable our marriage to take
place six or even twelve months earlier.

There is so much to say about all this which I will let you guess,
my darling. In any case I am heavily burdened with obligations

---

[1] Quotation from a poem by Uhland (see footnote 4 to letter of July 16, 1882).
[2] Allusion to a book planned only for fun. The title of Volume I was to be
"Poverty," that of Volume II "Riches."

toward other people, so much so that it rather oppresses me. But isn't it wonderful that normally parsimonious persons should be moved by the power of their and our true love to become warm and willing to make sacrifices? And isn't it wonderful again that a wealthy man should mitigate the injustice of our poor origins and the unfairness of his own favored position? And think how much happier and more capable of work I shall be when I have you at my side! And then I will work and earn so much that I shall no longer need to feel ashamed.

With this gesture Paneth is entitled to make a greater claim on my friendship—needless to say there was no question of the loan being provoked—and my only regret is that I seriously believe I won't be able to enjoy this new friendship very long.[3] I shall have to thank her verbally or in writing. And as in every novel there are always two or several couples and intrigues ("plots"), so something new happened this very day to Schönberg, too; something good and actually more honorable than that which happened to me. Bühler told him that Prof. Monier Williams[4] of Oxford wants to have Schönberg with him as soon as the middle of May, and so he has to take his degree in the greatest haste, for which Bühler is offering him every possible facility. I do think, though, that he will have to take a few weeks longer so as not to overwork himself. I understand that his salary will be anything up to £150, and that there is a possibility of his name being mentioned in the title of the dictionary on which he is collaborating. He will be seeing you all earlier than I. In any case Minna should be pleased and will realize that this rare stroke of luck is not falling on an ordinary person.

And now your letter, the best, most beautiful you have ever written me, the most valuable, a letter that puts an end to all my doubts. Let us love one another and work.

With fondest greetings
from your
Sigmund

---

[3] Paneth was suspected of having tuberculosis.
[4] Editor of a Sanskrit dictionary.

42 *To* MARTHA BERNAYS

Vienna, Saturday, at the *Journal*
April 19, 1884

My precious Marty

Most certainly you can take seriously what I said, and please don't believe for a moment that I am making any sacrifices for you which you cannot think about with a free heart. Believe me, it is only natural that I should object more than you to the protracted waiting; I just stand it less well; it is a general rule that brides are happier than bridegrooms. So it is more for my own sake that I have decided on a short-term career, and besides I am quite convinced that your eyes—the part must stand for the whole—that you, my darling, will compensate me for a great deal; you too must believe this. And what am I sacrificing for it? I haven't got particularly far, and in the two years we still have to wait nothing very decisive is likely to happen. At best a slight change in my position in society. It won't cost me any effort; on the contrary, I will be only too happy to give up what is unimportant, of uncertain value and prospect in favor of something so worthy, refreshing and rich as sharing life with a beloved who is going to be not only a housekeeper and cook but a precious friend and a cherished sweetheart as well. Add to this what I have often written to you, that in one field of science I am independent enough to make contributions without any further contacts or assistance, by which I mean my knowledge of the nervous system, and I am happy to think that you will be able to help me with it. So the world will not be allowed to forget my name just yet. The trouble is I have so little ambition. I know I am someone, without having to be told so.

By a German region[1] I was of course thinking of Lower Austria, Moravia or Silesia.

For the time being anyhow I am still quite ready to fight and have no intention of breaking off my battle for a future in Vienna. The "struggle for existence" still means for me a struggle for existence here. This past week the chances of being a *Dozent* next winter have, I must admit, seemed very remote. Owing to my

[1] Regarding the possibility of starting a practice there.

medical activities with Frau S., I have hardly been able to do any work at all. No doubt I shall be able to buy clothes with the fifty-odd gulden, but I would far rather have spent this time going short of things with more chances to work. Bettelheim[2] has brought the instruments; I myself bought another one today—i.e., paid the half; on Monday I expect the whole thing will start. But I am afraid the brain anatomy has been terribly neglected, nor has the preliminary work for my next publication got very far. Frau S. is better again today; I hope during the next week to improve her health sufficiently to stop treating her. The only unpleasant symptom is an old but suspicious catarrh of the lung which has affected one apex. If this gets worse, or if it proves to be connected with the general illness, then the outlook is bad. I don't anticipate this, however, and believe that the restored heart will hold out for a while until another fainting fit gets her down again. But this might not happen for years.

I must ask you to forgive me for taking so long to talk about your situation.[3] I am so sorry about it. Don't you at least take turns with Minna and go into the air a little? Marty, if you get ill over this I shall make a big fuss and you will see that it is not only patients but also lovers who can be very egotistic. The consultant's verdict doesn't impress me much; I cannot see any reason why this business should last forever or return. Nor why the consultant should appear without being called; dropping in is not considered etiquette.

Shall I send you a book by Fritz Reuter?[4] Reading aloud might perhaps help you over a few difficult hours?

Please write again soon, my darling, and tell me that you have been out—provided your weather is less awful than ours is here.

With affectionate greetings
Your
Sigmund

[2] Dr. Karl Bettelheim (1840-1895), Lecturer on Internal Medicine at the University of Vienna, with whom Freud planned to study the influence of electricity on the nervous system.

[3] Martha was nursing her mother, who was ill.

[4] Fritz Reuter (1810-1874), German author who wrote prose and poems in the Low German (*plattdeutsch*) dialect.

43 *To* MARTHA BERNAYS

Vienna, Monday, at the *Journal*
April 21, 1884

You will certainly be surprised, my darling, to hear that I am sitting here again after having written to you as recently as Saturday from the same spot; this is the result of my having been absent through being laid up so long, and rather awkward it is, too. I feel there is something altogether missing at the moment; I cannot work in the laboratory because of the prospering practice; work on the experiments, from which I expect a little recognition, is lying idle.—It gave me quite a turn today when the proofs of my paper on the Method arrived from Leipzig; since then, with the exception of two small discoveries, I have done no work whatever. But otherwise I am very well, feel fitter than ever, I also love you even more than during our best days here, and if I write to you so rarely it is because of the beastly combination of being on duty and work at the *Journal* during these past few days; even yesterday, Sunday, I was in harness. Paneth was here today and told me that I may *perhaps* be summoned to a nervous case in Schwechat.[1] Alois Schönberg[2] has mentioned the prospect of a job in Pest. All these are simply beginnings, which do not necessarily have to materialize, but they are nonetheless beginnings. Frau S. is much better now; I would be very pleased if nothing happened and I could stop treating her in a week's time. I would then advise her to go to the country at once.

I am also toying now with a project and a hope which I will tell you about; perhaps nothing will come of this, either. It is a therapeutic experiment. I have been reading about cocaine, the effective ingredient of coca leaves, which some Indian tribes chew in order to make themselves resistant to privation and fatigue. A German has tested this stuff on soldiers and reported that it has really rendered them strong and capable of endurance. I have now ordered some of it and for obvious reasons am going to try it out on cases of heart disease, then on nervous exhaustion, particularly

---

[1] Small town in Lower Austria.
[2] Ignaz Schönberg's brother.

in the awful condition following withdrawal of morphine (as in the case of Dr. Fleischl). There may be any number of other people experimenting on it already; perhaps it won't work. But I am certainly going to try it and, as you know, if one tries something often enough and goes on wanting it, one day it may succeed. We need no more than one stroke of luck of this kind to consider setting up house. But, my little woman, do not be too convinced that it will come off this time. As you know, an explorer's temperament requires two basic qualities: optimism in attempt, criticism in work.

Now that I have talked out everything concerning myself, I shall come to you, my precious girl. No, I am still here. I don't even consider seeing you in the spring, I would like to have achieved something really good before we meet again. And this is what I am looking forward to more than I can say.

I am expecting the mailman today with the parcel and money; it looks as though he isn't coming, but this doesn't mean that you will have to wait long for your visiting cards and your seal. It is so nice of you to have expressed a wish; the fact that you are taking walks in the woods also pleases me very much. All alone, my Marty? Yesterday Dolfi said how very nice it would be if one day you could say, proudly of course: "I waited four years for my husband." By the way, Marty, little Pauli has already fallen happily in love, what do you think of that? With the 28-year-old brother of her girl friend, Glaser, with whom she used to spend her vacations. He has taken his law degree and is a junior counselor-at-law in our town of Neutitschein in Moravia. Anyway a serious-sounding person. What do you think of it? Keep it to yourself, for I don't mean to say that the girl has definitely committed herself; but doesn't it look as though our silly girls are very much in demand? Dolfi is the only one still unattached. Yesterday—I had invited her to tea to get her to mend my black coat—she said: "It must be wonderful to be the fiancée of a cultured man, but a cultured man wouldn't want me, would he?" I couldn't help laughing at this category.

The mailman has just arrived, Marty; he brought very few nice things, but a letter containing twenty-eight florins. It does a man good to have some money, darling; now another ten florins are coming to you; I shall hold onto them for a while, for I have no other money, but they are yours. Now what plans have you for your wardrobe? A jersey jacket? Are they still in fashion?

I am holding onto the money for a while not because I am stingy, but because the cocaine will cost something and because I impoverished myself yesterday by paying ten florins for an electric apparatus.

Now the apparatuses are all there and we begin work tomorrow. I am seeing Frau S. only once a day. Schönberg is hard at work on Kant and Horace, but looks well and is in good form. Marty, doesn't all this together look rather like the second volume?[3]

Now please write to me as much about yourself as I have of myself. And also whether you are well, completely well. Whether the iron is doing you good and whether you are drinking any wine. I shall be angry if you don't say yes to both.

You wanted to return to something in my last letter. What was it?

Fondest greetings
Your
Sigmund

[3] See note 2 to letter of Apr. 15, 1884.

44 *To* MARTHA BERNAYS

Vienna, Thursday
May 29, 1884

My darling, my precious darling, so this is how you neglect me? Two long days without a letter, and only because I failed you for a similar time? Don't you grant me this little privilege? If only you knew how things have been accumulating here, things of which I am the center, how the world has been bringing pressure on me from all sides! And you, darling, surely you should have time for me, even if once in a while I have none for you. But I don't want to be angry; you probably have a reason too, but it mustn't be because you are not well, or because you don't love me.

Tuesday evening there was a Club meeting and Nothnagel invited me to accompany him. This was rather less than I had expected; I would have preferred to spend an evening in his room. What he said on the way also fell rather short of my expectations, or perhaps matched my earlier ones. I am afraid there won't be a golden snake, child, but it was a step forward all the same. You had better judge for yourself. It amounted to what we call "Etzes,"[1] and I would like to have told him so. He began by reminding me that I am engaged and twenty-eight years old, and that I surely don't want my engagement to last another five to six years? "No." *Nothn:* "Well, in that case I suggest the best thing for you to do would be to go to a provincial town, make a fortune there, and then—when, say, Breuer retires—come to Vienna. (May my friend Breuer, I thought to myself, go on working for many a year and not think of retiring.) You know how hard it is to get along in Vienna, how hard our colleagues work from morning to night and still barely eke out a living. Ah, there was a cousin of mine, for instance—he was my assistant in Jena and already a *Privatdozent*— he got engaged and so had to abandon his academic career and start a practice. 'Go to Nauheim,' I told him; 'a few years there and you will be a rich man.' No, he had to go to Berlin; there he is now, stuck, and can't get any further." *Fr:* "But, Herr Hofrat, an Austrian provincial town is not like a German one." *Nothn:* "All right, perhaps I don't know the conditions here; then it might be a good idea to go abroad. I could give you some recommendations to Buenos Aires, where a former assistant of mine has a practice; or to Madrid, where I have any number of connections. *Fr:* "Yes, I also consider emigrating, but first I want to try and see if I can't get along here. I have the capacity to work, and I am tied here by other things than the proximity of these beautiful buildings." *Nothn:* "How do you imagine going about it?" *Fr:* "In the winter I intend to apply for a *Dozentur;* most of my papers are theoretical; would that be a disadvantage?" *Noth:* "No, not as long as there are some clinical ones among them." *Fr:* "There would be; then I shall spend a certain time, say six months, seeing how the lecture courses develop and whether I can get the kind of practice that could be expected by someone with a minor reputation in the hospital; and if this doesn't work, then I'll get right out." *Nothn:* "Now, you must realize how difficult it is to get anywhere with these lectures. As you know, if a man is in an official position, people come flocking to him, whether he is a blockhead or not. But when a man without a position begins lecturing, he may well fare as I did when I started in Breslau. In my first course I had four students. You won't have any more, and will have to wait a long time before a course of this

---

[1] Yiddish—good advice.

kind pays its way. If you want to remain in Vienna you'll have to
count at best three years before being able to set up house. And
if you had to leave after all, you would have to start all over again
from scratch. How much do you think you'll need at first?" *Fr:*
"I think I could manage with three thousand florins." *Nothn:* "I see
you are well informed. Three to four thousand gulden, that's the
minimum, and even then think how economically you would have
to live!" *Fr:* "That doesn't apply to people who have been brought
up the way I have, Herr Hofrat." *Nothn:* "I wasn't brought up any
differently; I have known all kinds of privation; even as an assistant my lunch for the first six months consisted of two boiled
eggs and for a long time afterwards all I had for supper was a
piece of bread and a glass of water." (Then Nothnagel began
again): "Well, the electrotherapy idea isn't bad. There's room for
someone in addition to Rosenthal[2] and Benedikt,[3] as Weiss's example has shown. Eulenburg,[4] who might have been a competitor,
isn't coming here either, and I don't imagine you are afraid of
Weiss's brother. So I suggest you go on working as before, but
the papers you have done up to now won't be of any use to you;
general practitioners, on whom everything depends, are prosaic
people who will think to themselves: 'What's the good of Freud's
knowledge of brain anatomy? That won't help him to treat a
radialis paralysis!' You'll have to show them you can do that, too,
have to give lectures to the *Medical Association,* publish clinical
papers." *Fr:* "I intend to do all that and I'm in the midst of writing
these papers." *Nothn:* "Well, that's good; you'll have to go on
writing your other papers for other reasons, but you can build up
a practice only if the general practitioners send you patients for
electrical treatment."

By this time we had reached his house, where he said, "Goodbye, dear Freud"; I thanked him and went my way feeling rather
depressed. All this good advice I have known myself for ages; he
evidently thought he was dealing with a helpless person and even
admitted his surprise at my being so well informed. The one

---

[2] Dr. Moritz Rosenthal (1833-1889), Professor of Nervous Diseases at the University of Vienna.

[3] Dr. Moritz Benedikt (1835-1920), Professor of Neurology at the University of
Vienna, Director of the Neurological Clinic.

[4] Dr. Albert Eulenburg (1840-1917), Professor of Nervous Diseases at the University of Berlin.

valuable assurance, that he would send me patients, he failed to give me. But to judge from the whole situation, there is no doubt that he will do so, just as he talked of the *Dozentur* as something certain. He wasn't in the least patronizing. I explained the contrast between his former warmth and today's reticence by the fact that as a stubborn and trusting man he clings to what his colleagues have told him as well as to what Breuer and Fleischl have said about me. Hence the certain impersonal interest he showed in speaking to me. The meeting, however, has been an advance, if not in Fleischl's sense, then in that of Breuer, whose interpretation has proved to be correct. I have no personal relationship with him.

Afterwards I went to Breuer, who had not been at the Club, so as to recover from my disappointment. They are both dear, good, and understanding friends. We talked till 1 A.M. She always insists on my taking a small apartment before long and hanging out a sign, just a sign, a beautiful sign.

I have so much more to tell you, but I had better send this letter off; you have been waiting long enough already. Your precious letter arrived early this morning; it did me a world of good; do write soon again. . . .

It's all right, I won't bring you anything, but I expect you to ask for something when I am there.

Fond Whitsun greetings, darling. What memories this season brings back—precious, lovely ones, and some bitter ones as well. If only you had stayed here! Your leaving will have cost me part of my life. I shall be with you for your birthday, after all.

Once more a fond Whitsun greeting from

Your

Sigmund

45 *To* MARTHA BERNAYS

Vienna, Thursday
June 19, 1884

My beloved treasure

 I can't remember ever having been so rushed, otherwise I would have answered all your sweet, good letters with long pages of explanation; but as it is I have to be brief today, too; after all, I hope we will soon be able to talk.

Coca[1] wasn't finished till last night; the first half has already been corrected today; it will be 1½ sheets long; the few gulden I have earned by it I had to subtract from my pupil, whom I sent away yesterday and today. Now there is still the correction of a second paper;[2] in addition I have to give electrical treatments, read, and work at the *Journal*, but I am as strong as a lion, gay and cheerful, and you can well imagine that this isn't the mood in which to drop everything and become a male nurse to a mental case.[3]

My beloved girl, you must utterly banish from your mind gloomy thoughts such as that you are hindering me from earning a living. After all, you know the key to my life: that I can work only when spurred on by great hopes for things uppermost in my mind. Before I met you I didn't know the joy of living, and now that "in principle" you are mine, to have you completely is the one condition I make to life, which I otherwise don't set any great store by. I am very stubborn and very reckless and need great challenges; I have done a number of things which any sensible person would be bound to consider very rash. For example, to take up science as a poverty-stricken man, then as a poverty-stricken man to capture a poor girl—but this must continue to be my way of life: risking a lot, hoping a lot, working a lot. To average bourgeois common sense I have been lost long ago. And now I am supposed not to see you for three months—and this in addition to our uncertain circumstances, and with people as unpredictable as our families! In three months Eli may be in Hamburg, or the situation in my family may prevent me from leaving. In short, I know nothing about the future. I daren't count on it, but what I do know is that I need the refreshment of holding you in my arms again as urgently as I need food and drink; I know perfectly well that I have inflicted upon you enough worry and privation and mustn't rob you of our few happy weeks together, even if you were willing to renounce them yourself. I am going to follow my impulse and continue my venture; I want to strengthen myself through you and then with renewed strength go on trying to improve my position rather than tear myself away from all work for three months. The latter would have no great advantage; what I would save in money I would lose in time,

---

[1] "Über Coca" ("On Coca"), *Zentralblatt für die gesamte Therapie*, 1884, 2, 289.

[2] See note 2 to letter of Feb. 14, 1884.

[3] Breuer had offered Freud the well-paid job as companion to a critically ill patient on a journey of several months.

and not much money would be saved anyway. Could you imagine me having a thousand gulden in the drawer and letting Rosa and Dolfi go hungry? At least half of it I would give to them and the rest would be just sufficient to make up for the time I would have lost. It's true, they will be the losers, but I have to do the one thing that is right for my nature and our situation. I am completely of one mind about this.

Paneth came today, also convinced of course of the necessity of my accepting the job, but I possess the good quality of being able to believe in my own judgment. I have also found a number of people who agree with me. Anyhow, my darling, I know I shall be seeing you again before very long. Keep well; I must stop, for again there is a paper to correct.

Your
Sigmund

46 *To* JOSEF BREUER

Vienna
June 23, 1884

Dear and admired friend

The attack of conscientiousness during which I wanted to leave to you the decision whether I travel to see my girl or effortlessly earn a thousand florins has passed long ago. I marvel that I ever started to be so sensible; the explanation is that I wanted to act under the impression of bad news from outside. Now, you were magnanimous enough to refuse to make a decision, and now I know I will take the trip. I cannot travel earlier than I had intended, nor do I wish to return before the money has given out. I expect to be able to keep myself in Wandsbek for four weeks, and not return before August 15. If the position you are trying to procure for me is still open by then, it will tie me down till November. If I didn't take the trip, I could be free at the beginning of October. These six weeks are very important for the continuation of my work in the hospital; what's more, in November I won't have any daylight for my histological work. In short, the trip and the job with P. don't go together, or both together don't coincide with my projects in the hospital and in the town, with my "struggle to remain."

I could have spared myself this whole explanation; it doesn't correspond to my real motives. To be honest, things are different. The journey to my Martha belongs to a certain rash, frivolous life plan inconsiderate of others, including yourself. This plan I was willing to sacrifice for a while in order to live according to bourgeois timidity and caution. But from lack of talent for this mode of life I abandoned this course and am now determined to do nothing that contradicts the original plan. P.'s thousand florins belong to another plan.

I am so jealous of your good opinion, even of a nuance of your good opinion, that I cannot accept your remark about my mental epidermis. I know quite a number of people whom I respect highly —among them not only Meynert, but you, too—whose epidermis would have been torn open by the traumatic events of my life, whereas I may achieve in time quite a solid layer of scars. With you I don't have to protect myself against the suspicion of exaggeration.

Greeting you warmly and asking you to remember me to your *Hausfrau* in Gmunden.[1]

> Your
> Dr. Sigm. Freud

[1] Resort on Lake Traun in the Austrian Salzkammergut, where Breuer owned a house.

47 *To* MARTHA BERNAYS

Vienna, Sunday
June 29, 1884

My beloved sweetheart

You are quite right. From now on I too will write only about the journey. I can no longer think of anything else. If you really insist on meeting me at the station, I cannot stop you. I was against it because I don't want the station and the luggage to get mixed up with our first kisses. But if you are not embarrassed by the serious Hamburgers and will give me a kiss as soon as I see you, and on our way to Wandsbek a second one, and a third, etc., then I will give in. I won't be tired because I shall be traveling under the influence of coca, in order to curb my terrible impatience.

Do rent a little room for me, very close to you and very modest, otherwise I will grumble that you are not being economical; if possible a little attic, I give you unlimited powers to decide.

My wardrobe won't be very grand, but respectable. I have a gray suit which I am now wearing, and a dark one which is still at the tailor's, a new overcoat and hat. For shirts I am rather badly off; I was going to buy some here, but Father has suggested I buy them in Hamburg, where everything is better and cheaper, and what's more, you understand what one should buy.

I still haven't got my leave, will have to fight for it, if necessary with the threat to quit altogether, but I have no doubt that I shall get it. I hear that Anna is now leaving [Wandsbek] already on the tenth, in which case I shall come a few days earlier, for my pupil will probably release me on the tenth, certainly not before; perhaps even later if I stay longer, and since each day brings in three florins this cannot be despised.

Cocaine runs to twenty-five pages, ready only today; you will see it before you see me. You know what I have been working on today.[1]

Let us not worry about the weather. If it rains we can sit together and talk and read. I am going to bring a few books on neurology; apart from *this* branch of science I want to forget everything connected with Vienna in your presence. For you I will not bring anything, girl, but you will be having your birthday while I am there. I am very undecided whether to hand over the money and the bookkeeping to you or keep accounts myself. I think I shall save you the trouble and not let the control pass out of my hands too early. For two and occasionally three people the sum is not very impressive. On the contrary, if I can still scrape something together, I shall do so. I have to leave something for my family. Dolfi seems to need a little for herself. Yesterday I took her to the Prater; for the first time I was the rich man of the family. Rosa leaves today for Oberwaltersdorf[2] with Herzig for three weeks. Dolfi and Pauli have jobs starting on the fifteenth. Father is bearing up, but he has a lot of troubles. Oh, girl, I must become a rich man and then when they want something they will all have to come to you.

Your
Sigmund

[1] See note 2 to letter of Feb. 14, 1884.
[2] Village in Lower Austria.

48   *To* MARTHA BERNAYS

Vienna, Monday
June 30, 1884

My beloved girl

I am so glad we now see eye to eye and that you won't have to reproach yourself for anything while waiting for me. I am also so happy in anticipation of the beautiful days we are going to spend together. I know that at this point you want to interrupt: I must not anticipate anything so as to avoid disappointment. But, Marty, the beauty of these days depends on ourselves alone and not on the weather, not on the moods of other people, nor on the good or bad news that may come in the meantime. I want to bring back from the journey nothing but the certainty, the final conviction that you are utterly mine—in your attitude, great love, and all the little signs of affection. To take a retrospective glance as you do is quite justified; I really think I have always loved you much more than you me, or, more correctly: until we were separated you hadn't surmounted the *primum falsum* of our love—as a logician would call it—i.e., that I forced myself upon you and you accepted me without any great affection. I know it has finally changed and this success, which I wanted more than anything else, and the prolonged absence of which has been my greatest misery, gives me hope for the other successes which I still need.

Do you remember how you often used to tell me that I had a talent for repeatedly provoking your resistance? How we were always fighting, and you would never give in to me? We were two people who diverged in every detail of life and who were yet determined to love each other, and did love each other. And then, after no hard words had been exchanged between us for a long time, I had to admit to myself that you were indeed my beloved, but so seldom took my side that no one would have realized from your behavior that you were preparing to share my life; and you admitted that I had no influence over you. I found you so fully matured and every corner in you occupied, and you were hard and reserved and I had no power over you. This resistance of yours only made you the more precious to me, but at the same time I was very unhappy, and when at the corner of the Alser Strasse[1] we said goodbye for thirteen months, my hopes were very low, and I walked away like a soldier who knows he is defending a lost position. And whereas our being together threatened to estrange us, during the separation I received signs that I might be victorious after all—not from any abundance of tender words, not from the fulfillment of any wish—I myself don't know how, but I did notice that I was beginning to mean something else to you, that the stiffness and reserve which you yourself so often deplored would vanish the moment we were together. And since then I too have become another person, many wounds that went deeper than you knew have been closed, and I feel within me a gaiety and a self-confidence which for a whole year had been unknown to me. And for this reason I don't want to postpone strengthening myself for new tasks by the long-desired happiness of a close harmony with you. Only occasional whispered doubts crop up: if at the moment you love so fondly the me whom you haven't seen for such a long time and then you see me again, see the gesture, hear the voice and the opinions which invariably used to arouse your defiance, won't you discover that your fondness was directed at an idea that you made for yourself, and not at the living person who perhaps will have upon you the same effect he did a year or two ago? The "no" to this, my love, I can only experience and I am waiting so impatiently for it. Waiting is as much my fate as yours. To wait in peace and with resignation, or to wait in the midst of struggle and agitation—the difference is not so great, no greater than our different ways of facing the world. Another two weeks— but further I refuse to think. The years beyond are hidden from me as though by a screen. I love you so much and am longing to hear from you that you love me too, and I want to spend four weeks that are not sacrificed to the future, as all time hitherto has been, but which are the future itself.

I trust that you are well again, my sweet child? I have never felt better, and I now miss the work. I must think how best to spend the next two weeks. Probably writing reviews for periodicals and regular daily observation of patients. I am very much respected in the department.

[1] Street in Vienna in which the General Hospital stood.

Tuesday, July 1, 1884

My sweet treasure

The new month has begun with rain, but in my case with high spirits and good news. And I am hoping to experience so many lovely things before it ends.

My darling, I have been interrupted so many countless times, I must stop, which I do with fondest greetings and happiest expectations. It is even possible that something nice can be acquired for your birthday; a pupil in brain anatomy has been suggested to me; he wants to take a four-week course; I am going to offer him one of only two weeks. If he comes and accepts, it will mean quite a bit of money, and then we will go for a walk through Hamburg and look for something Marty wants. What's more, I may get a free ticket through Franceschini as far as the border, in which case I shall be able to leave a little something for my family. Coca appeared today, but I haven't seen it yet.

I hope there is nothing in this letter that offends you, my sweet Marty? If so, you must tell

Your
Sigmund

49 *To* MARTHA BERNAYS

Vienna, Sunday
August 3, 1884

My sweet Princess

Things are so unevenly distributed, I have my work if not to comfort me[1] at least to stupefy me, and you have nothing but my picture. I wonder what it tells you? How I wish I could let you know what I am thinking and hoping! Long, indescribably intense happiness if only you keep well. I am always hearing that you look pale. My pale little princess—and four weeks; but stop, I am not to talk about it.

I am writing today to send you two small bits of good news, or at least that is what I consider them. First, that I will get the specimen of my beautiful diagnosis which has caused considerable

[1] The trip to Hamburg had had to be postponed for four weeks.

sensation here, so I shall be able to write quite a nice paper[2] on it; second, that Dr. Heitler[3] has returned and that we have decided to take our money to the instrument-maker tomorrow to buy what we need and then work very hard so as to finish the electrical experiment. Aren't you pleased, my darling? Work, nothing but work; I myself am surprised at the amount of work I can get through. But I know what's driving me; the heart is well again, the giant strong again, gigantically strong. Are you laughing at me for calling myself a giant? Sometimes I have such a sense of power I feel there must be something I could still do to bring us together sooner. How I will love you then, laugh at you and scold you, and you won't say a word, because you are a silent darling.

I received the second very warm letter from Hammerschlag; he inquires among other things whether you have firmly decided to take a job here and says that apart from his interest for me he has an additional one—just what I cannot guess. He also says that if Tedesco's daughters have inherited a single characteristic from their mother, whom he knows, then a job with one of them would be very pleasant.

I will write to Fleischl sooner, tomorrow morning, and add that you won't be free till the end of September. If the lady hasn't found anyone yet, she will probably wait till she returns from the country. If this doesn't work, then we will find something else. Breuer, Fleischl, Schwab, Hammerschlag—my friendly guard will help.

August 4, evening

I broke off yesterday to finish the department's statistical report for July, and today I can respond to your sweet letter which reminds me of our most peaceful days. Oh, I could echo the words of Heine's shepherd boy: "It is a weary task to reign . . ." etc., except that my queen is still only a princess.

You will know by now that whatever happens I am all in favor of your leaving; once you are free we will live so happily, each of us serving and working, each in a state of constraint and renunciation, but happy nevertheless. I just cannot imagine what it will feel like not to be separated from each other by two days.

[2] "Ein Fall von Muskelatrophie mit ausgebreiteten Sensibilitätsstörungen (Syringomyelie)." ["A Case of Muscular Atrophy with Extensive Disturbances of Sensitivity (Syringomyelia)"] *Wiener Medizinische Wochenschrift* 35, 13-14, 389-392, 425-429.

[3] Dr. Moritz Heitler (1848-1923), Professor of Internal Medicine at the University of Vienna.

I received the specimen today and will soon have a drawing made of it. The publication will take three to four months because the microscopic examination has to be finished first. I am thinking that when I return from my trip I will be able to work better still and that when you come to see me we will shut the door, you will sit close beside me and lean against my shoulder while I go on working till I am tired and longing to kiss you. Just now a patient to whom I have been giving very successful electrical treatment for buzzing in the ear promised me some of the most beautiful fruit for "my young woman," the lady whom I certainly must have—she refrained from saying "sweetheart" out of respect. Actually, she is the fruit seller from The Three Ravens, where one turns off to the Seitenstättengasse.[4]

We don't have to worry about the cholera,[5] my darling; it is not likely to prevent my leaving here in September; instead, I think it will arrive here suddenly next year.

Rumor has it that Breuer has again applied for the vacant position as *Primarius* in the hospital; I would be so delighted if he succeeded, then I would do everything to become his *Sekundararzt* and learn a lot. But he won't get it.

How are you feeling, sweet darling? At least you are better off than I am. When you read this it will already be the sixth of August; I am writing this only on the fourth of this endless month.

With fondest greetings and kisses
Your
Sigmund

Please thank Minna very much.

[4] Street in the oldest section of Vienna, where the old synagogue stood.
[5] A cholera epidemic would have made it difficult for Freud to leave the hospital for a holiday.

50 *To* MARTHA BERNAYS

Vienna, Thursday
August 14, 1884

My severe little Princess

I am taking the liberty of pointing out that there are only seventeen more days and that between the time I write this and you read it half of this wretched month will have been slain. May we never have another like it! Amen.

I am pleased to detect from the tone of your letter a note of returning calm and well-being. Let us be wonderfully well when we meet in seventeen days. I am afraid this second journey won't be quite like the first; all kinds of little things are missing. First, the belief—I won't believe it till I am actually sitting in the train. Then I am rather more tired, my best suit is so worn I hardly dare travel in it, and I will have only the second one in which to parade in Wandsbek. It is really mostly stains that spoil it, and perhaps something can be done about this. My hat is no longer new, I have been obliged to buy shirts here, of an inferior quality, but my love is the same and the longing has increased. I am not working so hard nowadays, am almost lazy, although I am busy all day, but I am getting through things without any great effort; after all, I feel sure of new contacts when I return, and all my old papers have at last appeared. The most recent one I sent you was an amplification of a previous work,[1] the best I have ever done, although so far I haven't received one word of recognition for it, nothing but reproaches for the alleged lack of references to the relevant literature. At the moment I have nothing in print, but all kinds of things in my head.

Rosa was here yesterday, was seen by several colleagues and made a great impression with her beauty. Poor thing,[2] she is leaving in a few days for Frau Königstein in Gmunden. Tomorrow I am going home for once to take the family out. Today the Hammerschlags, the dear people, are celebrating their silver wedding. Yesterday I sent them a photograph of myself. As soon as I get some more, I will send one to Minna; who can resist flattery? It will interest her to hear that Frau P., with whom she was in Italy, has died.

Tomorrow there is going to be a simple church service to celebrate the centenary of this hideous, grandiose building[3] which was founded by Emperor Joseph [II]. I am not likely to go; it doesn't fit in with my time and I am not anxious to hear a Mass. Till the first of September I am wound up and then I will unwind, then

---

[1] See note 2 to letter of Feb. 14, 1884.
[2] Rosa had accepted a job as a lady's companion.
[3] The Allgemeine Krankenhaus (General Hospital).

I hope the visit to my sweet princess will supply me with a new spring.

> With fondest greetings
> Your
> Sigmund

51 *To* MARTHA BERNAYS

Vienna, Sunday
August 17, 1884

My girl

That was sweet of you; what a lovely scent these flowers have, what memories they evoke! They lift me out of all the litter of my bookish, bachelor life! And at the same time warn me that summer is coming to an end and that we will have to hurry if we are to enjoy the autumn. And this I will certainly take to heart and nothing is going to stop me. Roughly another two weeks! A week from today the chief[1] is coming back, and the price I shall have to pay to return to you and come to my senses depends on his mood. No price will seem too high. I have changed more than I myself realized. Today four young hospital colleagues insisted on my joining them on an excursion to Dornbach.[2] On our way back we stopped at the same inn where you and I once spent a quiet evening. I felt very out of place; I think I can feel happy nowadays only in your presence, and I cannot imagine that I shall ever be able to enjoy myself again without you. With the exception of the minutes brightened by your letters, I have experienced during the past fourteen months only three or four happy days—when some minor piece of work of mine happened to succeed. And that is too little for a human being who is still young and yet has never felt young. And because I know it is only your going away that has hurt me so much, I feel incapable of reconciling myself to her whose heartlessness and caprice I hold responsible for your departure. And this is something you will have understood long ago, my darling.

But only another two weeks after all; I will manage to doze

---

[1] The head of the department.
[2] Suburb of Vienna.

through the days, and then I will quench my thirst with your kisses and lead a very different life that will make both of us whole and young again, and when we have to separate once more we will do so with the confidence of never again taking upon us hardships such as we have endured these past fourteen months.

I have just had a letter from Mother, saying she is arriving at 4:30. Today, however, I am on duty.

> With fondest greetings & kisses
> Your
> Sigmund

52 *To* MINNA BERNAYS

Vienna
August 28, 1884

Dear Minna

I would like to say: you can be very nice—if this didn't imply that you are not always so. To avoid committing such an injustice, even indirectly, I prefer to say that you are very nice whenever you have the chance. Our relationship nowadays exists exclusively on little presents, one in which I am bound to appear as the less active party. But you shouldn't have to acquire my picture, the very thought shocks me, and that I have failed to send it before is entirely the fault of the tardy photographer. It isn't too late, I trust?

It is the picture of a dispossessed man of importance,[1] something I have been until today, but as of today I am once more a poor little devil and you will very soon be able to convince yourself of the truth of it. "Very soon" is what the historians and similar unfeeling people say. While it is the privilege of the Almighty that to Him millenniums are as moments, we poor human creatures indulge in the opposite: to us days are as millenniums. It seems an invention of the devil that suffering should prolong time, joy shorten it.

I have had several urgent letters from your beloved; since I answered him he is silent again. He probably thinks I lack the strength for an answer just now. I am really so weary it is no longer decent, and apathetic into the bargain. The engine must have some

---

[1] Freud had been in charge of the department during the chief's absence.

rest; I long to put into port and get my considerable damage repaired; this year's crossing has been a stormy one. I want to see no other friendly faces but yours and that of my Martha. Woe to them who try to spoil so much as one hour of my stay with their friendly intentions! Should you hear or know of anyone dying to meet me, please help me to scare him away. In return we will often take you with us and be very nice to you. By the way, don't you think it is just a lot of well-meant talk on the part of our Martha and that no one will dream of laying claim to my honored company?

If I had anything more to tell you it would be too late by now to put it on paper. Looking forward to a happy meeting next month, my dear little sister.

Pull yourself together and write to me at once what I can bring you, providing (1) it is not beyond my traveling expenses, and (2) it won't cause a conspicuous increase in my luggage.

Surely you can't resist such a friendly invitation?

With fond regards
Your brother
Sigmund

53 *To* MARTHA BERNAYS

Vienna, at night
September 30, 1884

My sweet little woman

So you managed to get here first! Thank you so much for your dear letter which has just taken me by surprise. The post office must know I am back again. Everything here seems so strange; to think I am really no longer with you! It was so lovely and it will be lovely again and remain so—no, even lovelier.

I have a whole bag full of news for you, big news and little news, but it would be useless to try and tell you it all now. It's 11:30 and I am very tired, but not mentally; in fact I am very fresh and very happy. No despondency whatever, I feel nothing but infinite gratitude to you, you dear, auspicious one! To my surprise I feel quite gay, more courageous, I would almost say more magnanimous, than before. I know I will work hard, put up with all sorts of difficulties, and that for a long time I will consider myself richly rewarded by the memory of our time together.

Just these few lines; don't consider this letter as completed, I just want to send you a sign of life. I had the most pleasant journey, have established myself again in the hospital, have been to Hammerschlag's and home, and now I am off to bed and send you my warm and fond greetings.

Your

Sigmund

Do I need to tell you that tomorrow I shall spend every free quarter of an hour writing to you?

Many thousand greetings. I am writing on a scrap of paper and an old envelope I have just come across in the confusion of my belongings.

54 *To* MARTHA BERNAYS

Vienna, Monday
November 17, 1884

My sweet little woman

My fondest greetings for the seventeenth. We will soon have belonged to each other for 2½ years; to think that we have been waiting for each other all that time! I would be sad if I didn't know the fact that you are mine is far more important than the other fact—that you are not here at my side. You are so sweet and you are so far away, but I intend to dwell on the former.

Dr. Leslie
Dr. Darling
Dr. Montgomery
Dr. Giles
Dr. Green
Dr. Campbell

Do you know what this means, my precious? Have you read *Don Quixote* and do you remember the condition the hero makes to all knights he has conquered? They have to walk to Toboso and kiss the hand of the incomparable Dulcinea. Now my six students are kissing your hand. Yes, my lecture course has become a reality. Today I read—i.e., I lectured in English for a whole hour and demonstrated on a patient, and the little box which I bought on the Speersort[1] contains a hundred self-earned gulden, of which I am sending you a sample. How good this tastes, girl! One of them, as a matter of fact, didn't pay; he is the organizer of the lectures, Dr. Leslie, whom of course I am only too pleased to take free of charge. But this evening he came to see me again and sang my praises, which I find suspicious. Do you think he is going to charge a commission at the end? I don't think so.

What am I going to do with the money? From now on Marty and Minna are going to drink port; there will be a regular monthly contribution to my family; a pair of winter trousers can be ordered; and if what is left will enable me to relieve Breuer in December, how wonderful! An isolated income of this kind doesn't make much difference to my budget, you know, but should the lectures continue regularly, girl, it will mean the end of the sponger's existence and the beginning of the end of the "Dalles."[2]

I am so busy at the moment. Just think: the department, the lecture course and the difficult preparation for it, the work on brain anatomy and the Ecgonin,[3] for which nothing has been done during the past week—how is all this going to work out? I am going to economize both in time and money, and work on vigorously and valiantly now that I see more chances of getting ahead.

Lustgarten is back and, what's more, as a great man with a great invention, but he is very friendly. He was only one day in Hamburg and very depressed at that time because he thought his discovery had miscarried; which is why he didn't come to see you. Oh, they have all outstripped me in fame, but not in happiness and not in contentment so long as you are going to be mine.

Your

Sigmund

---

[1] A street in the older section of Hamburg.
[2] Yiddish—penury.
[3] A cocaine derivative which Freud had been asked to examine.

1885

1886

55 *To* MARTHA BERNAYS

Vienna, Thursday
January 6, 1885

My precious darling

In the confusion of the past few days I haven't found a moment's peace to write to you. The hospital is in an uproar. You will hear at once what it is all about.

On Sunday Koller[1] was on duty at the *Journal*, the man who made cocaine so famous and with whom I have recently become more intimate. He had a difference of opinion about some minor technical matter with the man who acts as surgeon for Billroth's clinic, and the latter suddenly called Koller a "Jewish swine." Now you must try to imagine the kind of atmosphere we live in here, the general bitterness—in short, we would all have reacted just as Koller did: by hitting the man in the face. The man rushed off, denounced Koller to the director who, however, called him down thoroughly and categorically took Koller's side. This was a great relief to us all. But since they are both reserve officers, he is obliged to challenge Koller to a duel and at this very moment they are fighting with sabres under rather severe conditions. Lustgarten and Bettelheim (the regimental surgeon) are Koller's seconds.

I am too upset to write any more now, but I won't send this letter off till I can tell you the result of the duel. So much could be said about all this.—

Your pleasure over the little presents made me very happy; surely Minna wouldn't think that I would confine her to a calendar! The Eliot[2] is for her, I have reminded them again. As for the money, my

---

[1] Dr. Carl Koller (1858-1944), known for the introduction of the use of cocaine in ophthalmic surgery. See letter of Aug. 15, 1924.

[2] George Eliot (Mary Ann Evans, 1819-1890).

little woman, you keep it; Minna has a claim to part of the previous sum; it will be a long time before either of you get more.

Paneth has given me six bottles of very good wine, some of which will go to my family, but some will be drunk by myself and others here in my room. One bottle has gone off today to Koller to fortify him for the fight. I am considering a reckless purchase. For the forty-two florins' interest from Paneth I am going to buy myself a decent silver watch with a chronograph in the back; it has the value of a scientific instrument, and my old wreck of a thing never keeps proper time. Without a watch I am really not a civilized person. These watches cost forty florins.—I am too impatient to go on writing.

So far my neuralgia injections are working very well; the trouble is I have very few cases. Yesterday I went to see Prof. Weinlechner[3] and Standhartner,[4] who gave me permission to use the treatment on all cases of this kind in their department. I hope to learn more soon about the value of the procedure.

I must now go and see if they are back.

All is well, my little woman. Our friend is quite unharmed and his opponent got two deep gashes. We are all delighted, a proud day for us. We are going to give Koller a present as a lasting reminder of his victory.

Farewell, my sweetheart, and write again soon to

Your

Sigmund

[3] Dr. Joseph Weinlechner (1829-1906), Professor of Surgery at the University of Vienna.
[4] Dr. Josef Standhartner (1818-1892), professor at the University of Vienna.

56 *To* MARTHA BERNAYS

Vienna, Wednesday
January 7, 1885

My beloved darling

At last a letter from you again, which makes me laugh, for it informs me that you now possess three copies of the article[1] you wanted. Now you can send one on to Rosa.

[1] See note 1 to letter of June 19, 1884.

On one point I cannot agree with you, Marty. You say how sensible we are now and how foolishly we treated one another in the past. I gladly agree that we are now sensible enough to believe in our love without any doubts, but we would never have reached this point had it not been for all that went before. It was the very intensity of my misery brought about by the many hours of suffering you caused me two years ago and since, that convinced me of my love for you. Today, what with all the work, chasing after money, position, and reputation, all of which hardly allows me time to drop you an affectionate line, I could never reach that conviction. Let us not despise the times when for me a day was made worth living merely by a letter from you, when a decision from you meant a decision between life and death. I really don't know what else I could have done at that time; it was a difficult period of struggle and finally of victory, and only after it was all over could I find the inner peace to work toward our future. In those days I was fighting for your love as I am now for your person, and you must admit that I had to work as hard for the one as I am now for the other.—

During the past few days I have been feeling a bit seedy, engaged as I am in the twofold struggle which forms the content of *Auch Einer*:[2] struggle against a cold and against "the object." I have a combination of nose-throat-gums-ear catarrh and am correspondingly miserable. I suggest you read about it in Vischer.

My object has a specific name, it is called neuralgia—face-ache. The question is whether I shall succeed in curing it. I have already told you about one case which has very much improved; but now I am treating a second, a clear, much nicer case, at Prof. Weinlechner's. The result of the first day was very good. But what will the following days produce? I am so excited about it, for if it works I would be assured for some time to come of attracting the attention so essential for getting on in the world. Everything we hope for would be there and perhaps even Fleischl could benefit from it. And even if it isn't absolutely sensational, something is bound to come out of it.

I now have eleven subscribers for the lecture course, but wretchedly few cases, and I am continuously worried as to how I am to find the necessary material, but I will manage.

[2] Novel by the German author Friedrich Theodor Vischer (1807-1887).

Yesterday evening I went to see Breuer, where I met Fleischl, who was very talkative but not in a particularly pleasant way. If only I could relieve him of the pain!

Goodnight, my little woman. You are quite right, it is sad that we cannot exchange kisses, only letters.

Your
Sigmund

57 *To* MARTHA BERNAYS

Vienna, Friday
January 16, 1885

My sweet darling

A very affectionate greeting for the seventeenth; do you realize, by the way, that my lecture course also started on a seventeenth? And here, quickly, is my news, to make you happy at once. The die is cast. Today I had my wild beard trimmed and went to see Nothnagel, handed in my card: "takes the liberty of asking if and when the Herr *Hofrat* can be seen on an important personal matter." The usual crowd, usual anxious whispering among the people round me, whether I was a doctor and would be admitted before them who had already been waiting so long. The conversation I understood best was between a lady in mourning and her brother. Her feminine eye immediately diagnosed something suspicious about me, whereas the brother with a superior smile contradicted her suggestion that I could be a member of that pernicious profession. At last came the disappointment, for I was admitted ahead of them all to the man who had so often played a decisive part in my life, and behind him once more the picture of the thoughtful, serious, dead woman. I asked whether he would like me to state my request now or later. If brief, now, he said, otherwise it would be better to talk it over another time. I promised to be brief. "You once said you would be willing to assist me, and I believed it because it was you who said so. Now the opportunity has arrived. I would like to ask your opinion whether on the strength of my existing publications I should apply for the *Dozentur* or whether I should wait till I have more." "What are your papers on, Doctor? Coca—" (So coca is associated with my name.) I interrupted him to produce my collected writings, those from the pre-

Marty days and those of a later date. He just glanced at the number.
"You seem to have eight or nine," he said. "Oh, by all means send
in your application. When I think of the kind of people who get
the *Dozentur* . . . ! There won't be the slightest objection." "But
I have several more things to be published, two of them in the
immediate future." "You won't need them; these are more than
enough." "But there isn't much about neuropathology among them."
"That doesn't matter. Who knows anything about neuropathology
unless he has studied anatomy and physiology? You do want the
Dozentur for neuropathology, don't you? In that case three people
will be chosen to report—Meynert, Bamberger,[1] and probably myself. There won't be any opposition, and if any objection is raised
on the faculty, surely we are men enough to put it through?" "So
I may assume that you will support my application for the *Dozentur?* I know Meynert will." "Certainly, and I don't think there
will be any objections; if there are, we'll push it through just the
same." I added: "It's a question of legalizing an unauthorized
lecture course I'm giving. Actually, I'm only lecturing to some
English people in their language, but there's quite a run on it."
Then we shook hands warmly and off I went as the newest *Dozent*.
I will send in my application next week. This time you won't miss
your golden snake.

Let one fond kiss stand for many from
Your
Sigmund

1 Dr. Heinrich von Bamberger (1822-1888), Professor of Pathology at the Universities of Würzburg and Vienna.

58 *To* MARTHA BERNAYS

Vienna, Wednesday
January 21, 1885

My little sweetheart

All kinds of things have been happening, so forgive me if
this is a bit muddled. There has been a rumor that the German
Kaiser[1] has died. But now he is supposed to be alive and will certainly outlive us all.

Today I handed in my application for the *Dozentur* and talked

1 William I of Germany, born in 1797, was eighty-eight years old at this time.

to Prof. Ludwig and Meynert. The latter was decidedly optimistic and also mentioned very suggestively the neurological ward on which he is counting. I have an idea that if he gets it he will take me. There must have been a lot of talk about me this evening. Fleischl has been invited to Meynert's and is going to put in a good word for me, while Ludwig has gone to a restaurant to work on the dangerous Kundradt,[2] the pathological anatomist.

And now for your letter. There is a lot to answer in it. First, if I will allow you to skate. Definitely not, I am too jealous for that. I myself cannot do it, and anyway wouldn't have the time to accompany you, and accompanied you would have to be. So drop that idea. Then I insist on your buying a decent rug up to the price of twenty-eight marks, which I will send you out of the earnings from my next lecture—at the moment I am quite poor. If you still have any money left, please use it for this purpose and I will pay you back.

Third, I really cannot see why you should be cold. Is there no stove, no wood in Wandsbek? Explanation urgently required. I trust it won't again come to the point of your not being able to write to me in one room because it's too cold and in the other because you are too often disturbed. That was the most terrible letter I have ever had from you, and one I certainly will not forget even if I live to be eighty-five and you were to give me a kiss every day, which is perhaps asking rather much! Darling, is it possible that you can be affectionate only in summer and that in winter you freeze up? Now sit down and answer me at once so that I will still have time to get myself a winter girl.

And what else? That you will have to be pretty unlucky to miss the golden snake this time. Perhaps you don't know that the brides of *Dozents* are obliged to wear golden snakes to distinguish themselves from ordinary doctor's brides.

One other thing I want to say: Just because someone happens to cross our path, you mustn't call him a nasty character. Pfungen especially is quite within his rights, and there is nothing nasty about his intentions.—Anyway, at the moment the worst has been avoided.

I intend to have a number of books bound. Beginning tomorrow, I am going to eat supper in my rooms. Otherwise I will get out of the habit of working in the evening.

[2] Dr. Hans Kundradt (1845-1893), Professor of Pathology and Anatomy at the University of Vienna.

Goodnight, little woman, be very good and love me a little bit.
                          Your
                          Sigmund

59 *To* MARTHA BERNAYS

Vienna
March 10, 1885

My sweet darling

Woe to the day you become so rich[1] that I, like a character in a bad novel, have to ask you politely whether you wish to continue to be my betrothed as I don't want to stand in the way of your happiness, etc. I am already looking forward to writing the letter and getting your answer—but I seem to remember that we already indulged in this kind of fantasy about a year ago. Don't you know, by the way, that only the poor have difficulty in accepting presents, never the rich?

Otherwise, my darling, I feel splendidly well, I am like "Lucky Hans," today the last of my various affairs will be settled and then I shall have a frightening amount of time, all the afternoon except one hour to use the ophthalmoscope, and the light is so good. I feel no desire to be lazy, I am in quite an industrious mood.—Today the application[2] has been handed in, the case seems hopeless, although Lustgarten has put in a word for me with Prof. Ludwig, and perhaps it will lead to the new *Primarius'* forming a good opinion of me and allowing me to lecture in his department.

I brought the lectures to an end today; a lady arrived to sign up for the next, and I had to tell her there wasn't going to be one.

Today, you know, marks a clear dividing line in my life; all the old things have been finished, and I am in a completely new situation. But it has been a good time; I have only pleasant memories of the lectures; it wasn't just the money, but the learning and the teaching; and it has enhanced my reputation in the hospital.

Shall I go and see Breuer today and say goodbye for the time I am going to dig myself in? I think I will. It is a long time since

---

[1] An aunt had promised Martha a sum of money as her wedding present.

[2] Freud's application for the position of Senior *Sekundararzt* (see letter of March 31, 1885).

I have felt so well as during these bad days and I have hardly ever looked so fit. I am not seeing the family now, it is too painful for me to admit my lack of money. They are aware of it, anyhow.

I am keeping on my old apartment this month, but I have changed my charwoman and feel all the better for it. I am now eating supper in my own room, modestly, but I enjoy it, and I can make plans, read, and write reports to my heart's content.

I wrote to Fleischl yesterday, but did not insist on an answer because writing is so difficult for him. On Friday or Saturday, when I have come to the end of my money, I will go and see him. I wonder if he will lend me anything. . . .

You don't write a word about Minna? I hope she will be all right when we meet again.

Fond greetings
Your Sigmund

I give you the solemn promise that I will marry you even if you don't get the 1500 marks. If necessary, I will marry you with 150,000,000 marks.

60 *To* MARTHA BERNAYS

Vienna, Tuesday
March 31, 1885

My sweet darling

Apart from your two charming letters, quite a number of pleasant things have come my way during the past few days on which I will now report to you in detail. First, my second paper on coca[1] has been printed verbatim in a *Zentralblatt*; second, I have received from Dr. Pritchard, whom you all know, a nice letter which I shall not fail to answer and which I herewith enclose. I am now especially pleased that I told him to go to Wandsbek. But then the most important thing: a few excellent discoveries in brain anatomy, five to six in number, which are to adorn my next big paper. Some of these things I am discovering are actually being published piecemeal every week by someone else (from

---

[1] "Über die Allgemeinwirkung des Cocains" ("On the General Effect of Cocaine"), (1885) *Medizinisch-chirurgisches Zentralblatt*, 20, No. 32, 374.

Leipzig); but I am going to wait patiently till I have all the material together before starting my paper. I am not sure if I should count the following event among the pleasant ones. My successful rival's election for the position of *Sekundararzt* has not been confirmed by the local government because he is a Hungarian, and from now on Hungarians are to be treated as foreigners. It is generally considered possible that the local government will appoint me in his place. But at the moment I have little desire to become once more part of the hospital establishment. What I want, as you know, is to go to Paris via Wandsbek, have enough leisure to finish my work on the brain, and then the independence to find out what chances there are for us here. If I do accept the position I shall first of all not be able to finish the work on the brain; secondly, I will not be granted leave for the journey and so would have to abandon the position again in two months; this would simply have the effect of annoying *Primarius* Hein;[2] on the other hand, if I renounce the journey and continue with the hospital routine, I would soon lose my patience. It's true, of course, I haven't got the traveling grant yet; lots of people would say it is sheer folly to turn down a job I applied for a month ago. But a human being's demon is the best part of him, it is himself. One shouldn't embark on anything unless one feels wholehearted about it. What do you think? Let me know.

It is four years today since I got my doctor's degree, and I celebrated the occasion by doing nothing and paying a call on Breuer at noon. Work starts again tomorrow. I am very well and my Marty, too, I trust. If only I could see what she looks like. Would I recognize her in the street? Now and again I see a girl in the street who looks like her in one way or another, whereupon I invariably follow her for a while to convince myself she isn't here. She probably won't see Vienna again until she is my wife. If only this could be soon

is the wish of

Your

Sigmund

[2] Dr. Julius Isidor Hein (1840-1885), *Primarius* at the Hospital of the Rudolf Foundation, Vienna.

61   *To* MARTHA BERNAYS

Vienna
April 28, 1885

My precious darling

To be angry with you for demanding news in such an impatient fashion is difficult, and I am sure you are not serious about it. I am too happy in the knowledge that someone loves me, and that this someone is you. By now I hope you are reassured. You can always believe what I say; don't you know that yet? Oh, what a wicked girl you are! If I were feeling better I would try and take you to task, but I am so dreadfully tired that I would be very grateful for a few affectionate words from you. But they would take two days to reach me; by then I hope to be over my tiredness.

Now I want to tell you about my minor projects. By the end of this week I will be relieved of my duty to consider myself a public danger.[1] I expect the tailor to bring my summer suit on Saturday morning. If the weather is fine I shall at once board a train and ramble about on the Semmering[2] for at least three days. Alone, without you, it cannot be beautiful; in fact, it is not meant to be a pleasure, simply a medicine. Before this, on the thirtieth, the chemist is to pay me for the research on cocaine. Now we must wait and see if everything works out according to plan.

This has been a bad, barren month. How glad I am it is soon coming to an end! I do nothing all day; sometimes I browse in Russian history, and now and again I torture the two rabbits which nibble away at turnips in the little room and make a mess of the floor. One intention as a matter of fact I have almost finished carrying out, an intention which a number of as yet unborn and unfortunate people will one day resent. Since you won't guess what kind of people I am referring to, I will tell you at once: they are my biographers. I have destroyed all my notes of the past fourteen years, as well as letters, scientific excerpts, and the manuscripts of my papers. As for letters, only those from the family have been spared. Yours, my darling, were never in danger. In doing so all

---

[1] Freud had been ill with a mild form of smallpox.
[2] A mountainous region south of Vienna.

old friendships and relationships presented themselves once again and then silently received the *coup de grâce* (my imagination is still living in Russian history); all my thoughts and feelings about the world in general and about myself in particular have been found unworthy of further existence. They will now have to be thought all over again, and I certainly had accumulated some scribbling. But that stuff settles round me like sand drifts round the Sphinx; soon nothing but my nostrils would have been visible above the paper; I couldn't have matured or died without worrying about who would get hold of those old papers. Everything, moreover, that lies beyond the great turning point in my life, beyond our love and my choice of profession, died long ago and must not be deprived of a worthy funeral. As for the biographers, let them worry, we have no desire to make it too easy for them. Each one of them will be right in his opinion of "The Development of the Hero," and I am already looking forward to seeing them go astray.

This morning I went to see Fleischl; I had already been there twice, but he was asleep. His condition has not changed; he begins to give his lectures today, and I am wondering how he will stand it. He keeps a parrot in his room, a bird which means more to him than many a human being. It is a creature with plumage of outrageous colors to which he attributes all manner of subtleties, whereas I maintain that it is very stupid. The beast emits a kind of croaking sound which he interprets as *Bröckerl;** today to my surprise it actually uttered its name, Lore, several times. The animal does know one trick: it spreads its wings on order, allowing one to admire their beauty, but today Fleischl spent a good half hour imploring the bird to display them with the kind of ardor that normally a man might muster once in a lifetime for a girl; and the brute paid not the slightest attention. He couldn't help admitting that this behavior testified to a bad character.

I have heard that Leidesdorf[3] and Pollitzer[3] have been won over to my side. I am expected to call on the latter, but not yet; I must first feel human again.

---

* Scraps of language.—*Tr.*

[3] The two Professors, Max Leidesdorf (1818-1889), Director of the Psychiatric Clinic, and Pollitzer had promised to vote in favor of Freud for the award of the traveling grant to Paris.

I greet you fondly and thank you for your many letters.

Your
Sigmund

62 *To* MARTHA BERNAYS

Vienna, Wednesday
April 29, 1885

Highly esteemed Princess

If you keep your word, there should be a letter from you tomorrow instead of a card, my sweet darling! I promise you that once we have got through this terrible period of waiting, you won't have to touch a pen for ten years. What do you think, shall we get married in August after I have got my *Dozentur?* I will have to take an apartment anyway and we are surely both eligible. Which reminds me: you never respond to such suggestions, you just let me talk and occasionally laugh at me—won't you for once kindly let me know what you think it is all going to be like, how long, how much, under what conditions, and so on? So I expect a very detailed exposé about our future.

I might add that today I feel once again in a healthy frame of mind, something like the high spirits of convalescence. So don't be annoyed, my darling. To be healthy is so wonderful if one isn't condemned to be alone. Well, once summer comes the sluggish waters will begin to stir.

Greetings and kisses from
Your
Sigmund

63 *To* MARTHA BERNAYS

Vienna, Thursday
May 7, 1885

My precious Princess

Today my treasures arrived and gave me the greatest pleasure. I had expected the writing set to be much more complicated; it will certainly come in handy; the little prescription block is quite charming; I can hardly believe you managed it with so little advice, but I won't begin using it yet, it is too beautiful, not till I start my own practice. And finally the crackers, they have a most wonderful spicy flavor; quite incredible that crackers can taste like that. While enjoying the tender care which you have lavished on me, a number of thoughts have occurred to me which can be summed up like this: preparation for marriage is like the writing of a paper: one never finishes it; one just has to set oneself a deadline and break off somewhere. And I for my part have decided that an end of our misery must be made on June 17, '87, that by then we must be man and wife, whether things are going well or badly or not at all. Why shouldn't we be able to endure some hardships together, and anyway for the first year we would be safe with your money—mine by that time would have been spent. The carrying out of this plan, which is meant very seriously, depends entirely on the consent of one person only—you, Marty.

Today I went to see the family and apart from this took a very bold step—went to Tischer[1] and ordered the two suits I need so urgently. Do you approve? When they heard who I am, I was received with open arms.

Goodnight, my sweet darling. It's 1:30 A.M.; the day has simply fled by. Hope to work tomorrow. The flowers are for Minna, and I hope there will be a letter for me tomorrow.

Your
Sigmund

[1] A tailor who was a friend of the Bernays family.

64 *To* MARTHA BERNAYS

Vienna, Tuesday
May 12, 1885

My precious little woman

I am as delighted about the new yes[1] in your last letter as I was about the first one. Let us arrange things as you suggest if it is at all possible. I realize of course that the time of worry and trouble won't come to an end even then, but I think you realize

[1] Martha's consent to fix the day of the wedding.

this too and will appreciate the fact that we shall be going through it together and that our greatest desire will have been fulfilled. This long waiting certainly makes us neither happier nor younger and, as you admit, it doesn't bring us to the end of worrying about our future. I take it that your consent is seriously meant and not inspired by some whim of the moment, and I am deeply happy about it; I can't express it in other words.

Now let me tell you that today I lectured for an hour in the Club on brain anatomy—actually for one person only, Prof. Obersteiner,[2] because the others must have been very bored; but I enjoyed it and dealt not too badly with the difficult subject. I have also just written a letter to Prof. Mendel in Berlin, editor of a neurological journal, asking him to include a preliminary paper of mine; I am rather pleased with these things. But you mustn't get the idea that I am doing anything but brain anatomy these days.

I knew you would be pleased about Tischer; I did it only for your sake, for I am rather overawed by his high prices. I have so far received only one suit, which I wear in the mornings (ophthalmology is clean work) and take off in the afternoons when I go to the laboratory.

This business about Dr. R., which so horrified you, is nothing bad and certainly not new. Patronage it is, but not misplaced, for he really is an able man. There is no other way of becoming an assistant professor except by a recommendation from an associate professor.

What does it matter about the cross?[3] We are not superstitious or piously orthodox.

Are you all right again, my darling? I am in excellent health. That bit of success in my work also helps me along. Nothnagel was at the meeting today, but very impatient to get home. Either one of his children was ill or ten patients at ten florins a head were waiting for him.

I have been to see Fleischl three times, but each time he was asleep. It's impossible to get along with Meynert; he doesn't listen, nor does he understand what one says. I am giving him lots of my slides, so he rather likes seeing me these days.

[2] Dr. Heinrich Obersteiner (1847-1922), Professor of Psychiatry and Director of the Sanatorium for Nervous Diseases at Oberdöbling.

[3] A jocular reference to Martha's remark that she ought to "go to the cross"—i.e., humble herself.

Goodnight, my darling, good luck to us and may our dreams come true.

Your
Sigmund

65 *To* MARTHA BERNAYS

Vienna, Sunday
May 17, 1885

Precious darling

While you are taking such pleasure in the activities and management of the household, I am at the moment tempted by the desire to solve the riddle of the structure of the brain; I think brain anatomy is the only legitimate rival you have or ever will have. Let me tell you about this first: I have had some rather lucky ideas recently and enough points of view from which to go on working. I even hope here and there to draw an important conclusion from my discoveries. Yesterday I received from Berlin a very flattering letter promising to include my paper, provided I send it at once and that it doesn't run to more than a page, as promised. When the letter came I was suffering from migraine, the third attack this week, by the way, although I am otherwise in excellent health —I suspect the tartar sauce I had for lunch in Fleischl's room disagreed with me—I took some cocaine, watched the migraine vanish at once, went on writing my paper as well as a letter to Prof. Mendel, but I was so wound up that I had to go on working and writing and couldn't get to sleep before four in the morning. Today I am in fine fettle again, very pleased with my paper[1] which is short but contains some very important information and should raise my esteem again in the eyes of the public. It is to appear either on the first or fifteenth of June, depending on whether it can still be placed or has to take its turn.

Today I missed making one important find because the slide in which I had put great hopes turned out to be useless, so now I shall have to wait for a new one. Otherwise I go on working hard.

You, Marty, why are you keeping me on tenterhooks like this?

---

[1] "Zur Kenntnis der Olivenzwischenschicht" ("Concerning the Knowledge of the Intermediary Layer of the Olive") *Neurologisches Zentralblatt* 4, No. 12, 268.

If what you want to tell no one but me concerns Elise and it is only on her account that you are all worked up and preoccupied, I am perfectly happy to wait till we have nothing more important to talk about. But if it concerns you and me as well, as is possible, then I very seriously ask to be informed, and shall be quite nervous till I know what it is about. What could there be that you don't want to write to me about? If it is Elise's private affair, all you have to do is to say so, and I won't want to hear any more about it.

If I would like to see you again? Darling, what a question! Where the money is coming from? Darling, I don't know. If it doesn't come from the very doubtful traveling grant, then I am quite willing to take a job for the summer with a rich family and visit you in the autumn with the money saved. Like last year, when I was offered a job for two hundred florins a month. That time I did turn it down, but never regretted it, for otherwise I would have missed my most successful period.

How nice of you to have remembered Paneth. I quite forgot about it. I realize that people will love me only on your account. Tischer has turned up with the second magnificent suit; I wonder if you will see it while it is still new. God only knows what I owe him already! I will pay him in installments as several colleagues in the hospital do.

The weather is lovely again today. I didn't go out, I dislike any distraction that doesn't make me forget I am alone. Going for walks makes me melancholy.

I have just finished a letter to Schönberg; now I must ask you for his address.

For our thirty-fifth monthly memorial I send you fond greetings and (long to!) kiss you many times.

Your
Sigmund

66 *To* MARTHA BERNAYS

Vienna, Tuesday
May 26, 1885

My precious darling

It would seem that as a result of the sympathy existing between us, your Whitsun has been no better than mine; that would

be bad. Did you never wonder when you left Vienna how we should
ever meet again? Don't you remember how pleased I was when
you promised me to remain here? I was well aware what I should
have to thank your fond relations for. As yet I see no road leading
to Wandsbek this year. My American is good for only forty florins
a month, if he lasts, and altogether won't produce more than 120
florins. Of this sum at least half will have to be put aside for
Mother. The traveling grant wouldn't get me to you before October, but how glad I would be even of that! You know what we
will do? We will just sit still and be discontented, I think, and
that's all.—I am not so fond and capable of work as I should be;
perhaps lack of sleep is to blame. Whitsunday night I was at
Fleischl's again, and consider the tangible benefit of such a night,
the intellectual elation, the stimulation and the clarifying of so
many opinions, well worth the loss of sleep. But at 4 A.M. I am
afraid I fell asleep in his armchair, and when I woke at 6:30 his
face showed its tenderest, most suffering expression, and he was
writing a treatise. This magic world of intellect and unhappiness
contributes a great deal, of course, toward estranging me from my
surroundings; I have seldom felt so uncomfortable in the hospital
as I do now. My work is also proceeding very slowly. I expect the
paper to appear at the beginning of June.

I have unfavorable news about the traveling grant which I trust
will be decided on Saturday. When I went to introduce myself to
Dittel[1] not long ago, he told me that one of my rivals, the more
dangerous of the two, is going to withdraw on account of his
"youth." Now, my hope was that the Christian and to me hostile
votes would be divided between the two other applicants, so that
neither could equal my number of votes, as I am certain of more
than a third of the total. If one of them withdraws, then of course
the other is more likely to beat me. Perhaps it wouldn't be so good
if I were granted it, anyhow. The strange thing is, I have had several vivid dreams about it. Once I dreamed that I had got it, but
the details were very blurred, and last night I dreamed very clearly
that I was present at the announcement of the decision and learned
the grant had been divided and "Christian" (it's not his name at
all) Dimmer had received the one half, I the other, whereupon I
wrote an extremely rude letter pointing out that three hundred

[1] Dr. Leopold Ritter von Dittel (1815-1898), Professor of Surgery at the University of Vienna.

florins might be sufficient to get me to Wandsbek but certainly not to Paris, and that I should have to borrow double this sum, in which case I preferred to withdraw! Paneth as a matter of fact has offered to prolong my stay in Paris out of his own pocket.

The Paneths have invited me to go and see them on Wednesday. Darling, darling, I wonder how you are, if you are feeling really well? For if so, we are sure to get if not silk, then at least linen bedspreads soon. Do you know that I really hate this whole trousseau business—and why? It strikes me as a very worthy object for jealousy. Understand?

Fondest greetings from
Your
Sigmund

Minna's birthday!

67 *To* MARTHA BERNAYS

Vienna
June 6, 1885

My precious darling

Well, something is happening at last! Today I got the invitation for the oral exam which I am to take on Saturday the thirteenth before the board of professors. A mock exam, I think; nothing more. But the things that go with it! Top hat and gloves to be bought, and then what kind of coat am I to wear? I have to appear in a dress coat—am I to hire it or have it made? I have just been to see Tischer and ordered a frock coat, but I am not at all sure whether to stick to the order, for in that case I would have to hire a dress coat for the oral and for the trial lecture, but I also need a black coat as well; in fact I need both. How on earth is this to be solved? I am absolutely bewildered, and when I think of all the debts!

My leave has been granted, I have borrowed a traveling bag from Paneth and I have just had a letter from Obersteiner telling me that I cannot sleep there[1] till Thursday. This will facilitate the move, as I can take my things over there in several stages. I am taking books and brain slides along.

[1] At the Sanatorium in Oberdöbling.

I handed over fifty florins to Mother today. Breuer has again behaved splendidly in the Fleischl affair. By saying only good things about him one doesn't give a proper picture of his character; one ought to emphasize the absence of so many bad things.

I would welcome the traveling grant, my darling, more than anything; I was really quite ready to renounce it when in my mind I had also abandoned the trip to Wandsbek; but now that I am assured of a small sum (a hundred florins) for the latter purpose, I find it hard to dismiss the thought of such valuable assistance. One hundred florins is such a small sum for the visit, no matter how we economize and even if I don't give you any presents; I would hardly be able to stay ten days after deducting the train fare. What's more, my salary will have ceased, for I cannot ask for an extension of leave; I have to resign on September 1, which I do only too gladly. It is so appalling not to have any money, my darling. I cannot imagine who invented the tale about women's dresses being so expensive that a man simply dare not marry! I shudder at the thought of my tailor's bill!

The events[2] expected to take place on the next two Saturdays are bound to tide me over the probable boredom in Döbling. My paper on brain anatomy is due to appear on June 15. The anniversary of our engagement and Minna's birthday also fall in this period. Really an eventful month. If only everything goes well!

My American has paid his first twenty florins; they are lying here for you. He is going to pay once every two weeks and generally owes a tribute to my princess and my princess's sister. My only other source of income at the moment is Baron Sp., who has paid me two visits and perhaps will pay me two more this month.

One thing worries me. I am so terribly lazy I daren't think seriously what it is going to lead to. And the heat on top of it all! Marty, you will realize that I am not quite in control of things today.

Please go on writing to me at the old address.

Fondest greetings, my precious darling.

Your

Sigmund

[2] The colloquium and the decision about the traveling grant.

68 *To* MARTHA BERNAYS

Monday, Heilanstalt in Oberdöbling[1]
June 8, 1885

My sweet darling

Well, one lives and learns. Yesterday was a most amusing day; and today also seems strange and funny, although it is only 10:30. It would be easier for me to talk to you about it than describe it, but I will see what I can do. Have you ever seen this sanatorium? Do you remember the lovely park at the end of the Hirschen Strasse which continues towards Grinzing[1] where the road curves? On a little hill in this park stands the sanatorium, which consists of the two-storied big "house," the small house, and a new building; opposite is the so-called nursing home for the chronic cases.

I arrived here yesterday at 8 A.M. with a walking stick for luggage and became a member of this very mixed community. I must describe the people in greater detail. First, there's Prof. B., the overlord, whom I have known so far only by sight and reputation. An old gentleman, converted Jew, twisted features, a little wig and stiff walk, the result of gout or some nervous disease. He is associate professor of psychiatry, superintendent of the lunatic asylum, and was Meynert's teacher, but the pupil has elbowed the teacher out of all scientific jobs; only in medical practice has he been unable to get the better of him. He is not particularly talented, rather what is known as very shrewd, an old practitioner of most doubtful character, egotistic, completely unreliable, and in spite of being sixty-five or more, ready to indulge in any kind of pleasure. He runs the sanatorium with a Dr. Obersteiner, a step-brother of Minister Haymerle.[2] As B. had a single daughter and Obersteiner a single son, they got married and young Prof. Obersteiner, the son-in-law of B., is the man who really runs the place. Obersteiner is a friend of Breuer, Fleischl, Exner,[3] et al., pupil of Brücke, so I

---

[1] Suburban districts of Vienna.

[2] Heinrich von Haymerle (1828-1881), Austrian Foreign Minister.

[3] Dr. Sigmund Ritter von Exner (1846-1926), Professor of Physiology at the University of Vienna, later (1891) successor to Professor von Brücke, and Department Head at the Ministry of Education (see letter of Nov. 25, 1901).

have known him for quite a while. I have often been to see him to borrow the books I need for my publications. He is small, thin and insignificant, but of a very amiable disposition, extremely conscientious and decent. As a scientist he is industrious without having achieved anything outstanding, as a physician timid and modest. His wife is tall, pale, with pleasant features and an unmistakable resemblance to her father; she is in charge of the household management of the establishment, gets up early and shares in all the work. I have also seen two children. The elder, a boy, unfortunately half-paralyzed as the result of a brain disease. At meals I also made the acquaintance of the assistant, Dr. K., who has already been here twelve years, a good-looking and terribly boring Teuton. He is married, lives in with his family; his wife looks strikingly like my niece, Pauline.[4] Both left today. The staff consists of an inspector, an imposing fellow, a Fräulein Toni in the kitchen, and a Fräulein Marie as a companion for the ladies, both respectable, solid matrons; countless male nurses and housemaids, the latter very pretty and probably chosen by the old professor. There are sixty patients in the house, mental cases of every shade from light feeble-mindedness, which the layman wouldn't detect, to the final stages of withdrawal. The medical treatment is negligible, confined to their secondary surgical and internal complaints; the rest consists of supervision, nursing, diet, and noninterference. The kitchen of course is in the [big] house. The mildest cases lunch with the director, the doctor, and the inspector. Needless to say, they are all rich people: counts, countesses, barons, etc. The *pièces de résistance* are two highnesses, Prince S. and Prince M. The latter, as you may remember, is a son of Marie Louise, wife of Napoleon, and thus, like our emperor, a grandson of Emperor Franz. You cannot imagine how dilapidated these princes and counts look, although they are not actually feeble-minded, rather a mixture of feeble-minded and eccentric.

Now for myself. I have every reason to be pleased with the reception I was given. The old professor welcomed me with great friendliness, inquired about the oral and the traveling grant, for which he held out some hope. I was alone with the professors for breakfast and coffee both yesterday and today. Obersteiner of course showed me the ropes and is as nice as he always has been.

---

4 Daughter of Freud's half brother Emanuel.

It seems that I impressed him quite a lot with some diagnoses, and he praised my usefulness in connection with an American who arrived the day before yesterday, and my talent for remembering the names and faces of patients, although in this respect he is bound to receive some disappointments. The food is very good; there's a second breakfast at 11:30, lunch at 3 P.M. Yesterday before supper I went into town. As I am still homeless, Obersteiner has lent me his library, a cool room with views over all the hills round Vienna; there is a microscope and the walls are covered with a wealth of literature on the nervous system, which will make it hard to be bored. They have arranged a corner of B.'s salon for me to eat and write in. At the moment the table is being laid in there; I am writing this in the library. On Thursday a room will be put at my disposal where I will also have my meals alone. The working schedule is as follows: from 8:30 to 10 A.M. we make our rounds together; Obersteiner then drives to town and returns between 2 and 4 P.M.; during his absence I am in charge; once in a while there is a clinical job to be done, a young lady may have to be fed with a probang, as happened today, information given to visitors or official committees. From the morning rounds until lunch at three I am free, apart from the possibility of the above-mentioned interruptions, and then from 3 to 7 P.M. when there is another round of the sickrooms to be made. Provided one is neither the director nor the cook, there is very little to do, and one really could lead an idyllic life here with wife and child if it weren't for the lack of the challenging and stimulating element of the struggle for existence. It is actually something like a civil service job. But if things outside don't go well and I am absolutely determined to continue work on brain anatomy, I will inquire of my little woman whether such an existence, in which she wouldn't have to worry even about the kitchen, would suit her. It has its pros and cons, but I don't want to think about it now.

Apart from the idleness and the good food, I shall be able to make other uses of these three weeks: I am writing a case history and collecting the material for a new publication in connection with which some anatomical examinations will have to be made. I will also study the slides which I made last month.

Yes, write to me at this address: Dr. S. F., physician in the Oberdöbling Sanatorium, Hirschengasse 71. I want to spend as much time as possible here so that the people in return for their good treatment will also benefit from my presence. I am having my consulting hours only on Wednesdays and Fridays, when my American comes, otherwise I won't have much to do in town.

Fondest greetings to you

Your

Sigmund

69 *To* MARTHA BERNAYS

Vienna, Friday
June 19, 1885

My beloved little woman

I have been yearning for you recently more than I ever have since that first time we were forced to separate. This is the result of your sweet, tender letter, which I carry with me wherever I go. I am so boundlessly happy about it, but I realize that such complete satisfaction renders one speechless. All I can tell you is that had it taken not three but seven years—according to our patriarch's custom—for my courting to succeed, I would have considered it neither too early nor too late. How silly that sounds, and how annoying it is when one is accustomed to be in command of the language and all of a sudden it refuses to obey. I have always respected you highly for the very reticence of which I have so often complained: I could never trust the love that readily responds to the first call and dismisses the right to grow and unfold with time and experience—no, I just cannot find the right words. I would rather dwell on how fast the next 2½ months are going to pass, how happy we will be with one another then and how we shall work toward keeping the time we have set ourselves, or even curtail it. Then I will tell you everything and you will understand me better than I can hope at the moment. But I do hope that you doubt my love only very rarely, only in moments of great agitation. You know, after all, how from the moment I first saw you I was determined—no, I was compelled—to woo you, and how I persisted, despite all the warnings of common sense, and how immeasurably happy I have been ever since, how I regained all my confidence and so on —my beloved Marty.

Life in the sanatorium is far more pleasant than I dared to expect, largely on account of Obersteiner's great, unhypocritical amiability, which springs from genuine goodness. I also get long very well with the old man; once in a while when someone comes to be treated by him, I act as his private assistant, and he has also promised to send me patients for electrical treatment. Here and there he drops a bit of advice—for instance, to concentrate on nervous diseases among children; if only one could get an official call for this!

Tomorrow, Saturday, they are not only going to report and vote on my oral as well as set me a subject and date for my trial lecture, there will also be a vote on the traveling grant, which is terribly important to me, although no longer quite so urgent as at the time when my coming to you depended exclusively on this decision.

I dream about this grant every night; yesterday, for instance, I dreamed that Brücke told me I couldn't get it, that there were seven other applicants, all of whom had greater chances!

With these two pictures our album is now full; the moment a new photograph is taken—in September—I have decided to start a new Martha album.

I greet you and Minna fondly and will soon give you further news [about the decision] and . . .

Your
Sigmund

70 *To* MARTHA BERNAYS

Vienna, Saturday
June 20, 1885

Princess, my little Princess

Oh, how wonderful it will be! I am coming with money and staying a long time and bringing something beautiful for you and then go on to Paris and become a great scholar and then come back to Vienna with a huge, enormous halo, and then we will soon get married, and I will cure all the incurable nervous cases and through you I shall be healthy and I will go on kissing you till you are strong and gay and happy—and "if they haven't died, they are still alive today."

I wanted to send you a telegram to say that I got the traveling

grant by thirteen votes to eight, but then realized you would have
two whole days without any further details, and so perhaps the
card will please you more. Your presentiment about the 1500
marks[1]= 608 florins has come true. I expect a lot of good to come
out of this windfall. It also goes to show that I am not unpopular
with the board of professors. I am quite unspeakably happy. June
really is a kind month. At the same meeting my *Dozentur* was also
approved, by nineteen votes to three. At the first ballot I got
nineteen to one. So only two fiends joined the opposition. My trial
lecture is a week from today, the twenty-seventh, on a topic connected with brain anatomy, which suits me very well.

I send you fondest greetings and just cannot get used to the
idea that I am lucky, too. But did I not have the greatest bit of
luck on June 17 three years ago!

With 100,000 kisses, all of which are to be cashed

Your

Sigmund

[1] The amount of the traveling grant.

71 *To* MARTHA BERNAYS

Vienna, Tuesday
June 23, 1885

My sweet darling
I have just received your long-awaited letter with the sad,
anticipated news. I cannot entirely share your opinion, but there
is probably no great divergence in our views. He [Schönberg] cannot marry her now, this is clear for every possible reason; he will
not be able to marry her if he dies from his illness, and he ought
not to see her as someone else's wife if he remains alive. There just
isn't any decision to be made, is there? It decides itself. To break
off an engagement now in view of a probable event for which one
can only wait—surely this isn't necessary. As for Minna herself, she
won't want to do anything but stand by Schönberg, as long as
there is such a person. And you wouldn't behave differently,
wouldn't leave me before I died, if it looked as though I were going
to die. And I certainly wouldn't give up what is most precious to
me as long as I am alive.—But we are alive and are going to be

happy, and poor Minna, unless India helps, will have a lot to forget. So for the moment do let her cling to the shred of hope that remains.

It is terribly hard, my beloved little woman. Do you realize that all happiness is hemmed in by limitations which one only has to think of in order to become very unhappy? Let us cling to one another all the closer so as to be of some support to all those around us.

Fondest greetings. The time is passing very slowly for me now. There is so much to wait for.

Your devoted
Sigmund

Do write Anna Hammerschlag a few words on a postcard, darling.

72 *To* MARTHA BERNAYS

Vienna, Thursday
June 25, 1885

Sweet darling

You are quite right to talk about Schönberg and Minna rather than ourselves. I feel the same, except that I lack the direct impression. If your observations are correct, Marty—and, alas, I don't doubt it—then we cannot count on his recovery and we will have to save our energy for the time when Minna needs to be comforted. He is lost, a good man, worthy of a loving memory. I wouldn't dare say this to a patient or a relative of his, for there are winners among the blanks, but how few! In my opinion the prognosis has been made.

What we can do, my darling? Not much, alas. You are right, we cannot change human society even for the sake of our patient, we cannot turn the man who has to work into one who can afford simply to enjoy life and take care of his health. This we cannot do, and there's the rub. It is not the disease that is incurable, it is a man's social standing and his obligations that become an in-curable disease. Keep going, my darling, you will have to be more than ever Minna's elder and only sister and at the same time not take anything away from me, for I will lend you to no one—no one. Take good care of yourself and of her. If I arrive and find

you both pale and sick I will go straight on to Paris—no, I will
stay and spend the entire six hundred florins in Wandsbek. I will
write to you about the project with Fleischl after supper.

Your

Sigmund

73 *To* MARTHA BERNAYS

Vienna, Friday

June 26, 1885

My sweet treasure

Nothing new except your dear letter. I am brooding over my
trial lecture[1] in the most appalling heat. There are certain difficulties, for after all one does want to say something intelligent,
and this is not so easy; on the other hand, it isn't much fun slaving
away for nothing and spending one's energy on things one may
not be able to use. One has about twenty minutes to talk in; if only
I knew exactly how much I can pack into that time; as it is, I am
afraid what I have prepared may come to an end before my allotted time, or that from behind the folding screen of my introduction—nothing will appear. Fortunately I have already formulated
the whole thing in my mind, and I intend to write it out this
evening.

Dolfi's birthday is on July 23, three days before yours.

I do wish I could bring you the beautiful roses, my darling, which
I so often find in my room. I share this favor with most of the
patients. You shall be compensated when I am in Wandsbek. I
do wish I could get some more money from somewhere; the two
months ahead must produce something so that instead of economizing we can live as we did last year. A little happiness does
everyone good, especially after so much bad luck. I must tell
you something: I feel I can hardly survive another two months;
when it was six months and there was no certain prospect of our
meeting, it was easier. You know, when one is traveling to America
the journey as far as Stockerau[2] passes very quickly, but from there

---

[1] A trial lecture had to be delivered and approved of in order to become a
Dozent (Lecturer).

[2] A railway station near Vienna.

time begins to drag. And the return journey is the same, the last bit from Stockerau seems so slow. I don't want to fall into the habit of putting everything off by saying, "We will talk about this when we are together." Just remember what things were like two months ago; I had smallpox, and that was all. Now so much has changed. I even think of my stay in Döbling as a lucky episode; Breuer pointed out to me that I may owe the grant to it, as it brought me the votes of B. and his friends.

I am just about to be called to supper—till then I will keep on writing. I think after all I would prefer the room in your house, for then I will get up early and startle you out of your sleep every morning with lots of kisses. One can love one another properly only when one is close. What is a memory compared to what one can behold!—

I am wondering who will turn up at the lecture tomorrow. I haven't invited anyone. Strange to think that I shall be standing in Brücke's auditorium where I did my first work and with an enthusiasm I have never known since, and where I had hoped to stand at least as an assistant beside the old man. Could this be an omen suggesting that I may after all return here for scientific work and teaching? Do you believe in omens? Since I learned that the first sight of a little girl sitting at a well-known long table talking so cleverly while peeling an apple with her delicate fingers, could disconcert me so lastingly, I have actually become quite superstitious. Do you remember, you unsuspecting worm?

If the energy I feel within me remains with me, we may yet leave behind us some traces of our complicated existence. I don't think I am ambitious, although not exactly unsusceptible to recognition. I want to have you all to myself, some freedom, and a few possessions; I want to keep my nervous system intact and to be left in peace by the rest of my body.

I have been reading off and on a few things by the "mad" Hoffmann,[3] mad, fantastic stuff, here and there a brilliant thought. Once, for instance, a fairy presents a bride with a necklace which has the power of preventing her from ever being annoyed about a greasy spot on her dress or a spoiled soup. Isn't that amusing?

Saturday, June 27

So now I can tell you that the trial lecture is over, too. At last!

[3] E. T. A. Hoffmann (1776-1822), German author.

It went off quite well, only at the end something funny happened. I finished before the Dean had shouted his "Sufficit" (Enough), bowed and went off, which is against the rules; one is supposed to go on talking until interrupted by the "Sufficit." Brücke, Meynert, and Fleischl were there. My friends didn't turn up till toward the end because the Dean had made me change places with someone else. It is only from now on that I am considered a real *Dozent*.

I had to have lunch with Fleischl afterwards, and didn't get home till four. I will stay here another two days; on Tuesday I will be in the hospital. My interest is of course concentrated on the St. Gilgen business.[4] I have an idea I will go, although I won't have the opportunity of doing any work, and for this reason I would rather it didn't come about. On the other hand I see clearly that I could be of some use to him. The whole story will remain undecided for several days, because on Sunday and Monday Breuer is going to join his wife in Berchtesgaden. I on the other hand am still two months away from my darling, to whom go fond greetings from her

privatdozierenden
Sigmund

It has been frightfully hot, and I am exhausted.

[4] As Fleischl's doctor, Freud was to accompany his friend to St. Gilgen, a lakeside resort in the Salzkammergut.

74 *To* MARTHA BERNAYS

Vienna, Sunday
July 5, 1885

My sweet darling

Look here, I really don't understand you at all. To be quite so good-natured as to let people get away with everything and to become incapable of taking offense really ceases to be a virtue. I am not prudish and respect you all the more for not being so, either; but how you, after all that has happened in connection with Elise, and above all after the last incident, could think of honoring her with your visit—this is beyond me. I will spare myself the sermon which you can preach to yourself, but it all reminds me so vividly of our bad times, the conditions for which I had thought were gone for good. A human being must be able to pull himself together to form a judgment, otherwise he turns into what we Viennese call a *guten Potschen.** It is just the kind of thing that almost persuaded me to leave you to Herr Fritz Wahle when he insisted he had older claims on you and you failed to find word or gesture to turn them down. *Pfui,* I don't want to think about that and what might have happened as a result, and you really shouldn't do anything to remind me of it. What is the good of your feeling that you are now so mature that this relationship can't do you any harm? A girl doesn't intentionally lower herself to irresponsible behavior such as your friend has always suggested and finally quite openly displayed. I am not worrying about the question of decency, which doesn't seem to worry Elise, but about the utter weakness and lack of principle. Let her by all means be the poor girl who looks for a man, no matter where, and let us be glad she has found one. But don't put yourself on the same level by keeping up friendly relations. Don't say I'm too hard; you are far too soft, and this is something I have got to correct, for what one of us does will also be charged to the other's account. You have given me a bad day, Marty; please let me know soon that you feel a bit sorry about it. I know that with you all this springs from pity, but human beings, apart from feeling pity for others, must have consideration for themselves as well.

I just couldn't let all this pass without reproaching you, and I don't think I shall regret it. But, my poor sweet darling, are you so unhappy at home that you want to leave under any conditions? Please let me know what is going on. What is it that has come between us? Am I still not your confidant? Can you separate your confidence from your love? Just you wait, when I come you will soon get used to having a master again. And a severe one, too, but you couldn't have one who loves you more or who could be so deeply concerned about you. This you know yourself. Oh, how I curse the wretched time till September, till I can snatch a kiss from my sweet, good little princess who is incapable of being cross, which makes me cross.

Your
Sigmund

* A doormat.—*Tr.*

75 *To* MARTHA BERNAYS

Vienna, Sunday, at night
July 5, 1885

My sweet darling

I have never written to you from this place before. I am
sitting at Fleischl's desk while he is asleep next door and I don't
know how long I shall be able to write. I am writing to you because I so deeply regret the angry words I wrote to you this afternoon. I don't know how you will take what I said, and that's why
I am adding this. I am so terribly sorry if I said something which
sounded unloving, and especially from this distance. When we are
together a serious or severe word can be followed at once by a
tender reconciliation, and the little storm only testifies to the
soundness of the structure. But when we are so far apart each
word has time to engrave itself on one's memory, and there is no
friendly hand to smooth it out. I don't know what to do. I just
couldn't accept what you wanted to do without making serious objections, and yet—I realize how one can offend by love the person
one loves most—and yet I am now, afterwards, inclined to let anything happen rather than have brought you to the verge of tears.

I do hope you won't take it too much to heart. You are so good,
and sweet, and full of compassion and kind interpretations, you
surely won't consider the influence I am trying to exert upon you as
the result of an unloving and unjust disposition. Do you remember
how once, after we had parted in anger, I soon came back to you,
and you said you would never forget? In the same way I am not
ashamed to come back to you now and ask once more for a kind
word, a friendly glance. You are my precious little woman and
even if you make a mistake you are none the less so, but you must
be able to take some criticism and return it if you feel like doing
so. There was a time when you did me a gross injustice and caused
me great suffering and I think you really had to hear about it often
enough, but believe me, it affected me more deeply than anything
else in my life. But if you are in my debt I am proud of it, and if
my love were not so strong I would have been less violent in my
accusations and not so sensitive to the memory of them now.

If you are still keeping something from me, I promise I won't say another word about it, but please clear it up soon and don't do it again, my beloved girl. After all, it is so unnecessary to make me feel I needn't know everything that concerns you; you know this is just what I demand of you, and I do it myself. You have also often promised me to do it and I have always been so happy when I have been able to think of you with absolute trust. Now we will soon be seeing one another again and have a lovely time together, and then it looks as though I can expect a swift promotion and I will fetch you at the appointed time and then we will live in such unclouded happiness and undisturbed closeness, and for this time, which is soon to come, you must give me some credit and you won't be disappointed. But you know all this, my sweet child, and I am sure you are glad we are so much closer to our goal, and when in an hour of yearning you write as you did recently for our anniversary, I fall silent with joy and am so grateful to have you, but I cannot put it into words.

So what are you doing, precious Martha, my deeply loved little bride; why do you want to leave home, what can I do to make the journey possible? After all, we have money enough at the moment. Please unburden your heart to me once more. Could I possibly have become too inattentive to understand your subtly implied wishes by reporting day after day on the favorable and unfavorable things that occur to me on my way to you? Just demand what you want, just tell me, don't make me feel too much how far away I am from you. Remain for me as fresh and gay as I can only imagine you to be; we will survive this slice of time, too—and then I will never write to you again, nothing good and nothing bad, for I will never leave you till I realize I have become a bit too much for you, and then I will go away and wait and see if you will call me back.

Your devoted
Sigmund

76 *To* MARTHA BERNAYS

Meidling,[1] 12:45 P.M.
July 23

My little princess

Your card received early this morning; I am so sorry I didn't understand your Hamburgese. Very sweet of you to send me the

[1] A suburb of Vienna.

five marks, girl; why are we already anticipating the bad times?
Please send me a financial report. For your information, I already
have a trunk and a traveling bag; Moritz[2] has left both with me.

Now to explain the situation. We—Dolfi and I, of course—are
lunching here on our great Semmering excursion. At 1:30 we are
taking the train to Payerbach,[3] then we walk part of the way, spend
the night somewhere and return early tomorrow morning. Quite
an experience for the little one.

Semmering, 10 P.M.

Everything has gone very well, most glorious weather, excellent
butter, honey, and a quarter *Gespritzter,** everything at its best.
Having thought at the outset that any pleasure without you must
be a torture, I ended up by enjoying it myself. We took the road
from Klamm into the Adlitzgräben, then on to Semmering. In the
Adlitzgräben we came upon a charmingly situated *Gasthaus* with
a dear, tiny little waitress, and Dolfi, with her common sense,
suggested spending the night there. But I insisted on going on to
the Archduke Johann on the Styrian border.

Now come the adventures. We arrive there; despite hints I fail
to ask if we may spend the night. We take another walk, return
late—and they won't have us. Now by faint moonlight over the
mountain to the hotel, then to the tourist house, to the dairy, no
room anywhere; we inquire the way to the *Gasthaus* but they refuse
to believe we can find it in the dark. Finally the innkeeper con-
sents to make up some beds in his little dining room, and here
we are eating our supper in peace. Dolfi is holding out very well,
she marches like an old soldier, has no fear of "dark forests," is
continually gay and happy and doesn't reproach me in the least
although she has every reason for doing so. I have given free play
to all the fountains of my irresponsibility: I haven't even brought
enough money along, and she has to help me out! I know how you
would scold me under these circumstances, but in that case I
could kiss you, and then I would have deserved it. I really have
arranged things very stupidly, but I am enjoying myself very much
all the same. Oh, if only I had you with me, my sweet princess! I
will really have to pull myself together during the next forty days

---

[2] A cousin, Moritz Freud, later married to Freud's third sister, Marie (Mitzi,
1861-1942).

[3] Small mountain resort at the foot of the Semmering.

* White wine and club soda.—*Tr.*

and reduce my careless way of living. I can understand very well why you don't like my counting the days. I wanted to give my poor little sister a nice birthday and this has been a success. But from now on till I see you there will be only serious, hardworking, and thrifty days so as to feel I have earned my luck.

I tore the sheet of notepaper I brought along in two, gave one half to Dolfi, and am writing this on a scrap of paper I happened to have on me. We return tomorrow morning; I want to spend your birthday in solitary contemplation.

Goodnight, my little woman.
Your
Sigmund

77 *To* MARTHA BERNAYS

Vienna, Wednesday
August 5, 1885

My precious darling

You too are surprised at Schönberg's traveling plans, aren't you? As a matter of fact he isn't here; please have a look at the card and see whether he says the "middle of next month" or "week." Perhaps it is just a slip of the pen. Yes, we shall be very happy in Wandsbek; anyway for me Wandsbek is not what it is for you. More like what Vienna is for me.

I must give you a report on last night's jollification with our chief. It was very genial, particularly in his company; only on our way home we had to admit that the evening had left a bad taste in our mouths. But from this kind of remark one doesn't learn much; I must describe it to you in detail. Well, he—do you know L.?

I will tell you his history. Once upon a time there lived in Pressburg[1] a certain Moritz L. (certainly not the only one of that name), a man as poor as only a Jew can be. Or rather, his father was as poor as that; I am not sure whether he was a peddler or a *schammes*[2] or a dealer in secondhand clothes, I think the second of the three. But this didn't prevent the young Moritz from at-

---

[1] Small town on the Austro-Hungarian frontier.
[2] Yiddish—caretaker of a synagogue.

tending Pressburg grammar school, and there it transpired that he
had come into the world with a so-called Jewish brain. At school
he was almost invariably top, very industrious, although he had
to spend most of his time giving private lessons; a typical little Jew
with sly features, a boundless vivaciousness and the gift of the gab,
otherwise quite an honest fellow. Needless to say, Moritz L. had
to go to Vienna to study medicine; there he suffered months and
years of starvation, finally became a tutor and as a result could
take his doctor's degree in peace. Then he joined the hospital, for
a while he was assistant physician at Loew's sanatorium, and there
he made the acquaintance of the professors. The day came for him
to be made *Sekundararzt* and his application got him the position
with J., which was more or less the same as being his assistant. At
this time J. was writing his great work and needed a stenographer,
for which he employed his *Sekundararzt* who before long became a
collaborator. Thus Dr. Moritz L. grew to be an intimate of the J.
household, in which considerable brutality and crudeness existed,
but also a beautiful daughter by the name of Marthe. Suddenly he
became J.'s son-in-law; rumor had it that this was considered necessary because the child of the Viennese professor had grown too
fond of the intelligent Jewish boy. But there is no need to believe
this—or, if one does, to assume any evil intention on the part of
either. In short, L., son of the *Schammes,* turned into L. the
Dozent, associate professor, and finally, since 1881, J.'s successor,
who by the way is very fond of his wife and invited us only because
she is in the country and he was feeling lonely.

And there he sat, the same man despite his forty-eight years,
the same sly features and incessant talk. We were his guests: old
Dr. Ricchetti,[3] Lustgarten, and I, the two *Aspirants,* and an American who has been attached to the clinic for a long time. He treated
us to fish, meat, poultry, cheese, beer, wine, champagne, and
cigars and never for a moment stopped talking, mostly about himself; as a true parvenu he indulged in memories of his poverty-
stricken youth when one coffee a day was all he had to sustain
him; he revealed the whole series of lucky incidents to which he
owed his rise to the position of J.'s assistant; he was very full of
himself and very happy. The atmosphere was unrestrained, for

---

[3] Dr. Ricchetti lived and practiced in Venice. Freud met him again in Paris (see
letter of Nov. 19, 1885).

the character of this vain, rather transparent man has aspects which one cannot help respecting, among them the lack of any trace of pompousness and conceit, no shame at being a Jew. There isn't very much to be said against him except that he seems to consider he has fulfilled his task in life by having been so fortunate. Since becoming a professor he hasn't made a worthwhile contribution, and instead of assuming the leadership in his field, he lets everything pass him by and amuses himself by cultivating the role of a wordly-wise, rather senile observer.

That in spite of all my moderation the banquet has not contributed to making me feel any better is, I suppose, to be expected. Yesterday and today have been bad days despite Karlsbad salts and a cold bath. The truth is, my energy has been spent again; every inch of me is longing to be happy with you, to kiss you, and to be compensated for this endlessly long year.

Please thank Minna warmly for her friendly letter; I am too lazy and feel too disagreeable to answer now. But don't you worry, I will revive as though touched by a magic wand the moment I am with you, and then everything will be so beautiful.

Fondest greetings and kisses
Your
Sigmund

78 *To* MARTHA BERNAYS

Vienna, Thursday
August 6, 1885

My lovable little woman

I have just got back from Baden. As you have seen him [Schönberg] recently I won't describe to you what he looks like now—without blood or flesh, without voice or breath. One of his lungs is completely destroyed, and the other probably riddled with disease. I consider him a lost man; how fast or how slowly the miserable remains will take to burn themselves out, I don't know; in Vienna he would probably last three months; what could be achieved by a better climate, care, and quiet remains to be seen.

For us in any case he is lost. His wretched soul is weary; enthusiasm for an aim, passion, the halo with which one surrounds

the woman of one's choice, all these are the products of health.
When the breath comes short, interest narrows, the heart abandons
all desire, nothing remains but a tired, resigned philosopher; he
has found his way back to the once despised family, is incapable
of holding anything against them, grateful for any attention they
show him, and above all is in need of peace, only peace. What a
lot there would have been to talk about in the past if one of us had
seen our girls! When I broached the subject, he said: "You agree,
don't you, with my having broken off the engagement?" I then
began repeating to him what I had written to Minna: that the
breaking off is not important since their feelings remain the same
and that otherwise they were dependent on circumstances over
which they had no control; but he said "no," and I suddenly realized
I was wrong and that his love had died before him. What has
brought him to the point of renouncing everything he has clung
to for so long—work and position, independence from his brother,
and his own willfulness—I don't know. Is this the end of a long,
hard struggle or a symptom of the psyche going to sleep? He had
difficulty in speaking, and said little of importance beyond warding
off my anger with his family.

Tomorrow or Tuesday I will go out and see him with Dr. Müller.
For I have made it quite clear to him that I am acting simply as
his friend and have also told Geza[1] a few bitter truths, but that
chap is too stupid and conceited. As a matter of fact they now
seem to realize he is in need of being taken care of and agree to
everything, the silly fools.

Schönberg let one word slip out which hurt me very much. He
told me how good Martha had been to him and that she looked
well but had "dark rings under her eyes"—Why has my little
woman dark rings under her eyes? From that moment my spirits
fell and all the selfishness of a human being revealed itself by my
being much more shaken by your dark rings than by the poor man's
deplorable condition.

When I come I am going to fatten you up and kiss some color
into your cheeks; and just you wait, I am not going to allow myself to be sent away on October first.

Your

Sigmund

[1] Schönberg's brother.

79 *To* MARTHA BERNAYS

Vienna, Wednesday
August 12, 1885

My wandering princess

Fancy, Lübeck! Should that be allowed? Two single girls traveling alone in North Germany! This is a revolt against the male prerogative, the beginning of the realization that one doesn't have to be lonely without a man. Haven't you had any adventures? I would rather have enjoyed that. So there is nothing left for me to do but express my pleasure that you got along so well in Lübeck, which I do herewith.

You won't expect any great change in my circumstances since last night, but something new I can report is that I received a summons to appear tomorrow at the police station; but don't be alarmed, it is obviously connected with my *Dozentur*. The government wants to know whether I haven't some vile deed to confess that would render me unworthy of the noble title. I am determined not to divulge a thing.

Eliot's *Middlemarch* in four volumes is lying in front of me, and my handkerchiefs are coming to an end, but not so my cold. And now I am off to luncheon and when I get back I will report on my financial misery.

I have to tell you about it because to my own deep regret I must reveal the fact that I won't be able to bring you any present, something I had been looking forward to so much, probably more than you. So listen. I have asked Paneth to send me three hundred florins, and I intend to borrow a further ninety from Breuer, which will bring my debt to him to fifteen hundred. Now, this is to be distributed as follows: one hundred florins to Tischer, the first and probably the only installment for a long time to come; two hundred florins for September in Wandsbek, journey included, rather too little than too much, for last year we went through more than that and still had to economize toward the end. This leaves only one hundred and seventy for my stay. Now, of the ninety florins from Breuer, seventy-five goes to the bookseller, seven to the shoemaker, five to the French teacher (I have decided to take only five more lessons). Trunk, box, packer—will thirty florins be too much for all this? (A hat—but no, there will be time for that in Hamburg.) My charwoman, five to eight florins; in short, I realize I shall be very lucky if I am left with twenty florins, of which something must go to those at home. If I am left with thirty, which would pay for the journey, then I would bring a full two hundred florins to Wandsbek, which is what I would like to do. Remember, we won't be getting another penny before October 1. Thus there won't be anything left for you, my darling; I feel the loss of the forty florins caused by the departure of my last patient. That I won't be able to bring you anything I have foreseen ever since then, and it doesn't contribute to raising my spirits, for I would love to have given myself this pleasure. May I point out that if I leave on the thirtieth, there are still eighteen days to be endured? A long and difficult time, and one not likely to make me feel any better.

Let us take some French lessons in Wandsbek; can you find a decent and not too expensive teacher? Life is so full of little worries.

Fondest greetings.
Your
Sigmund

80 *To* MARTHA BERNAYS

Vienna, Friday
August 14, 1885

My precious little woman

I went to Baden yesterday afternoon and didn't get back till this morning. Astonishing how much better he [Schönberg] looks; most of the symptoms have subsided; I don't think the end will come for some time. He is so satisfied with the way his brother and sister-in-law are looking after him that I too find myself reconciled to this couple. In conversation the brother tried to clear himself of all responsibility for the neglect, and in fact it turns out that it was mainly because of Schönberg's irritability while here, his permanent secretiveness and his protestations about his improving health while in England, that the family didn't make it their business to look after his health before. We went to the theater in the Baden Arena, to a box where one could smoke!—and during the performance of *The Beggar Student*[1] I questioned him much as a father confessor questions his penitent.

I spent the night with Schönberg to give him a chance of confiding in me once more; he has a beautiful room in the Frauenhof, and we talked for a long time before going to sleep. I asked him outright why, when we sounded the alarm, he put up such a resistance against coming here. His answer seemed to be sincere. For diplomatic reasons, he said. At that time he wasn't ill enough to expect any help from his family and was afraid that if he came here of his own accord he would again be without any financial support; he added that he had the greatest difficulties in obtaining the money for the journey to Oxford. When it comes to guilt, human beings seem to be quite willing to share!

I also asked him if he had any particular wish, and he answered that his greatest, indeed his only wish at the moment was to know for certain that his engagement to Minna had been broken off. He said he had informed his brother that he had been engaged but that he wasn't any longer, because he had felt an overwhelming desire to establish in the presence of one person that he is not engaged. He added that he felt an actual aversion to Minna, hoped her vivacious temperament would soon help her to forget him and that she will behave as an unattached girl; I promised to do everything to make things easy for him. "Don't you think that the psychic burden of such a relationship is too great for me? My egotism is beginning to assert itself. All I really want is to keep going for a few years." The poor man, I told him that of course the dislike he had taken to everything was the result of his exhaustion. During the evening, after he had bowed to a lady, his sister-in-law remarked: "It's also a sign of improvement that he is beginning to be more polite to women again." I leave it to your judgment, which in Schönberg's case has always turned out to be so correct, how much of this you will feel like passing on to Minna. The final prognosis of his condition remains the same, in spite of the present improvement.

In the evening I kept thinking that if serious illness made it impossible for us to marry, we two would behave differently. I have been looking upon you for a long time as my own, and I would

[1] Operetta by Karl Millöcker (1842-1899).

never set you free; I would accept the fact that you suffered with
and about me, and I doubt that you, my little woman, would do
otherwise. A human being is so miserable when all he wants is to
stay alive.

I am going to see Hammerschlag tomorrow morning, spend the
night there, and on Sunday go on to Herzig. I am delighted to hear
the news from Segeberg.[2]

Fondest greetings from
Your
Sigmund

[2] Spa in Holstein where Martha had gone for a holiday.

81 *To* MARTHA BERNAYS

Paris
October 19, 1885

My beloved darling

My lazy life could have come to an end today. I went to the
Salpêtrière, which is just as big and has as many courtyards as our
hospital, to introduce myself to the medical assistant and ask when
Charcot[1] is expected. The assistant, however, was not there; he
has already been replaced by a new one, and Charcot was in the
wards. I could have gone in there, but I had left my introduction at
home, and so this step upon which so much depends has to wait
till tomorrow. The *Consultation Externe*—i.e., the consultation for
outpatients—takes place at 9:30. So perhaps by tomorrow I shall
already be occupied with work. The lectures at the École de
Médecine don't begin till November 5, but if things with Charcot
work out well I shall hardly have much to do there. The medical
library is on the first floor of the École de Médecine; it contains a
great number of magazines, including German and English ones,
and here I shall probably spend many an hour.

One of Charcot's books, which I already own in the German
translation, I have now bought for four francs so as to learn French
from the translation. My laziness is beginning to worry me terribly;

[1] Jean Martin Charcot (1825-1893), Professor of Pathological Anatomy at the
University of Paris and Director of the Clinic for Nervous Diseases at the *Salpêtrière*.
Famous neurologist, best known for his work on hysteria.

for days now my sense of guilt has not allowed me one calm hour. Apart from some subjective and scientific profit, I expect so little from my stay here that in this respect I cannot be disappointed.

I can barely remember what I did yesterday. My evening at the theater on the seventeenth gave me migraine. Believe it or not, performances here last from eight till midnight! and the heat is appalling. I went with John;[2] the lowest (i.e., highest) seat costs one franc; we paid 1.50, *quatrième loge de côté,* really disgraceful pigeonhole boxes, in a corner of the highest balcony, where one is aware of being alone, but not much more. Just think of our evening in the theater in Hamburg! I was struck by the total absence of elegant dresses; I suppose they keep these for the Opera. There is no music, no orchestra, and the signal for the play to begin consists of three blows with a hammer behind the curtain. The plays were *Le Mariage Forcé, Tartuffe* and *Les Précieuses Ridicules,* all by Molière, and although I couldn't understand a word of what the women, and only half of what the men said, I enjoyed enormously the brilliant acting. *Tartuffe* I knew, of course, and what was remarkable about the last play was not so much the dialogue as the high comedy of the two Coquelins.[3] During *Tartuffe* the audience applauded every speech of any length. My migraine is rather discouraging me from frequent visits to the theater; I really went in the hope of learning French, for I have no one to talk to and every day I seem to get worse at uttering these wretched sounds. I don't think I am mistaken if I say already that I shall never achieve a tolerable "accent," but it must at least be possible to construct a sentence correctly.

The walk I took three days ago, of which I owe you a description, led along the Quai d'Orsay, where the ministries are, past the Dôme des Invalides, across the Seine and on to the Avenue des Champs Elysées, the most stylish part of Paris, as John would say; here there are no shops at all and people travel only by horse and carriage. Elegant ladies walk here with expressions suggesting that they deny the existence in this world of anyone but themselves and their husbands or are at least graciously trying to ignore it; one side of the avenue is formed by an extensive park in which the

---

[2] The artist John Philipp, a cousin of Martha's.

[3] The brothers Benoît Constant (1841-1909) and Ernest (1848-1909) Coquelin, famous French actors.

prettiest children spin their tops, ride on merry-go-rounds, watch
the Punch-&-Judy show, or drive themselves about in little carriages
drawn by goats. On the benches sit wet nurses feeding their
babies, and nursemaids to whom the children dash screaming
after they have had a quarrel. I couldn't help thinking of poor
Mitzi[4] and grew very, very furious and full of revolutionary
thoughts. Walking on, one comes to the Place de la Concorde, in
the center of which stands a real obelisk from Luxor. Imagine, a
genuine obelisk, scribbled all over with the most beautiful birds'
heads, little seated men and other hieroglyphs, at least 3000 years
older than the vulgar crowd around it, built in honor of a king
whose name today only a few people can read and who, but for
this monument, might be forgotten! The Place de la Concorde
leads into the Tuileries Gardens, which you can think of as being
very like the square between the two *Burgtoren* in Vienna (including the *Volksgarten* and both museums). Then comes the
Louvre. Now I remember, of course—yesterday I went to the
Louvre, at least to the antiquities wing, which contains an incredible number of Greek and Roman statues, gravestones, inscriptions,
and relics. I saw a few wonderful things, ancient gods represented
over and over again, as well as the famous armless Venus de Milo
to whom I paid the traditional compliment. I remember that old
Mendelssohn (the father in *The Family M.*)[5] reported on it from
Paris as a new acquisition without any great show of enthusiasm.
I believe the beauty of the statue was not discovered till later, and
that it has become fashionable to think so. For me these things have
more historical than aesthetic interest. What attracted me most
was the large number of emperors' busts, some of them excellent
characterizations. Most of them are represented several times and
don't look in the least alike. Many of them must have been produced in factories according to the prevailing fashion. I just had
time for a fleeting glance at the Assyrian and Egyptian rooms,
which I must visit again several times. There were Assyrian kings
—tall as trees and holding lions for lapdogs in their arms, winged
human animals with beautifully dressed hair, cuneiform inscriptions as clear as if they had been done yesterday, and then Egyptian

---

[4] Freud's sister Marie (see footnote 2 to letter of July 23, 1885), who at this time
had accepted a position as a governess.

[5] *The Family Mendelssohn*, by Sebastian Hensel, Berlin, 1879.

bas-reliefs decorated in fiery colors, veritable colossi of kings, real sphinxes, a dreamlike world.

Today I walked in an arc similar to that of three days ago, but away from the Seine and off the map which I sent you the day before yesterday. I found myself surrounded by the most frantic Paris hubbub until I worked my way through to the well-known Boulevards and the Rue Richelieu. On the Place de la République I saw the gigantic statue with the pictorial presentations of the years 1789, 1792, 1830, 1848, and 1870. This gives some idea of the poor Republic's interrupted existence. Yesterday the bye-elections took place in France (and Paris); all the Republicans got together, for as a result of the split between the Opportunists and the Radicals, almost only Monarchists got in at the first election. The yelling of the newspaper vendors was deafening; some papers appeared in four and five editions, and I myself bought two copies. Needless to say, these bye-elections are now Republican.

Do you like the way I write from Paris? It strikes me that with all the news and descriptions I hardly ever have time for anything personal.

It is now a week since I saw you, and every day I still think I am going to see you. Again I can hardly imagine what you look like! Would it have been better for me to have gone to Berlin? I could have left there every Saturday evening and spent Sundays with you. The great benefit of my stay in Wandsbek, the physical fitness and mental calm, are still with me, but I can't really enjoy myself; I am too much in love and feel too much out of place.

News in brief; the coffee here is delicious everywhere, and the children wear the same kind of blouses as yours from San Francisco. Just think, for three toilet articles (some talcum, tar, and mouthwash) I had to pay 3.50 francs. And then one is expected to economize!

You don't mention the tooth trouble. Please tell me everything. Your last letter was not even signed; but it was from you all right, for who else would write to me so affectionately?

> Your devoted
> Sigmund

82 *To* MARTHA BERNAYS

Paris
October 21, 1885

My beloved treasure

Your letters know how to find me by now, so there is no
need for a more precise address. . . .

Today you may miss the note of melancholy to which you will
have grown accustomed in my letters from Paris. The reason is
that I spent yesterday and today in the Salpêtrière, where everything went off better than I had expected. I am already in the
midst of work and full of hope. For a deposit of three francs I was
given the key of a closet in the laboratory and a *tablier* (apron)
by the hospital administration. On the receipt I am described as
"M. Freud, élève de médecine." Let me tell you about it in detail.
When I arrived in the Salpêtrière yesterday morning the *Consultation Externe*—i.e., for outpatients—was being held. In one room
sat the patients, in the other, a small one, several guest doctors, the
internes, and the *Chef de Clinique* M. Marie,[1] who examined the
patients as they were admitted one at a time. At ten o'clock M. Charcot arrived, a tall man of fifty-eight, wearing a top hat, with dark,
strangely soft eyes (or rather, one is; the other is expressionless and
has an inward cast), long wisps of hair stuck behind his ears, clean
shaven, very expressive features with full protruding lips—in short,
like a worldly priest from whom one expects a ready wit and an appreciation of good living. He sat down and began examining the
patients. I was very much impressed by his brilliant diagnosis and
the lively interest he took in everything, so unlike what we are accustomed to from our great men with their veneer of distinguished
superficiality. I gave my card to the *Chef* who handed it to Charcot. The latter fingered it for a while and after the consultation
asked where I was. I came forward and gave him my introduction.
He recognized Benedikt's handwriting, stepped aside to read it,
said "Charmé de vous voir," and invited me to accompany him.
He advised me to make my working arrangements with the *Chef*

[1] Pierre Marie (1853-1940), Charcot's assistant, later his successor at the
Salpêtrière.

de Clinique, and without any further ado I was accepted. He then proceeded to show me everything in the laboratory and the lecture hall, passed through several wards and explained a great many things to me. In short, although fewer formalities were exchanged than I had expected, I soon felt very much at ease, and I realized that in the most inconspicuous fashion he was showing me a great deal of consideration. I asked his permission to show him some of my slides, which I did briefly today.

Today was the day for the ophthalmological consultations. The clinic has its own ophthalmologist, whose consulting room is as accessible to me as everything else. Altogether the atmosphere is very informal and democratic. Charcot lets fall quite casually any number of the most brilliant remarks, is constantly asking questions and always good enough to correct my wretched French. As long as he is present I try to keep near him, and already feel quite at home. The *Chef* Marie is an excellent fellow and my only regret is that he is leaving in ten days. His successor hasn't yet arrived. Without a moment's hesitation Marie gave me the material I need for my self-chosen work (have you ever heard of secondary degeneration?), and today Charcot wrote a letter to another professor to get me some children's brains. The morning is devoted to patients, the afternoon to study. So I have good reason to be satisfied. Today I went to the clinic in the afternoon as well. The people here are not very busy yet; lectures haven't even started. The evenings I intend to spend reading in the library or studying at home, as I did today. I don't think I shall see the boys[2] so soon again. I am completely happy to be back at work.

My new shoes arrived today, with laces and English soles, but twenty-two francs! Altogether, you would hardly believe the amount of money one needs for the most ordinary things, and how poor I am already! Needless to say, I have already written to Paneth. But my stay here is going to be well worth it, this I can see clearly. If I didn't have to think of the misery at home, I would feel quite all right. But I am so old and so weak or so wicked that I cannot deny myself a thing. I eat my fill and I smoke and I cannot do anything but—be sorry. Whenever I think of them it upsets me, but this doesn't do them any good.

[2] Martha's cousins Julius and John Philipp.

You, my darling, do write as much as possible about yourself. Couldn't I for once have you and the work at the same time?

With many kisses

Your

Sigmund

83 *To* MARTHA BERNAYS

Paris, Wednesday
November 4, 1885

My beloved darling

Well, the great news is that yesterday on arriving rather late at the *Consultation Externe* (I am getting lazy and easygoing these days), I noticed in the audience a narrow, pale skull covered with thin, fair hair which nodded to me in recognition and which turned out to be that of my friend *in cerebro* Darkschewitsch[1] from Moscow. Let me tell you the previous history of our relationship: when I entered Meynert's laboratory to do research on the gold method, I found there an American, Mr. Barney Sachs, a particularly amiable and intelligent man (I learned later that he is a Jew), and my Russian Darkschewitsch. The latter attracted my attention by his melancholy disposition, typical of Ruthenians and Little Russians, but I got to know him better only after I had discovered my Method. Sachs translated my paper into English for *Brain*, or rather he corrected my translation, and D. offered to translate it for a Russian journal, which he did. He slowly began to confide in me and I discovered in him a quiet and profound fanatic. He was averse to all distractions and his soul was absorbed in the motherland, religion, and brain anatomy. His ambition was to write the first book on brain anatomy in the Russian language. Dissatisfied with Meynert, he went to Leipzig, to my rival Flechsig.[2] He wrote to me once from there, but I never got an answer to my

---

[1] See note 5 to letter of Jan. 28, 1884.

[2] Paul Flechsig (1847-1929), Professor of Psychology at the University and Director of the Psychiatric Clinic at Leipzig. Founder of the ontogenetic method of analysis of the internal structure of the central nervous system (see letter of Mar. 31, 1885).

177

reply. Since March, 1884—when he left Vienna—I have read several interesting papers by him on brain anatomy. So now he is here, to spend his last year abroad in Charcot's clinic. His government has promised him a professorship on his return. After the consultation he came over to me and gave me his address; I accompanied him at once and found him unchanged and in his own quiet way very friendly. He still remembered my engagement, inquired after the health of my fiancée and expressed the hope that I won't have to keep her waiting much longer. He had also once met my father in Vienna and asked how he was. I liked all this very much and in the evening called for him, we dined together, then drank tea in his room, and I began to feel less isolated. In one paper he showed me he mentions that my Method had given him by far the best illustrations for his investigation, and his drawings are copied from these slides. He told me my Method had created a sensation in Leipzig, which I was glad to hear. He described Flechsig to me as an insignificant man who doesn't know how to make proper use of his own discovery. When I referred jokingly to his melancholia, it turned out that he is just as much in love as I am and waiting for letters in the very same way, and this brought us that much closer. As he is not looking for any form of social life or pleasure, he is just the right kind of company for me. On Sunday we have decided to go to Versailles together. Of course I am not indifferent to the news of the effect my minor scientific achievements are having on others. His book is well advanced; with his Russian diligence and great sobriety, he works very hard. I am very pleased to have met him.

> Fondest greetings and kisses.
> Your
> Sigmund

84   *To* MARTHA BERNAYS

Paris
November 8, 1885

My beloved darling
    Dimly aware that I have not written to you for ages and reminded by a card from you that by now you may have once more

got used to Mama and would like to hear from me, I am writing to you again. All kinds of minor things have happened, but the most important fact for me is that my work is now proceeding smoothly and I am just reaching the proper pitch of enthusiasm, which is another reason why I haven't written. But I haven't made any discoveries as yet.

Yesterday my failure to write had another cause. My head was reeling; I had been to the Porte St. Martin theater to see Sarah Bernhardt.[1] I am still rather tired and ravaged by the heat and the blood-&-thunder melodrama, which lasted from 8 to 12:30, but it was worth it. How shall I begin to tell you about it? I am so clumsy today at arranging things. First the minor details. We (I was with one of my Russians) paid four francs and for this were given seats in the *stalles d'orchestre,* which I suggest should be translated simply as the orchestra stable. One could see and hear perfectly, but I think I will have more room and be more comfortable in my grave; at least I will be stretched out. The play started at 8 P.M., had five acts or eight scenes (*Theodora,* of course); after the first act the excruciating heat gradually increased until toward the end it was neither describable nor bearable. And on top of that the wretched megalomania of the French for insisting on 4½ hours of theater as they do on five- or six-course meals. To enjoy one's way through something quickly, allowing interest to help conquer fatigue, is too plebeian for them; so they prolong a 2½-hour play by two hours of entr'acte during which one can, it's true, go out into the beautiful evening and drink beer, smoke a cigar and eat oranges; but if one returns too early (as one invariably does), one suffers the ghastly tortures of anticipation in the oven. I really cannot praise the play, Victorien Sardou's[2] *Theodora* (he has already written a *Dora* and a *Feodora* and is said to be busy on a *Thermidora, Ecuadora,* and *Torreadora!*). A pompous trifle, magnificent Byzantine palaces and costumes, a conflagration, pageants of armed warriors and so on, but hardly a word anyone would want to commit to memory, and as for characterization, it leaves one completely cold. Theodora herself, Justinian's famous empress, originally a ballet dancer who, as history has found worth noting, once appeared in public *toute nue,* is in this

---

[1] Sarah Bernhardt (1844-1923), French actress.
[2] Victorien Sardou (1831-1908), French playwright.

play simply a *femme qui aime*. The French love such simplifications —think of *Donna Sol*.[3] Theodora is deeply in love with a young patrician who has ideals and republican sympathies; her whole complicated early life, which needless to say her lover throws in her face at the end, has to be imagined; it is certainly not shown in her behavior. But how this Sarah can act! After the first words uttered in an intimate, endearing voice, I felt I had known her all my life. I have never seen an actress who surprised me so little; I at once believed everything about her. She was almost never off the stage. In the first scene she is seen giving audience lying on a "throne sofa" with a bored, arrogant expression and receiving back in favor the disgraced Belisar. In the second scene she visits in disguise her wet nurse, who is a keeper in the circus menagerie; she plays with a tiger concealed behind some straw and seems to be rather enjoying life while helping to peel onions and sharing the nurse's meal. In the third scene, this time disguised as Myrtha, she visits her lover in his garden; in the fourth she has a little squabble with her husband, the emperor, a stiff and cowardly tyrant whom she reproaches for being a bigger hypocrite than herself. From now on the play moves a little faster; the lover and a friend enter the palace at night with intent to murder the emperor. Theodora however slams the door behind the friend, thus helping her Andreas to make his escape, and when the conspirator is caught and about to be tortured to make him divulge the name of his accomplice, she begs for a chance to speak to him, reveals to him her relationship with Andreas and orders him to think of a means of preventing him from talking under torture. He can think of only one: she has to kill him and, with the threat that he will otherwise tell the whole story, he forces her to pierce his heart (having first of all shown her the precise spot) with a golden pin which she wears in her hair. In Scene V she again visits her lover, who is in the midst of celebrating the funeral rites of his dead friend and who swears to wreak the most terrible vengeance on the murderess, Theodora. In Scene VI the emperor and empress are about to enter their box in the circus to watch the games when a man shoves his way through the crowd and yells an insult at her, whereupon he is seized and commanded to kneel in front of her before being executed. Needless to say, the man is Andreas. Pleading for him to

[3] Heroine in *Hernani*, drama by Victor Hugo (1802-1885).

be spared, she casts her wrap over the shackled prisoner. In Scene VII we see Justinian in the palace waiting in fear and trembling for the outcome of the battle which has broken out in the town. Andreas has made his escape and organized a rebellion. But Belisar wins, the prisoners are brought in, the gates are thrown open, and the town is seen in flames. The emperor has grown suspicious of Theodora, who is being told by her wet nurse that Andreas is lying wounded in the circus. In Scene VIII she visits him there, is compelled to listen to his reproaches and contempt and offers him a magic potion which the wet nurse has brought her to make Justinian submit to her will. The potion, however, is poison, the wet nurse's son has been executed by the emperor, and this was to have been her revenge; Andreas dies and while Theodora is mourning him several courtiers appear and present themselves with a silent bow. "Ah, je comprend," she says, glancing up. "L'empereur—le bourreau!" And then: "De quelle manière?" The hangman shows her a silk noose, she frees her neck, says "Now I'm ready to die!" and is throttled.

I have never seen a funnier figure than Sarah Bernhardt in Scene II, where she appears in a simple dress, I am really not exaggerating. And yet one was soon compelled to stop laughing, for every inch of this little figure was alive and bewitching. As for her caressing and pleading and embracing, the postures she assumes, the way she wraps herself round a man, the way she acts with every limb, every joint—it's incredible. A remarkable creature, and I can well imagine she is no different in life from what she is on the stage.

For the sake of historical truth let us add that I again had to pay for this pleasure with an attack of migraine, and so have decided to go to the theater only rarely and to pay not less than five or six francs for a seat.

The Russian I went with was Dr. Klikowicz, Botkin's[4] assistant, a vivacious, shrewd, and amiable young man to whom I owe several practical hints. He pointed out to me a *crêmerie* where one can get for thirty centimes what costs sixty in a cafe and took me to a new restaurant where one can eat *à prix fixe* and yet choose one's dishes, get twice as much to drink and more to eat than at Duval's, and yet pay twenty centimes less per meal. I would save more if I drank wine instead of beer, would pay 1.60 francs instead

[4] Russian doctor, physician to the Czar.

of 2.00. Today, Sunday, an excursion to Versailles had been planned, but I have decided to give my head and my pocket a rest.

I hope to go to several lectures with my other Russian, the "scientific" one, who has invited me to tea this evening. On Friday we actually went to one given by M. Hallopeau,[5] a young *Dozent,* where we got a glimpse of a French amphitheater, etc. I introduced myself to the gentleman, but there is no question of being received; all one hears is "charmé" (which is not true), then "I have been in Vienna, too, and made the acquaintance of Herr So-&-So," and in future I shall go where I want and save my cards. Other foreigners I have met here feel the same as I do about the so-called civility of the French.

That is enough of this chronicle. I must add that I had a very nice letter from Dolfi. Now that I have finished my report I will write to you again more affectionately and on more personal topics and hope meanwhile to hear from you at some length.

Fond greetings from
Your ever devoted
Sigmund

[5] François-Marie Hallopeau (1842-1919), Professor of Dermatology, Paris.

85 *To* MARTHA BERNAYS

Paris, Thursday
November 19, 1885

My lovely darling

You are so superhumanly good I really don't know how to thank you. There are things that are tied to geography, and because of the distance between Paris and Hamburg I cannot take you in my arms and kiss you as I would like to. The newspaper was an incomparable treat for me, especially the Viennese sections with the delightful passage by Spitzer.[1] My description[2] is altogether most one-sided and should be taken with a grain of salt because I always try to speak the truth as far as I can and dare!

You are right, my darling, in saying that I have even more to

[1] Daniel Spitzer (1835-1893), Austrian writer and journalist. His "Strolls through Vienna," published regularly in the *Neue Freie Presse,* were widely read.
[2] Refers to the accounts of his strolls through Paris which Freud sent to Martha.

tell you than before, and usually there is something I even forget to tell you, for instance my visit to Notre Dame de Paris on Sunday. My first impression on entering was a sensation I have never had before: "This is a church." And I looked about for Ricchetti, who knows the churches of Italy. There he stood, deeply lost in wonder. I have never seen anything so movingly serious and somber, quite unadorned and very narrow, which is no doubt partly responsible for the general impression. I really must read Victor Hugo's novel[3] here, for this is the place properly to understand it.

I hope you feel reassured about the overcoat; as a matter of fact, the cold has abated and we are having the most beautiful autumn weather. The anatomical work is very hard going; I am resigning myself more and more to my inability to work out the many stimulating ideas. Today Charcot gave me permission to embark on the clinical study for which I had an inspiration; but it looks as though Marie wants to withdraw, so I don't know what I shall be able to do. Probably nothing, but subtract all this and there is still enough to make my stay here valuable.

It is not very nice of you to lay so much stress on Frau Ricchetti's ugliness. You know well enough that however beautiful she might be, she could not be compared to Martha. In any case she is an exceptionally distinguished and nice woman, speaks four languages fluently, has read a great deal, doesn't boast about it, but rather irritates one by asking too many questions. She is very quiet when the men are talking but takes an interest in everything, is kindness itself toward her husband who cannot do without a certain amount of selfish comfort, with the result that he is always saying to me: "There's only one Louise." She allows herself to be sent wherever he wants her to go, stays with him when he wants her to, looks after him and enjoys things with him. I play a bit of the cavalier toward her, and she repays me with little attentions such as producing some doughnuts the day after I complained about the lack of dessert. She is fat, an indefatigable walker and usually walks alone. Her genealogy is as follows: she comes from Frankfurt, of a very rich family, but was herself not rich. The great Behrend in Manchester is her uncle; her dowry was sufficient for them to have lived on it very carefully, but only a few years ago she inherited a lot of money, so she is now worth some 340,000 francs. Since that

---

[3] *Notre-Dame de Paris*, 1831.

time they have kept their accounts separate; he says he doesn't want to be rich through her. His own fortune, earned from his practice, amounts to 250,000 francs. But they both live so frugally that on this money they are very well off. They make no effort to cultivate friends. Our relationship consists in his taking me to the Salpêtrière in the mornings, then we return to his house, pick up his wife, and go to Duval's. In the afternoon I return alone, come back to them at six, at seven o'clock we go out to dine, and then go for a walk or I spend another hour with them in their apartment.

I do wish we were already in the position to accept the invitation to stay with them in the Palazzo Buffo in Venice. You would get along splendidly with these two excellent people. By the way, she used to know Michael Bernays in Frankfurt—thirty-eight years ago. She was a bit confused about the time and asked if you are a sister of his, but soon realized you must be a niece.

Today I had a letter from Paneth and an interesting paper by him. So much news that there's hardly a chance for an affectionate word. My precious, beloved darling, when I think that you are my bride and have been so for 3½ years, I see myself in turn as worthy of envy and pity. Like the wind, you say? It is half the engagement time of our patriarchs, but they lived to an incredible age and God was on their side. Fare thee very well and thank you again for your amusing enclosures.

Your

Sigmund

86 *To* MARTHA BERNAYS

Paris

November 24, 1885

My precious sweetheart

. . . Both Mama's and Minna's letters are very nice and require an answer. I am not at all averse to writing and the Republic is pleased that I am in Paris because I spend so much money on stamps. Quite agree about the Christmas present for Frau Gehrke.[1] It is to come from Paris, isn't it?

I am really very comfortably installed now and I think I am changing a great deal. I will tell you in detail what is affecting me.

[1] Charwoman in Wandsbek.

Charcot, who is one of the greatest of physicians and a man whose common sense borders on genius, is simply wrecking all my aims and opinions. I sometimes come out of his lectures as from out of Notre Dame, with an entirely new idea about perfection. But he exhausts me; when I come away from him I no longer have any desire to work at my own silly things; it is three whole days since I have done any work, and I have no feelings of guilt. My brain is sated as after an evening in the theater. Whether the seed will ever bear any fruit, I don't know; but what I do know is that no other human being has ever affected me in the same way. Even old Ricchetti, who has known all the important men of his time, is absolutely bowled over by him. When I get home I feel completely resigned and say to myself: the great problems are for men between fifty and seventy; for young people like us there is Life itself. My ambition would be satisfied by a long life spent learning to understand something of the world, and my plans for the future are that we get married, love each other and work with the object of enjoying life together instead of exerting every ounce of my energy trying to pass the post first, like a race horse—in other words, trying to build myself a home that would involve such effort and privation that I couldn't expect to be granted more than two or three years of mental health. Or am I under the influence of this magically attractive and repulsive city? If so, it would be quite an indirect one. Have you anything to say to this, my darling?

Yesterday I committed a great psychologically interesting blunder. I wanted to buy a *Mémoire* by Charcot, which cost 5 francs. But it was out of print, and to own it I would have had to take a volume costing 12 francs from Charcot's Archive. The extra 7 francs struck me as too much. Then the man told me that if I subscribed to Charcot's Archive, I could have all the published works for 60 francs instead of 140. The annual subscription itself is 20 francs. This I did and thus spent 80 francs, ostensibly to save 80 francs. For me the Archive is of the greatest value, but I shall feel the loss of 80 francs. So if you find it so touching that I put away every day a little something for Christmas, just remember this wild extravagance of mine.

Thursday, November 26

My sweet darling

As a result of my monumental laziness and in spite of my persistent desire to write you a long letter, this page has been lying around for two days and you will now receive two letters by one mail, but this is something I don't intend to repeat.

As long as you are well, my darling, I am only too glad not to be the King of Spain.[2] He is the first sovereign of my own generation whom I have outlived, and his death has made a great impression on me. The utter absurdity of the hereditary system will once again be proven when, under the rule of a five-year-old queen following the death of one man, the whole country will rise up in arms.—I prefer to rejoice at hearing that Dr. Cohn[3] has done you so much good, for which achievement I consider no price too high. But I feel sure he won't overcharge you. It is about time Minna went to him, too; please insist on it.—

Now I must tell you about a second visit I paid yesterday. They wrote me from home that the wife of our family physician, Dr. Kreisler, is in Paris; I should have called on her long ago. I have just been there, Rue Bleue in the Faubourg Poissonière, next to the Conservatoire. The unfortunate woman has a ten-year-old son who, after two years in the Vienna Conservatorium, won the great prize there and is said to be highly gifted. Now instead of secretly throttling the prodigy, the wretched father, who is overworked and has a house full of children, has sent the boy with his mother to Paris to study at the Conservatoire and try for another prize. Just imagine the expense, the separation, the dispersal of the household! Needless to say the poor woman, who is giving up everything for the boy, is bored to tears. Little wonder that parents grow vain about their children, and even less that such children grow vain themselves. I will have to go and see them at least once a week, first because I feel under an obligation to him, secondly because she is rather in need of my medical attention, and then it seems good diplomacy to remain on pleasant terms with a Viennese colleague. The prodigy[4] is pale, plain, but looks pretty intelligent.

Meanwhile the Archive has arrived and contains highly valuable material, so I don't regret the expense.—Today Charcot announced he was not going to appear, and so instead of the Salpêtrière we went to the Louvre, or rather we drove there, for it was raining cats and dogs.

Yesterday evening John came to see me; he sends you his kind

[2] Alfonso XII of Spain (1857-1885).
[3] Martha's dentist.
[4] This "prodigy" was Fritz Kreisler, born 1875.

regards. He is quite a decent boy. What amused me was that he was evidently trying to sound me out about whether I keep a mistress here. I trust not on your instructions, my little woman? Perhaps it was Mary's[5] curiosity. Or I misinterpreted his rather naïve talk.

Now don't go and pay me back, my darling, and write to me soon again. But you are so good, I needn't have mentioned it. I couldn't possibly regret leaving Paris, for after all I am coming to Hamburg.

With fondest greetings and many kisses, which will be cashed in their own time

Your
Sigmund

[5] Martha's aunt, wife of Elias Philipp (see letter of July 7, 1882).

87 *To* MINNA BERNAYS

Paris
December 3, 1885

My dear Minna

How unnecessary of you to begin a correspondence between us with a formal introduction! Does my answering you require anything more than your writing to me? I would have answered before had I not been waiting for the opportunity to pick up the gauntlet thrown to me so gracefully, but Herr Otto Wolff[1] (or Wolf-fff) still hesitates to honor Paris with his presence or has settled elsewhere than in the Rue d'Hauteville.

I am under the full impact of Paris and, waxing very poetical, could compare it to a vast overdressed Sphinx who gobbles up every foreigner unable to solve her riddles. But I will save all this for verbal effusions. Suffice it to say that the city and its inhabitants strike me as uncanny; the people seem to me of a different species from ourselves; I feel they are all possessed of a thousand demons; instead of "Monsieur" and "Voilà l'Écho de Paris" I hear them yelling "À la lanterne" and "À bas" this man and that. I don't think they know the meaning of shame or fear; the women no less than the men crowd round nudities as much as they do round corpses in the Morgue or the ghastly posters in the streets announcing a

[1] Martha's cousin. Minna had corrected Freud for spelling Wolff with one f.

new novel in this or that newspaper and simultaneously showing a sample of its content. They are people given to psychical epidemics, historical mass convulsions, and they haven't changed since Victor Hugo wrote *Notre-Dame*. To understand Paris this is the novel you must read; although everything in it is fiction, one is convinced of its truth. But don't read it till you are in a perfectly calm frame of mind and in Paris.

If you do come here you will probably first of all be attracted by what captivates most people exclusively—the brilliant exterior, the swarming crowds, the infinite variety of attractively displayed goods, the streets stretching for miles, the flood of light in the evening, the over-all gaiety and politeness of the people; but to bring all this into harmony with the rest one has to know a great deal.

As you realize, my heart is German provincial and it hasn't accompanied me here; which raises the question whether I should not return to fetch it, like that boy from Lauterbach[2] and his stocking. Any good advice to offer?

Apart from books and stamps (1.50 francs a week), I have so far bought nothing here. If you want something, please tell me, for the diversity of the choice here makes me more than ever incapable of shopping. *Chocolat Marquis*—good, that you shall have.

Whenever John comes to see me I am embarrassed, for he keeps admiring my French, whereas I can't speak enough to ask for bread in a café. The French language is terribly poor in vowels, every whisper means a dozen different things and, somewhat modified, a dozen other things. For instance, when I said "du pain" to the *garçon,* I have no idea what he made of it; he just shook his head till I got so annoyed I refused to go to any café. I did at last manage to say "croissants," since when I always get *Kipfel** with my coffee.

Oh, those wonderful Wandsbek days! Paris is simply one long confused dream, and I shall be very glad to wake up.

Please write to me again soon; I do nothing here except allow myself to be wound up by Charcot in the mornings, and in the afternoon I have time to unwind and to write letters in between.

Fond greetings from

Your brother

Sigmund

---

[2] Allusion to the German folksong "I've Lost My Stocking in Lauterbach."

* Austrian: crescent-shaped roll.—*Tr.*

88   *To* MARTHA BERNAYS

Paris, Saturday
December 12, 1885

Dearest highly esteemed little Princess

Does your Highness really believe it is so easy to tear one-self away from Paris? Don't be alarmed, I am arriving in Hamburg on the morning of the twenty-first; this is certain; but I shall hardly see Berlin; I shall return to Paris instead. "What on earth has happened, you crazy man?" Nothing, my little woman, except that Charcot took me aside today and said: "J'ai un mot à vous dire." And then he told me he would gladly consent to my translating his Volume III into German—what's more, not only the first section, which has already appeared in French, but also the second, which hasn't yet been published. Are you pleased? I am. This is again something very gratifying. It is bound to make me known to doctors and patients in Germany and is well worth the expense of a few weeks and several hundred gulden, not to mention the few hundred gulden it will bring in. It will be of great advantage to my practice and moreover will pave the way for my own book[1] when that is ready for publication.

Ricchetti thinks that this would not be the moment to leave Charcot, when one has just begun to establish a contact with him, and I really do think he is right. But this has nothing to do with my being away from here for ten days, except that when I kiss you I will be richer by one prospect. And you really do deserve, my sweet treasure, to be a little pleased with me, considering the times you have had reason to be sad on my account. Well, this was a good day, similar to that which brought me the traveling grant, and I trust you will not advise me against coming to see you before I plunge into my new work in Paris.

Today so far has been like a scene from a comedy in which everything happens at once. Charcot's permission, a good letter from home; Rosa writes that she is terribly busy—my winter coat, my shirts and shoes! What the afternoon has in store I don't know—but I would like to know what happened in your life today.

[1] The plan for this book was dropped later on.

I shall probably give up my apartment, but I can surely get it back again; my books I will pack away in a box and store it at Ricchetti's. . . .

I feel like shouting and jumping for joy and what I would like best is to be with you already today, my dear, good darling. I hope you agree with my return under these circumstances.

I kiss you many thousand times.
Your
Sigmund

89 *To* MARTHA BERNAYS

Paris
December 18, 1885

My precious darling

Just one more short letter which may perhaps arrive at the same time as myself. I am glad you have abandoned your resistance to my coming. Do you still remember the first compliment I paid to you, the unsuspecting girl, more than 3½ years ago? I said that roses and pearls fall from your lips as with the princess in the fairy tale and that one is left wondering only whether it is goodness or intelligence that has the upper hand with you. This is how you acquired the name of Little Princess. And now that I know you so well I can but uphold the compliment, the aptness of which I could at the time only divine. May things always remain like this between us.

I leave here on Sunday morning at 7:30, and arrive in Hamburg at 6:18. Nothing has happened to change the original arrangements. I am taking with me a traveling bag belonging to the Ricchettis, probably a rug of theirs as well, the small English handbag, nothing else. Box and trunk are packed and will be stored with them.

I am bringing nothing but some candy for the children and some tiny trifles for the rest of you. In Cologne I will also buy a bottle of eau de cologne for Mama. . . .

Yesterday I had one more interview with Charcot, during which he very obligingly yielded to all the demands of the publisher. The whole thing is now safe and settled. I have written to the publisher and expect to receive his offer of a fee in Wandsbek.

Tell Minna from me that whenever we entertain friends, there
will always be a place laid for her.

I must stop, darling, it is midnight. The Russian has been here
and I have read my opus to him. May love and science never desert
Your
Sigmund

90-91 *To* MARTHA BERNAYS

Paris, Sunday
January 17, 1886

My sweet little Princess

In spite of my recent declaration of war on you for making
parcels instead of writing me long, affectionate, and flattering letters, I don't feel entitled to conceal from you that I love you excessively, in case you have already forgotten. Furthermore, that
I am having some very good days and so far as it is possible I am
quite contented, as will be seen by what follows:

First because a paper on brain anatomy has been concocted here
in the following manner: I am sitting beside my sick Russian friend
(grumbling about Ricchetti, who continues to pick his brains)
telling him of a nice discovery which I made in Vienna and which
I didn't publish because I was afraid there might be still more
behind it. Whereupon he shows me his notes on precisely the same
subject and tells me that he has divulged the result to a colleague
and rival, who is going to use it in a publication. "No, dear friend,"
say I, "we will publish that together—and, what's more, at once. I
have brought several slides along, you have any amount of them,
we will study them together, then you make a drawing, I will do
the text, and then we will get it off together."—Agreed.—The first
day we spent 5½ hours together studying the slides by day- and
lamp-light, searching and doubting endlessly till he fell back on the
bed and I left with such a buzzing in my head that I couldn't get
to sleep, so I walked the boulevards from ten to midnight to air
myself. The whole problem still seemed rather doubtful to me, but
next day in the microscope I glimpsed a few auspicious things,
got all excited, everything cleared up, and today we have all we
want. It works out very well, I am mulling over the text, he is

brooding over the drawings, and in three weeks you may be able to add to your collection of scientific articles in your old portfolio a new little work by Dr. F. and Dr. Darkschewitsch entitled "Zur Kenntnis des Hinterstrangskerns und der Bestandteile des Strickkörpers"[1] (eh!) or something like that. It is quite a nice paper which gives me great pleasure.—Needless to say, I had to interrupt the translation and my studies for three days on account of it.

Second pleasant event: Jules Bernays[2] arrived last night at theater time to drag me along to the Comédie Française, where they are playing *Le Mariage de Figaro* by Beaumarchais.[3] Being in a good mood I gladly went with him and enjoyed myself enormously. We took six-franc seats (I at last succeeded in not being his guest again), a splendid performance, a remarkable play sparkling with wit, and now some hundred years old. You know the story from the opera we heard together. But apart from the libretto for the opera there is tucked away within it another brilliant play. Now and again I missed the wonderful music, for instance when Susanna sits down to write the letter or when the Count shows his delight at the prospect of meeting her in the garden. The spirit of the play is extremely revolutionary and its performance is generally considered to be a precursor of the great Revolution. Of course I cannot describe to you how these actors move and speak; you really should have been there. Susanna was charming, so was the page, then there was an additional comic figure, the Judge, who is not in the opera, where everything relating to Figaro's lawsuit with Marcellin has been very much reduced. After the theater we (B.'s friend from last week making a third) went to a *brasserie* and drank beer, and I didn't get to bed till 2 A.M. No migraine today.

Third, and not least interesting event:

Monday, 11 P.M.
January 18, 1886

My sweet Princess

Last night after supper I went on writing the outline of my

---

[1] "Über die Beziehung des Strickkörpers zum Hinterstrang und Hinterstrangskern nebst Bemerkungen über zwei Felder der Oblongata" ("On the Relation of the Restiform Body to the Posterior Column and its Nucleus, with Some Remarks on Two Areas of the Medulla Oblongata"), 1886. *Neurologisches Zentralblatt* 5, No. 6, 121.

[2] Another cousin of Martha's.

[3] Pierre Augustin Caron de Beaumarchais (1732-1799).

anatomical paper till I just couldn't keep my eyes open. Today your sweet letter arrived at last, and I must now send off my answer in this mutilated form, otherwise you will have to wait even longer. I have so exhausted myself writing that I can barely hold the pen. Well, the third interesting event was that yesterday I spent more than an hour with Charcot, who handed me a further ten sheets. I would love to give you a description of his home; but this must wait for another day. What's more, he invited me (as well as Ricchetti) to come to his house tomorrow evening after dinner: "Il y aura du monde." You can probably imagine my apprehension mixed with curiosity and satisfaction. White tie and white gloves, even a fresh shirt, a careful brushing of my last remaining hair, and so on. A little cocaine, to untie my tongue. It is quite all right of course for this news to be widely distributed in Hamburg and Vienna, even with exaggerations such as that he kissed me on the forehead (à la Liszt). As you see, I am not doing at all badly and I am far from laughing at you and your plans.

My fondest greetings and I would love to be your dentist, who I am sure doesn't know how to value it, only how to charge.

Your
Sigmund

92 *To* MARTHA BERNAYS

Paris, Wednesday
January 20, 1886

My beloved little woman

I acknowledge with pleasure the open admission of the cardinal baseness of your actions, for tradition has always interpreted a semiconfession on the part of a "lady" in this way, and even a beloved fiancée retains enough of the unassailable and unchangeable character of a "lady" for the man who loves her. You will allow me quietly to observe that you have been wrong on almost every point and without good reason, and have paid less attention to the whole business than you usually do to things concerning us both. Furthermore, you must allow me to point out that you could easily have changed everything according to your desire if only you had informed me about it. It is already quite some time now since I have been unyielding toward you, especially in small matters; a man will always get annoyed if his little woman appears to be trying to get her own way by any other than straight means. If she is frank with him he will usually give in. But all this probably concerns only the future, for this time I am willing to believe you didn't do it on purpose. Thus renouncing any further explanations on your part about this matter, unless you feel the need for it, I herewith close this diplomatic incident and return to the familiar *du* of more intimate relations and to the carrying out of my duties as a reporter.

I had meant to write to you at midnight, but couldn't find the matches, and so had to take off my elegant clothes and go to bed by the light of the moon. So let us begin at the beginning. On Saturday Charcot came up to Ricchetti and invited him to dine at his house on Tuesday before leaving. Startled, R. declined, and finally accepted to go after dinner. Then Charcot turned to me and repeated the latter form of invitation, which I accepted with a bow, feeling delighted. What's more, he decided on Sunday at 1:30 as the time to discuss the translation. (I have already told you that I have been to see him and was given ten sheets to start with.) I just want to add what his study looks like. It is as big as the whole of our future apartment, a room worthy of the magic castle in which he lives. It is divided in two, of which the bigger section is dedicated to science, the other to comfort. Two projections from the wall separate the two sections. As one enters one looks through a triple window to the garden; the ordinary panes are separated by pieces of stained glass. Along the side walls of the larger section stands his enormous library on two levels, each with steps to reach the one above. On the left of the door is an immensely long table covered with periodicals and odd books; in front of the window are smaller tables with portfolios on them. On the right of the door is a smaller stained-glass window, and in front of it stands Charcot's writing table, quite flat and covered with manuscripts and books; his armchair and several other chairs. The other section has a fireplace, a table, and cases containing Indian and Chinese antiques. The walls are covered with Gobelins and pictures; the walls themselves are painted terra cotta. The little I saw of the other rooms on Sunday contained the same wealth of pictures, Gobelins, carpets and curios—in short, a museum.

After Charcot had reminded us once more of our appointment on Tuesday morning, we spent all the afternoon preparing for the evening. Ricchetti, who hitherto had been going about in the most incredibly shabby clothes, had been persuaded by his wife to buy a new pair of trousers and a hat; his tailor is said to have told him that for a party it is quite unnecessary to wear a tail coat and that he could go in a redingote, with the result that he was the only guest not in full evening dress. My appearance was immaculate except that I had replaced the unfortunate ready-made white tie with one of the beautiful black ones from Hamburg. This was my tail coat's first appearance; I had bought myself a new shirt and white gloves, as the washable pair are no longer very nice; I had my hair set and my rather wild beard trimmed in the French style; altogether I spent fourteen francs on the evening. As a result I looked very fine and made a favorable impression on myself. We drove there in a carriage the expenses of which we shared. R. was terribly nervous, I quite calm with the help of a small dose of cocaine, although his success was assured and I had reasons to fear making a blunder. We were the first after-dinner guests and as we had to wait for the others to come from the dining room, we spent the time admiring the wonderful salons. But then they came and we were under fire: M. and Madame Charcot; Mlle Charcot; M. Léon Charcot; a young M. Daudet, an unattractive youth, son of Alphonse Daudet;[1] Prof. Brouardel,[2] doctor of forensic medicine, a manly, intelligent head; M. Strauss, an assistant of Pasteur and well known for his work on cholera; Prof. Lépine[3] of Lyons, one of France's most distinguished clinicians, a small sickly man; M. Giles de la Tourette,[4] former assistant to Charcot, now to Brouardel, a true Provençal; a Prof. Brock, *membre de l'Institut*, mathematician and astronomer who at once started talking German and turned out to be a Norwegian; then came Charcot's brother, a gentleman who looked like Prof. Vulpian[5] but wasn't, and several others whose names I never learned; also an Italian painter, Tof-

[1] Alphonse Daudet (1840-1897).

[2] Paul Camille Hyppolite Brouardel (1836-1907), Professor of Forensic Medicine, later for many years doyen of the Faculty of Medicine in Paris.

[3] Professor Raphaël Lépine (1840-1919), member of the French Academy of Science.

[4] Giles de la Tourette (1857-1904).

[5] Professor Edmé Felix Alfred Vulpian (1826-1887).

fano.[6] And now you will be anxious to know how I fared in this distinguished company. Very well. I approached Lépine, whose work I knew, and had a long conversation with him; then I talked to Strauss and Giles de la Tourette, and accepted a cup of coffee from Mme Charcot; later on I drank beer, smoked like a chimney, and felt very much at ease without the slightest mishap occurring. Indeed, one couldn't help feeling at ease, for the whole atmosphere was so informal and a great deal of attention was paid to us foreigners. Lépine suggested I should join him in Lyons, which I wouldn't mind doing; he asked me a great many questions about the Vienna hospital staff and at one moment I became the center of attention. R. had been paying court to Mademoiselle and Madame and the latter suddenly became full of enthusiasm and announced: "qu'il parle toutes les langues. Et vous, Monsieur?" asked Madame Charcot, turning to me. "German, English, a little Spanish," I replied. "And French only very badly." She found this sufficient, and Charcot added: "Il est trop modeste, il ne lui manque que d'habituer l'oreille." I then admitted that I often don't understand what has been said until half a minute later, and compared this failing to the symptoms of tabes, which went very well.

These were my achievements (or rather the achievements of cocaine), which left me very satisfied. I also received permission to attend Prof. Brouardel's course in the Morgue, where I have been today. The lecture was fascinating, the subject matter not very suitable for delicate nerves. It is described as the latest murder in every Paris newspaper.

You will probably be as interested in the personalities of Madame and Mlle Charcot as in my achievements. The former is small, rotund, vivacious, hair powdered white, amiable, in appearance not very distinguished. The wealth comes from her; Charcot started out quite poor; her father is rumored to be worth countless millions. Mlle Charcot is something else: also small, rather buxom, and of an almost ridiculous resemblance to her great father, as a result so interesting that one doesn't ask oneself whether she is pretty or not. She is about twenty, very natural and amiable. I hardly talked to her as I kept to the old gentlemen, but R. spoke to her a lot. She is said to understand English and German. Now just suppose I were not in love already and were something of an adventurer; it

[6] Émile Toffano (1838-1920), whose picture "Enfin Seuls" (exhibited at the Paris Salon in 1880), was reproduced and sold all over the world.

would be a strong temptation to court her, for nothing is more dangerous than a young girl bearing the features of a man whom one admires. Then I would become a laughingstock, be thrown out, and would be the richer from the experience of a beautiful adventure. It is better as it is, after all.

I very much wonder, by the way, whether this invitation will be the last. I believe it may, for in fact I owe it to Ricchetti.

And now something else. Do you know the old song which goes: "And a little bit of falseness is always involved"?

Do you think I don't know that the parcel is being sent now because it was to include something for Eli, and that you did not tell me about it because you were afraid I would object? And am I to be pleased that you allow me to grow increasingly suspicious and that you won't tell me this simple story? Is that nice of you, my child? Not at all. And now as a punishment, don't leave that little jersey jacket till I return, but buy it at once as I always intended you to. You are not very lucky with subterfuges. One of these subterfuges kept annoying me for almost two years, and I would rather think of you with false teeth in your mouth than one dishonest word.

And now enough, you know I am always inclined to think longer and more intensely about such things than they deserve.

With a fond kiss
Your
Sigmund

93 *To* MARTHA BERNAYS

Paris, Wednesday
January 27, 1886

My sweet treasure

I am deeply happy that you have forgiven me; the idea that you were not thinking of me as affectionately as usual gave me a strange feeling of forlornness, a feeling I couldn't have stood for long, the less so because I had no one but myself to blame. Well, you have forgiven me, and I am deeply grateful, and yet I am not quite satisfied, for I believe that when one has quarreled one ought to love the other more than ever, otherwise the relationship is no longer what it was. For the first time in the 3½ years of our engagement I have the uneasy feeling that perhaps not everything has been straightened out, and if you can think of anything I could do to eradicate any bitter memory in you, please tell me. Then I shall feel happier again.

One thing has really surprised me: not that you have forgiven me so quickly—I knew you would do that even if you no longer loved me—but that such thoughts should pass through your head, such bitter thoughts which one recognizes immediately as alien to one's nature but which one just cannot prevent from cropping up. Some people are good simply because nothing bad enters their heads, others are good because they manage—always or frequently—to conquer their bad thoughts. I had counted you among the former. But it is probably my own fault that you have lost this guilelessness. And perhaps it doesn't matter so much; anyone frequently colliding with life is bound to lose it and to acquire a character instead.

If I could kiss you now, my precious girl, you would realize how much everything has remained unchanged, although of course I don't know what you may feel has changed.

I don't deserve your well-meant reproaches about my behavior towards my family, as you will know by now.

The Ricchettis left last night. I saw them off at the station and in saying goodbye I was happy to forget my many minor objections to him which—compared with his interest in me and the two things I owe indirectly to him, the translation and the invitation to Charcot's—seem negligible. They sent you their kindest regards. She asked me to go on sending her a Paris newspaper containing a novel she is interested in. So I am alone again.

The days are not entirely uneventful. Yesterday I finished the first section of my own book.[1] I am now going to lay it aside for a while and take up the translation, which I had dropped really from fear of acquiring writer's cramp. In fact, my nerves are somewhat frayed and I could do with a rest.

I am now the only foreigner at Charcot's. Today, when a number of small separate offprints (English) arrived, he handed me one and soon afterwards I had the chance of making a certain impression on him. He was talking about a patient, and while the others

[1] See note to letter of Dec. 12, 1885.

were laughing, I just said: "Vous parlez de ce cas dans vos leçons," and quoted a few words of his. This seems to have pleased him, for an hour later he said to his assistant: "Vous allez prendre cette observation avec M. Freud," then turned to me and asked if I would like to "prendre une observation" with M. Babinski.[2] Needless to say, I had no objection. It is a case that Charcot seems to find interesting; personally I don't think much of it, but I shall probably have to write a paper about it with the assistant. Anyway, the point of the incident is that Charcot singled me out at all, and since then the assistant's behavior toward me has changed. After Charcot had left at 11 A.M. I tackled the case and realized to my surprise that I could actually communicate with a Frenchman. After we had decided to postpone the final observation till 4 P.M., the assistant invited me (!!) to lunch with him and the other hospital doctors in the Salles des Internes—as their guest, of course. And all this in response to one nod from the Master! But how hard this little victory has been for me, and how easy for Ricchetti! I consider it a great misfortune that nature has not granted me that indefinite something which attracts people. I believe it is this lack more than any other which has deprived me of a rosy existence. It has taken me so long to win my friends, I have had to struggle so long for my precious girl, and every time I meet someone I realize that an impulse, which defies analysis, leads that person to underestimate me. This may be a question of expression or temperament, or some other secret of nature, but whatever it may be it affects one deeply. What compensates me for all this is the devotion shown to me by all those who have become my friends—but what am I talking about?

So the day was spent entirely in the Salpêtrière. From 4 to 7 P.M. we were busy with the patient. During the interrogation I jotted down the answers of the man, who is from the south of France and incapable of concentrating on anything for more than a moment or of organizing his statements. Then the assistant, who probably had no great desire to compete with me in the investigation, went off, and as I am not a novice like him, I found everything I wanted in a quarter of an hour and passed it on to him. As a matter of fact he treated me very decently throughout. Tomorrow we are going to present the case to Charcot.

As a result of this work I have decided to abandon my own paper

---

[2] Professor Joseph François Félix Babinski (1857-1932), assistant to Charcot.

on brain anatomy. In any case my slides are not sufficient to allow me to solve certain problems, and I am full of ideas and projects which I intend to work out and turn into decent papers when I get to Vienna.

Several books have arrived, I have bought an instrument, a dynamometer, to study my own nervous condition. In other words, work, science, everything is going well; if only I could see you again for one day! Isn't there anyone who could bring you to Paris? Today I am venturing again to send you the ten marks and hope soon to hear that the little jacket suits you well. Greetings to Mama and Minna and please explain to them why I cannot write just now.

With fondest kisses

Your
Sigmund

94 *To* MARTHA BERNAYS

Paris, Tuesday
February 2, 1886

My beloved sweet darling

You write so charmingly and sensibly and every time you speak your mind about something I feel soothed. I don't know how to thank you; I have recently decided to show you a special kind of consideration (you will laugh): by making up my mind not to be ill. For my tiredness is a sort of minor illness; neurasthenia, it is called; produced by the toils, the worries and excitements of these last years, and whenever I have been with you it has always left me as though touched by a magic wand. So I must aim at being with you very soon and for a long time, and since this is hardly possible except by marrying I must try soon to earn the famous 3000 florins a year; and as I am not unindustrious and the prospects aren't bad, I am not unhappy either and am not concerned about my nervousness.

I am delighted that you credit me with having remembered the fee.[1] It really wasn't out of thoughtlessness but of decency that I

---

[1] The fee for Freud's German translation of Charcot's book, *Leçons sur les maladies du système nerveux*, Tome 3ᵉ (*New Lectures on Nervous Diseases, Particularly on Hysteria*). Besides translating it Freud had written a preface and provided explanatory remarks. The book was published in Vienna in 1886.

made the mistake. There is hardly anything to add to what you say, my darling—that we are young and have to pay for our experiences. The bookseller still hasn't answered my letter. At first I felt rather ashamed of telling you the story, and I did so in the end only because the whole thing annoyed me so much.

The news of the day is that I received a very friendly letter from Obersteiner, in whose good will I place, as you know, a certain —though still rather vague—hope. Among other things, he tells me of the scientific scandals going on at the moment in Vienna. To think of the Viennese circle of highly respectable people does me good, even from a distance. One really must try not to become as wicked as people make one out, but one should learn to be careful. The reason he wrote is that he wants some information about the statutes of the Paris society of physicians, which I shall probably be able to get for him this evening. For it is now 6 P.M. and at 9:30 I am going, as you know, to Charcot's, not without the fear of having a most unamusing evening. Needless to say, I have fewer preparations to make than for the first time, but I have felt so out of sorts all day that I haven't done any work.

The bit of cocaine I have just taken is making me talkative, my little woman. I will go on writing and comment on your criticism of my wretched self. Do you realize how strangely a human being is constructed, that his virtues are often the seed of his downfall and his faults the source of his happiness? What you write about the character of the family Bernays is of course quite correct. But I have no reason to grumble about it. It is to this exaggeration (to which you yourself so charmingly admit) that I owe my luck, for otherwise I would never have found the courage to court you. Whether it was luck for you too, we won't go into. But if today were to be my last on earth and someone asked me how I had fared, he would be told by me that in spite of everything—poverty, long struggle for success, little favor among men, oversensitiveness, nervousness, and worries—I have nevertheless been happy simply because of the anticipation of one day having you to myself and of the certainty that you love me. I have always been frank with you, haven't I? I haven't even made use of the license usually granted to a person of the other sex—of showing you my best side. For a long, long time I have criticized you and picked you to pieces, and the result of it all is that I want nothing but to have you, and have you just as you are.

Do you really find my appearance so attractive? Well, this I very much doubt. I believe people see something alien in me and the real reason for this is that in my youth I was never young and now that I am entering the age of maturity I cannot mature properly. There was a time when I was all ambition and eager to learn, when day after day I felt aggrieved that nature had not, in one of her benevolent moods, stamped my face with that mark of genius which now and again she bestows on men. Now for a long time I have known that I am not a genius and cannot understand how I ever could have wanted to be one. I am not even very gifted; my whole capacity for work probably springs from my character and from the absence of outstanding intellectual weaknesses. But I know that this combination is very conducive to slow success, and that given favorable conditions I could achieve more than Nothnagel, to whom I consider myself superior, and might possibly reach the level of Charcot. By which I don't mean to say that I will get as far as that, for these favorable conditions no longer come my way, and I don't possess the genius, the power, to bring them about. Oh, how I run on! I really wanted to say something quite different. I wanted to explain the reason for my inaccessibility to and gruffness with strangers, which you mentioned. It is simply the result of suspicion due to my having learned that common or bad people treat me badly, but this is bound to disappear to the extent to which I grow stronger and more independent, and don't have to fear them any more. I always comfort myself with the fact that people subordinate to or on a par with me have never considered me unpleasant, only superiors or people otherwise above me. One would hardly guess it from looking at me, and yet even at school I was always the bold oppositionist, always on hand when an extreme had to be defended and usually ready to atone for it. As I moved up into the favored position of head boy, where I remained for years and was generally trusted, people no longer had any reason to complain about me. You know what Breuer told me one evening? I was so moved by what he said that in return I disclosed to him the secret of our engagement. He told me he had discovered that hidden under the surface of timidity there lay in me an extremely daring and fearless human being. I had always thought so, but never dared tell anyone. I have often felt as though I had inherited all the defiance and all the passions with which our ancestors defended their Temple and could gladly sacrifice my life for one great moment in history. And

at the same time I always felt so helpless and incapable of expressing these ardent passions even by a word or a poem. So I have
always restrained myself, and it is this, I think, which people must
see in me.

Here I am, making silly confessions to you, my sweet darling, and
really without any reason whatever unless it is the cocaine that
makes me talk so much. But now I must go out to supper and then
dress myself up and do some more writing. Tomorrow I will report
to you quite truthfully on how the evening at Charcot's turned
out. You of course must tell everyone that I had a wonderful time,
and I shall write the same to Vienna. The truth is for us alone.

12:30 A.M.

Thank God it's over and I can tell you at once how right I was.
It was so boring I nearly burst; only the bit of cocaine prevented
me from doing so. Just think: this time forty to fifty people, of whom
I knew three or four. No one was introduced to anyone, everyone
was left to do what he liked. Needless to say, I had nothing to do;
I don't think the others enjoyed themselves any better, but at
least they could talk. My French was even worse than usual. No
one paid any attention to me, or could pay attention to me, which
was quite all right and I was prepared for it. I bowed to Madame,
who clearly didn't expect to be amused by me and told me that
her husband was in the next room. The old man wasn't very agile,
sat most of the time in his chair and seemed very tired. Of course
he didn't fail to offer me some refreshment, which was the only
thing I got from him. Mademoiselle wore a Greek costume, and since
your jealousy probably won't have lasted very long I can tell you
that she looked quite attractive; she shook my hand as I came in
and never spoke another word to me. Only toward the end I embarked on a political conversation with Giles de la Tourette, during
which he of course predicted the most ferocious war with Germany.
I promptly explained that I am a Jew, adhering neither to Germany
nor Austria. But such conversations are always very embarrassing
to me, for I feel stirring within me something German which I
long ago decided to suppress.—At about 11:30 we were ushered
into the dining room, where there was a lot of drink and something to eat. I took a cup of chocolate.—You mustn't think that I
was disappointed; one cannot expect anything more from a *jour
fixe,* and all I know is that we will never have such a thing. But

please don't tell anyone how boring it was. We shall always talk about the first evening only. And now goodnight, my sweet darling, fondest greetings from

Your

Sigmund

95 *To* MINNA BERNAYS

Paris

February 7, 1886

My dear Minna

I had finished writing you a short, gay letter on Tuesday evening, was too lazy to mail it the same day, and by the following morning this became impossible on account of the news I received from Rosa. Your sad romance has come to an end, and when I think it over carefully I can only consider it fortunate that the news of Schönberg's death should reach you after such a long time of estrangement and cooling off. Let us give him his due by admitting that he himself tried and succeeded in sparing you the pain of losing your lover, even though it was less his high-minded intention than the moral weakness of his last years that prompted him to do so.

You will soon have forgotten the grievance his memory still evokes, and then you will tell yourself that you lost a good, noble and affectionate man without it being his or your fault. Oh, his rigidity and whatever it was that seemed to impede the intimate and undisturbed development of your relationship—none of this would have occurred had he been allowed to enjoy good health. In my opinion the most terrible aspect of this disease is that it destroys a human being before it makes him suffer. Can you visualize the change he must have undergone before he could misunderstand his condition as much as he did during his last months? Everyone suffering from this disease lives in the certain hope of recovering in the immediate future. Whenever a patient in the hospital asked to be discharged because he was feeling so well, we knew that he would be dead within twenty-four hours. Nature isn't always so charitable toward her victims.

I have no desire to conjure up before you again the memories that must in any case be with you now, nor to tell you how everything could have taken a different and better turn had it not been for powers beyond our control. I write about it only because I cannot bring myself to ignore it entirely and because I want to tie it up with a request. You haven't been made as unhappy as you so easily could have been; fate has grazed you only lightly; but you have suffered a lot, you have had little joy and a lot of worry, and in the end a great deal of pain from this relationshp. You were hardly out of your teens when you took upon you tasks normally faced in life only by adults. Now all that is over and I would like to ask you one thing: try to regain something of the lost youth during which one is meant to do nothing but grow and develop, give your emotions a long rest and live for a while quietly with the two of us who are closest to you now. You will have guessed that in' my recent joking letter I was trying to say the same thing.—

I well understand your desire (expressed in Martha's letter today) to make a gesture of friendship and ask for a memento. But I would advise both you and her against doing anything of the kind. We are dealing no longer with him but with his family, and from what we know of them it is best to break off all relations. You may as well be prepared within a few weeks or months to hear the whole family saying (and believing) that either your love or my medical treatment or Mama's relationship was the cause of his illness. Human beings are always glad when for some irrevocable event they can find a reason which is not entirely impersonal but tinged with some kind of emotion. As a matter of fact I doubt very much that he left any instructions concerning objects or any other possessions we might want as mementos, since he was unaware of his condition. One thing I am going to do first thing tomorrow is write to the brother with whom at the end I was on some kind of terms, and ask him to send to Mama's address the little case with your photographs, if it still exists. The man is fairly susceptible to the duties of a gentleman, and I hope to attack him from this side. I advise you to burn your letters while it is still winter, clear your ahead of all this, and think what a long life we have ahead of us, and what wonderful and extraordinary things may still happen to our little circle.

Warmest greetings in the hope of a brief, early meeting and a few lines in advance.

Your devoted brother
Sigmund

96 *To* MARTHA BERNAYS

Paris, Wednesday
February 10, 1886

My lovable darling

What a magic city this Paris is! Shall I start by talking about yesterday or shall I answer your many, many questions? Let me start with yesterday. It was the pleasantest evening I have spent here so far. I arrived very early and at the same time as Charcot himself, but he put me at my ease at once by saying that I had been invited not by him but by Madame. My early arrival gave me the advantage of having Mlle and then Madame to myself. Mlle was very friendly but, as you will be glad to hear, rather inaccessible. More details later. Madame was soon called away by sounds in the background and said by way of explanation: "C'est lui, il ne sait pas se mettre la cravate lui-même"! I was delighted to share this failing with the great man. He soon appeared and I had him to myself for a quarter of an hour during which I had the chance of mentioning a number of things: first, the news about the child outpatients,[1] to which he said: "Mais c'est quelque chose"; then about my departure, then about a little theory I have evolved around the case he put at my disposal, which he liked very much; then about the translation, and so on. He remarked that Paris was doing me good and that I had "engraissé." Gradually the guests arrived and we sat down to dinner. Apart from the Charcot family (four in number), there was the sculptor of the recently unveiled statue of Claude Bernard,[2] then Charcot's chief assistant Richet[3] and his wife—the latter in a somewhat denuded state for which, considering her beauty, one could hardly blame her and who, as a matter of fact, sat as silent as a statue—then a M. Mendelssohn, a Jew from Warsaw, who had been Charcot's assistant as well as a pupil of the Berlin physiologists and is now working under enviable

---

[1] Reference to the First Public Institute for Child Patients, which had been opened in Vienna. Dr. Max Kassowitz (1842-1913), Professor of Pediatrics at the University of Vienna, was the Director.

[2] Claude Bernard (1813-1878), Professor of Physiology, Paris.

[3] Charles Robert Richet (1850-1933), Charcot's senior assistant.

conditions with patients at the Salpêtrière; a M. Arène,[4] journalist and art historian whose articles I read every day in the papers; M. Toffano, an Italian painter whom I was meeting there for the third time, and myself. I sat next to Mlle Charcot; I enclose my place card for our archive. We were not given very much to eat, but each dish was exquisite and accompanied by different wines. The conversation was carried on mostly by Madame; Charcot himself was in a gay mood and the family remarked that he was "aimable." Well, now for Mlle. She is twenty years old and, despite her smallness, very pretty, moves with great ease of course, and her interests seem to be divided between her father and brother. "Si j'étais garçon," she said. She obviously takes a serious interest in medicine. I tried my best to be attentive and suggested we carry on our conversation in English, but when she told me that English had been her first language I soon stopped. She has a much older sister but not from the same father; the rest of the meal was taken up by a lively dispute between her and the young Charcot which the old man brought to an end with a good-natured "Assez, Mademoiselle!" When the meal was over I had the honor of leading Mlle back to the drawing room, as M. Richet was too far away. Since the dinner had made me feel at ease, I enjoyed myself very much and talked a lot more with Charcot himself, from whom I also borrowed a book and a magazine. A particularly pleasant event for me was the arrival during the evening of M. Ranvier,[5] the famous histologist, who had given me such a friendly reception in the Collège de France. I think he spoke to Charcot about me, and I myself had a pleasant talk with him later. My confidence as a judge of human nature received a considerable boost when he confided in me that he would have liked best to be a professor in a small German university—for instance, Bonn—for in a letter to Paneth I had described him as a "German university professor badly translated into French." The party grew larger and larger; the later guests included Cornu,[6] the famous optician, with a truly inspired expression; M. Peyron, director of the Assistance Publique, against whom the students recently instigated a huge scandal, no one knows why; and (prepare

[4] Emanuel Arène (1856-1908), journalist.

[5] Louis Antoine Ranvier (1835-1922), Professor of Anatomy in Paris and member of the French Academy of Science.

[6] Marie Alfred Cornu (1840-1902), Professor of Physics in Paris, known for his experiments on the speed of light.

for a surprise) Daudet himself. A magnificent face! Small figure, narrow head with a mass of black curly hair, a longish but not typically French beard, fine features, a resonant voice, and very lively in his movements. Mme D. was also there, and never left her husband's side; she is so unbeautiful that one cannot imagine she has ever been any better-looking; a worn face, prominent cheek bones. She was dressed like a very young woman, although her eighteen-year-old son, friend of Charcot's son, was present. Daudet doesn't look a day over forty; he must have married very young.

In short, the evening was very amusing. I left with M. de la Tourette and as late as 12:30 went up to his apartment to fetch a paper he had promised me.

The following day I couldn't help thinking what an ass I am to be leaving Paris now that spring is coming, Nôtre Dame looking so beautiful in the sunlight, and I have only to say one word to Charcot and I can do whatever I like with the patients. But I feel neither courageous nor reckless enough to stay any longer.

Next day—yesterday, Wednesday—I had another adventure. The Viennese, a truly dreadful fellow, called for me and we went together to the Salpêtrière. The man is a hydrotherapist at Winternitz's,[7] and as a result considers himself a great neuropathologist and made all kinds of condescending remarks which I took in my stride, confident of an imminent revenge. He had a letter of introduction to Charcot containing some outrageous flattery: that he had arrived to meet the greatest living physician. From this he had evidently expected I don't know what kind of reception, but I knew it would be rather cool. And indeed, after having received the letter, all Charcot said was "À votre service, Monsieur!" and added: "Vous connaissez M. Freud?" to which we both lowered our heads, he rather taken aback, I silently pleased. Then something else happened.

For a week now there has been a foreigner in the Salpêtrière, a definitely Germanic type and yet somehow different; I can't quite make him out. Wednesday is the day we go to the ophthalmological room, and there this foreigner suddenly began behaving with some authority; when he exchanged cards with Charcot's ophthalmologist, the latter became very polite and expressed the hope that

[7] Dr. Wilhelm Winternitz (1834-1917), Director of the Hydropathic Establishment in Kaltenleutgeben, near Vienna.

Monsieur would often return so that he could learn something from him. Whereupon we all began wondering who he might be. Before leaving he came over to us Viennese and said: "I heard you speaking German. I'd like to introduce myself." My *bête noire* exchanged cards with him first and I was still trying to find mine when the foreigner said: "I am a German, but I emigrated to America long ago." At last I gave him my card, but it happened to be one without an address. He glanced at it and said: "Could you be Dr. F. from Vienna? I've known your name for a long time, from your publications, especially the one on cocaine." I was a little surprised and inquired after his name, which turned out to be Knapp.[8] Now, Knapp is the foremost ophthalmologist in New York, who has also written a lot about cocaine and to whom I once wrote a letter in Koller's name. I greeted him accordingly and my *bête noire* stood there looking rather sheepish, first of all because he had failed to recognize the man, and second because he had again managed to make a fool of himself. When he heard the word *cocaine* mentioned he asked: "Have you also written about cocaine?" Whereupon Knapp replied: "Of course he has, it was he who started it all." This morning my Viennese was much more malleable and talked exclusively of the great practice that awaited me in Vienna.

I have had letters again from both the bookseller and Kassowitz. The former much more affable. K. writes only to say that he doesn't want to influence my choice between Breslau and Berlin, but asks me if it is to be the latter not to mention his name because he is on bad terms with the Berlin pediatricians. I am kept very busy with the case Charcot passed on to me; our association continues to be very satisfactory.

But now it is time to answer your questions, my darling. I don't know anything about the funds of the enterprise; I think it is run privately like the other polyclinic and very likely exists on voluntary contributions. There is no question of remuneration for a director of the department, which doesn't make the position less valuable. The consultations take place in a special room containing among other things electrical equipment; there will be one or two students to keep the records; the consultations are held two or three times a week, unpaid, but in return one has material and if one is a *Dozent* one can lecture on this material, if not at once, then in the

[8] Dr. Hermann Knapp (1832-1911).

winter. Do you understand it now? The chief advantages lie in having access to the material and in the reputation one can acquire in this way as a specialist.

I have never told you about my uncle in Breslau because I never think of him. I have seen him three times in my life, on each occasion for a quarter of an hour. He is a younger brother of my father, a rather ordinary man, a merchant, and the story of his family is very sad. Of the four children only one daughter is normal, and married in Poland. One son is a hydrocephalic and feeble-minded; another, who as a young man showed some promise, went insane at the age of nineteen, and a daughter went the same way when she was twenty-odd. I had so completely forgotten this uncle that I have always thought of my own family as free of any hereditary taint. But since I have been thinking about Breslau it all came back to me, and I am afraid the fact that one of the sons of the other (very unhappy) uncle in Vienna died an epileptic is something I cannot shift to the mother's side, with the result that I have to acknowledge to a considerable "neuropathological taint," as it is called. Fortunately, of us seven brothers and sisters there are very few symptoms of this kind to report except that we, Rosa and I (I don't count Emanuel), have a nicely developed tendency toward neurasthenia. As a neurologist I am about as worried by such things as a sailor is by the sea. But you, my darling, must realize that it is your duty to keep your nerves in good condition if the three children, of whom you have been prematurely dreaming, are to be healthy. And if the thought of medicine makes you shudder, darling, I can't blame you, but love me you must nevertheless, and if we get married soon we will be very happy, won't we? These stories are very common in Jewish families. But now that's enough about medicine.

The money situation is easily explained. The three hundred florins which are still outstanding for the translation and which I have meanwhile accepted from Paneth as an advance, were for January and February. So you see that money for traveling and for the month of March will have to come from elsewhere. What you say about the detour round Hamburg is quite right; but are you suggesting that I don't want to see you? I will stay only one day, and for everyone in Vienna and Hamburg I have traveled through direct. Has your Highness taken that in? Assian,[9] the only one likely to notice my presence, probably won't betray me. The ready money is a very unpleasant chapter, little Princess, but the prospect of having you sit beside me for the whole of one day is a very pleasant one, and on that day I refuse to pay any calls, nor will I allow you to do anything but chatter with me.

Fondest greetings and kisses, my little woman; I am ending on a different day from that on which I began, and tomorrow I expect another letter from you.

Your
Sigmund

[9] A florist in Wandsbek.

97 *To* MARTHA BERNAYS

Berlin, Wednesday
March 10, 1886

My sweet darling

What strange things you have been experiencing and what fascinating letters you are able to write! I take a very warm interest in the fate of the silver snake, and your comment on the sensation of receiving telegrams is really excellent. Fortunately nothing like that is happening here. I am going to take the liberty of being as boring today as I was yesterday; so far no adventure, no thrill, no éclat, as we used to have in Paris. Just quiet work. I have plunged into the translation with something akin to passion because I am afraid of not finishing it, no doubt a quite unreasonable fear. This morning I roused myself to the point of going to the Royal Museum, where I had a brief look at the antique fragments, with deep regret at not being able to understand more about them and with nostalgic memories of the Louvre, which is so much more magnificent and substantial. The most interesting things are of course the Pergamene sculptures, all of them fragments representing the battle between the gods and the giants—tense, dramatic scenes. As my colleague Dr. Türkheim used to say, one can't always be a doctor.

What appeals to me more than the stones, however, are the children in the clinic, who on account of their small format and because

they are usually well washed I find more attractive material than
the large editions of patients. As long as their brains are free of
disease, these little creatures are really charming and so touching
when they suffer. I think I would find my way about in a children's
practice in no time. A few more months' preparation wouldn't do
any harm, but I am afraid this is out of the question; the days of my
reckless daring have evaporated. Vienna weighs upon me and perhaps more than is right. I am afraid I am sinning nowadays against
my otherwise loyally adhered to principle of not tormenting myself
with new situations until I am in the midst of them. But I will
conquer my present mood, and then I won't worry about anything
till I see with my own eyes the detestable tower of St. Stephen's.

And now once more I have come to the end of my material and
I ask you to be very patient with me; I cannot after all initiate you
into the secrets of children's diseases, and even Baginsky[1] doesn't
seem to me a sufficiently important figure to warrant a detailed
description. I am secretly counting the days, but you don't have
to know the number I have got to.

With fondest greetings and kisses
Your
Sigmund

[1] Dr. Adolf Baginsky (1843-1918), Professor of Pediatrics and Director of the
Kaiser Friedrich Hospital in Berlin.

98 *To* MARTHA BERNAYS

Berlin, Friday
March 19, 1886

My sweet darling
        Nothing new has happened; I am annoyed that I am not
on my way to you. They[1] have left me in peace so far, and it may
turn out that I was afraid of an unloaded gun. Now I shall have
to endure it for another week, but then there will be a few lovely
days; perhaps by that time it will be possible to travel without the
danger of snowdrifts.

I am so industrious, organized, brave, and sober that I am almost
worried; no adventure of any kind comes my way. What else have

[1] Some of Martha's relatives who had announced their visit to Berlin.

I to tell you except that I had a letter from L.[2] in Breslau asking me
to look up his sister-in-law and a *Sanitätsrat*° who is also related to
him. Well, I suppose I must. But I am rather stingy with my time;
I have never enjoyed work so much. I have been left with such
a friendly, edifying memory of Charcot, in its way not unlike the
one I had after the ten days with you. I feel I have experienced
something precious that cannot be taken away from me. I feel increasingly self-confident, more urbane, more adroit in dealing with
colleagues. It is a pity I cannot stay here to take part in the so-called
holiday course which begins on the twenty-second. Oh, my little
darling, you have but one minor fault: you never win the lottery.
Just now I feel so ready to be very happy. I would stay here a few
more weeks, then I would get an apartment in Vienna, and we
could still get married in the spring, and then together we could
practice the skill in traveling which I have acquired during the past
seven months. What a pity this must remain a dream! And now I
am regretting all the kisses I could have had tomorrow and the day
after! But just you wait, I won't let you off these two days, and
if you are naughty I will stay longer. A return ticket to Hamburg
is valid for more than five days.

I hope Assian came to congratulate you on Saturday morning.
He was supposed to announce my arrival, but as I couldn't come
I didn't want to call everything off. I wonder if you have been
racking your brains as to who sent the bouquet. And why? Perhaps
you thought it was Hugo Kadisch[3] trying to bring about a pleasant
end to an old relationship.

Threatened by the illustrious visitor, I have decided to get my
already rather shapeless French beard (which, by the way, is
generally envied here) trimmed. Filled with understandable suspicion of the Berlin scissors, I went to the most fashionable barber,
the Court Hairdresser in the Unter den Linden, paid one reichs-
mark, and the man (who looked like a cabinet minister) did a bad
job. I don't think that's nice.

On Saturday or Sunday I am going to the theater out of gray,
grim despair. Not even a library is open on Sunday. And although
translating is quite a nice way of spending a Sunday, I have grown
quite stupid; I no longer know how to spell or how to insert relative

---

2 Presumably Professor Lichtheim, Hammerschlag's son-in-law.

° Member of the Board of Health; a title conferred on German physicians.

3 A friend of Martha's father who had asked for Martha's hand years ago.

clauses; in my ears nothing seems to sound German or French any more.

"How different it was in France!" I sighed like a Maria Stuart[4] among neuropathologists.

A human being can really go a long way in the art of self-persuasion. If I had had to travel from Paris to Vienna, I think I would have died en route. And now I have reached the point of actually looking forward to Vienna, especially to one thing: to the hours I am going to spend finishing my paper in Obersteiner's lovely library in Döbling.

Do you know what I have just come to realize very clearly? That if life in Vienna is to be at all humanly possible, I must have one or two thousand florins, and that to acquire this sum people to whom one pays interest will have to be found. Not usurers, needless to say. But where can semi-unselfish capitalists be found willing to lend money at a normal rate of interest with no more guarantee than a human head and two hands? This is the great problem that has to be solved, not at once but in two months' time when my present wretched thousand florins have come to an end.

I dare say greater revolutions have taken place in the world than that a man who has nothing now, later acquires a few thousand gulden. I have very little fear of the future. In any case, and this comes first among the achievements of this year of travel: I am coming to fetch you like an overdue bill of exchange on June 15, 1887—if this hasn't been possible before. Are you absolutely determined to be ready by then, my little girl?

Answer verbally at the end of the month to

Your

Sigmund

[4] The quotation erroneously ascribed by Freud to *Maria Stuart*, by Schiller, is to be found in Schiller's *Don Carlos*.

99   *To* MARTHA BERNAYS

Berlin, Tuesday
March 30, 1886

My sweet darling

If I ever have to leave Wandsbek again without you, I will certainly let you see me off at the station. I very nearly turned about at Schadendorf and came back to spend another night, but I felt a little ashamed, for up to now we really have managed to be pretty sensible at our partings. But this was a very difficult one for me, my darling; and you too, my treasure, shed some tears, and this hurt me to see. But now I really will go ahead, as you suggest, so that you may be spared many more partings and a long period of waiting.

And here I am again in Berlin and we are already as far apart as if I were in Vienna. A new era is beginning, a good one I hope, one that will bring good things. My sweet treasure, I just can't imagine what it will be like when I don't have to leave you any more, and yet I know that I won't be in the least surprised and will feel as if we had always been together.

This time no letter had arrived before me; the journey was very pleasant; I slept nearly the whole night under cover of my knitted rug. I am not in the least tired and am just off to the Café Bauer. Don't forget, my little Princess, that you have promised me to keep well and write again soon. I too have more time again now that the translation has come to an end.

With lots and lots of kisses, for which a new account begins today.

Your
Sigmund

Fond greetings to Mama and Minna.

100 *To* MARTHA BERNAYS

Vienna, Thursday
May 6, 1886

My sweet girl

Warmest thanks for your dear letter and for the parcel,[1] the contents of which I knew so well I felt I had been present at the purchase. I have always wanted a clythia[2] and I realize that you knew it. But that you apologize for your gift, dearest, really is unnecessary, and I feel quite ashamed when I think that I am now in your debt instead of competing with you in giving presents.

I am now so old, as you know, and yet on the eve of the fourth

---

[1] Martha's presents for Freud's thirtieth birthday.
[2] An evergreen plant.

anniversary of our engagement we still don't know when the married state which we have so often visualized will become a reality. But although still as far as ever from the goal, we are less far from the certainty. In a few weeks the money—which I still haven't touched—will have come to an end, and then we shall see whether I can go on living in Vienna. I would like to think that the next birthday will be just as you describe it, that you will be waking me up with a kiss and I won't be waiting for a letter from you. I really no longer care where this will be, whether here or in America, Australia or anywhere else. But I don't want to be much longer without you. I can put up with any amount of worry and hard work, but no longer alone. And between ourselves, I have very little hope of being able to make my way in Vienna.

I am continuing this in the evening, darling. There were two old patients of Breuer's at my consultation today, no one else. I usually have five: two for electric treatment, one for nothing, one *schnorrer*[3] and one—*schadchen*.[3]

Then came the congratulations: Pauli and Dolfi brought me a beautiful little brush box, Mitzi a large photograph of herself and two Makart bouquets,[4] Mother a cake and Rosa a lovely framed blotter for my writing table. There were written congratulations from Willenz, Schani, Kleinenberger[5] and Uncle Elias, whom I want you to thank very much for me. So they celebrated me like a prince; I have every reason to be tired and am going to bed early.

Work in the laboratory is giving me great pleasure. I certainly have time enough. I also have another therapeutic idea, which I shall try out very soon. But it is rather doubtful that it will prove as valuable as the coca one.

Goodnight, my little woman. Here's to next year!

Your
Sigmund

I'll write to Mama and Minna separately tomorrow.

---

[3] Yiddish—the former meaning: beggar; the latter: marriage broker.

[4] A bouquet consisting of dried palm branches, reed, bamboos, a peacock feather, etc., named after the Austrian artist Hans Makart (1840-1884).

[5] Relatives of Freud on his mother's side, except Schani (Freud's younger brother Alexander, 1866-1943).

The Freud family, 1876. Back row, left to right: Pauli, Anna, Sigmund, unidentified, Marie (Mitzi), Simon Nathanson. Second row: Dolfi, unidentified, Amalie, Jacob. Front row: Alexander, unidentified boy.

Sigmund Freud and Martha Bernays during their engagement, 1885. They were married the following year.

Jean Martin Charcot. "I sometimes come out of his lectures as from out of Nôtre Dame, with an entirely new idea about perfection" [Letter 86].

Dr. Josef Breuer. "By saying only good things about him one doesn't give a proper picture of his character; one ought to emphasize the absence of so many bad things" [Letter 67].

Lou Andreas-Salomé. "After all, you are an 'understander' *par excellence*" [Letter 172].

Marie Bonaparte with her chow Topsy. "She surpassed herself in tender care and attention" [Letter 299].

The Committee, Berlin, 1922. Back row, left to right: Otto Rank, Karl Abraham, Max Eitingon, Ernest Jones. Front row: Freud, Sandor Ferenczi, Hanns Sachs. "A group of men who were united in their devotion to psychoanalysis" [Letter 236].

Freud reading the manuscript of An Outline of Psychoanalysis, July 1938.

Nov. 16th 1938

20 MARESFIELD GARDENS.
LONDON. N.W.3.
TEL: HAMPSTEAD 2002.

To the Editor of "Time and Tide

I came to Vienna as a child of 4 years from a small town in Moravia. after 78 years of assiduous work I had to leave my home, saw the Scientific Society I had founded dissolved, our institutions destroyed, our printing Press ("Verlag,") taken over by the invaders, the books I had published confiscated or reduced to pulp, my children expelled from their professions. Don't you think you ought to reserve the columns of your special number for the utterances of non-Jewish people less personally involved than myself?

In this connection my mind gets hold of an old french saying:

Le bruit est pour le fat,
La plainte est pour le sot,
L'honnête homme trompé
S'en va et ne dit mot.

I feel deeply affected by the passage in your letter acknowledging "a certain growth of anti-Semitism even in this country." Ought this present persecution not rather give rise to a wave of sympathy in this country?

Respectfully yours
Sigm. Freud

Photostat of Letter 310.

Freud's study at Maresfield Gardens, London.

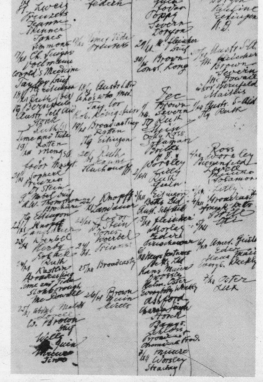

Photostat of a record of correspondence. Records of incoming and outgoing letters were kept through many years.

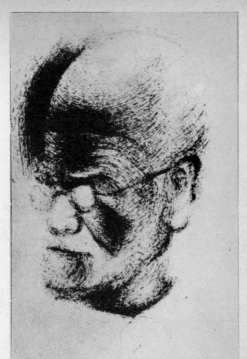

Sketch by Salvador Dali, July 19, 1938. "The young Spaniard, however, with his candid fanatical eyes and his undeniable technical mastery, has made me reconsider my opinion" [Letter 302].

Figure by Nemon, done in 1936; it is now at the New York Psychoanalytic Institute.

Arrival in London, June 6, 1938.

101 *To* MARTHA BERNAYS

Vienna, Thursday
May 13, 1886

My beloved darling

I won't be able to write to you any more during my consulting hours because there is too much going on. The waiting room is full of people and I shall hardly be finished by three o'clock. The takings are not yet very brilliant, but the patients who avail themselves of my services are quite numerous, though there aren't many paying ones among them: Frau Prof. M, who gives me a lot of trouble, the sciatica case which is almost cured, and the two police officials who come once a week. Tomorrow T. is coming. Today my earnings amounted to eight florins: three from a police official and five again through Breuer, who sent Frau Dr. K. She came to ask for some advice to relieve her husband.

I realize that for a doctor work and income are two very different things. Sometimes one makes money without lifting a finger, at other times one slaves away without reward. The day before yesterday, for instance, an American physician came to see me with a nervous complaint—a complicated case which interests me so much that I took it on without getting anything out of it. His case is complicated by his relationship with his beautiful and interesting wife with whom I also have to deal and on whose account I am going to see Prof. Chrobak[1] tomorrow. I am too tired to describe to you in detail all the delicate aspects. It seemed weird to me that on both the occasions she was here your photograph, which has otherwise never budged, fell off the writing table. I don't like such hints, and if a warning were needed—but none is needed.

And a doctor is supposed to economize! Here I am counting every gulden and I am called to go and see a distant acquaintance in the Stadtgutgasse, no remuneration of course, and two hours out of the day gone, for I cannot afford a cab. Today the same. When I get home I find an urgent message to go and see the man again. This time of course I am obliged to take a cab, and what I have saved in suppers during the past three days has to be spent on it.

[1] Dr. Rudolf Chrobak (1843-1910), Professor of Gynecology at the University of Vienna.

On Tuesday I gave a lecture in the Physiological Club on hypnotism; it went off very well and received general applause. I have announced the same lecture for two weeks from today in the Psychiatric Club, and during the next three weeks I will give another lecture on my Paris experiences before the Medical Association. So the battle of Vienna is in full swing, and if you were here I would tell you with a kiss that I haven't abandoned the hope of calling you my wife in six months.

I think I will have to arrange for a second consulting hour three times a week from three to four for nonpaying patients and those in need of slight electrical treatment. In spite of everything my position here is a strong one, as I can see from several indications.

Goodnight, my sweet darling.

Your
Sigmund

What do you think of a collective present for Mama again this year?

102 *To* JOSEF BREUER

Olmütz[1]
September 1, 1886

Dear Friend

I won't try to describe the extent of my pleasant surprise on hearing that you both went from Hamburg to visit my little girl and were so *nett** to her, as the locals say. May the Lord in return grant you the most beautiful holiday, the least unpleasant weather, and an undisturbed enjoyment of the happiest mood.

I am lying here on a short leash in this filthy hole—can't think of any other way of describing it—painting flagpoles black and yellow.[2] The few lectures I have given on medical service in the army were fairly well attended and have been translated into Czech. So far I haven't been locked up.

The only remarkable thing about this town is that it gives no

---

[1] Town in the Austrian province of Moravia where Freud had to present himself for military service.

* Nice.—*Tr.*

[2] Colors of the flag of the old Austrian Empire.

idea how remote it can be. Frequently one has to march for three
or four hours before reaching it, and at moments, not memorable
for anything else, I find myself miles away from it. As Paul Lindau[3]
once said in a review of a novel laid in the Middle Ages: "Most
readers will hardly remember that something like the middle of
the fourteenth century ever existed," so I would like to ask what decent, self-respecting human being would consider transferring one
of his activities to the hour of 3 to 3:30 in the morning. We continually play at war, once we even enacted the siege of a fortress, and I
am acting the role of the medical officer doling out handbills showing
gruesome injuries. While my battalion makes an assault, and I lie
with my medical orderlies on some old stones, and a battle is being
fought with blank cartridges, the general rides up as he did yesterday and says: "Men! Men! Where would you be if those things had
been loaded? Not one of you would be alive."

The only thing that makes Olmütz bearable is a citified cafe
which has ice creams, newspapers, and good pastry. I might add
that whenever I sing the praise of ices my fiancée always insists
that they are made with water instead of cream, and warns me
against them. The service, however, like everything else, suffers
from the military atmosphere. When the two or three generals—
who keep reminding me of parrots, I can't help it but mammals
just don't dress in such colors, with the exception of the reddish-blue
of mandrils' posteriors—whenever these generals sit together anywhere, a whole swarm of waiters buzzes around them to the exclu-
sion of everyone else. Once in despair I was forced to indulge in
some serious boasting. I seized one of these waiters by the coattails
and shouted: "Look, I too may be a general one day; go and get
me a glass of water."—It worked.

An officer is a miserable creature; he envies his equals, he bullies
his subordinates, and is afraid of the higher-ups; the higher up he
is himself, the more he is afraid. I deeply dislike having my value
written on my collar, as though I were a sample of some material.
And yet the system has its loopholes. It wasn't until recently when
the commanding officer from Brünn arrived here and paid a visit to
the swimming pool that I learned to my surprise that bathing
trunks show no signs of rank.

But it would be ungrateful not to admit that military life with

[3] Paul Lindau (1839-1919), German journalist, critic, and dramatist.

its hopeless "must" is a very good cure for neurasthenia. Every vestige is as though blown away, and this within a week.

This whole military interlude of mine is fizzling out, as Meynert would say. Only ten days more and I shall be fleeing northwards and have forgotten the crazy four weeks.

From the scientific point of view nothing of interest has come my way except that the strange case of paralysis agitans which I mentioned recently suddenly turned up here and maintains that he is deriving unmistakable benefit from the arsenic injections I have been giving him.

I hope you will excuse this stupid chatter which has slipped off my pen I don't know how, and I very much look forward to visiting you in Vienna for the first time with my wife.

Your devoted
Dr. Sigm. Freud

1887

1912

103 *To* EMMELINE AND MINNA BERNAYS

Monday, Vienna
—actually Tuesday, 12:30 A.M.[1]
October 16, 1887

Dear Mama and dear Minna

I am terribly tired and have so many letters to write, but
the first must go to you. You will have heard already by telegram
that we have a little daughter. She weighs nearly seven pounds,
which is quite respectable, looks terribly ugly, has been sucking at
her right hand from the first moment, seems otherwise to be very
good-tempered and behaves as though she really feels at home here.
In spite of her splendid voice she doesn't cry much, looks very happy,
lies snugly in her magnificent carriage and doesn't give any impression of being upset by her great adventure. She is called
Mathilde, of course, after Frau Dr. Breuer. How can one write so
much about a creature only five hours old? The fact is I already love
her very much, although I have not yet seen her in the daylight.
She was born at 7:45.

Perhaps you are more interested in the mother. She has been so
good, so brave and sweet all the way through. Not a sign of impatience or bad temper, and when she had to scream she apologized
each time to the doctor and the midwife. When she woke at 3 A.M.
with the first labor pains, we decided to stick it out till 5 A.M., then
I fetched Dr. Lott and the midwife who lives nearby; during the day
we refused all visitors, Eli and Ditta[2] among them; luckily it was
Sunday, no consultations, no visits to pay. At first things went very
slowly. Lott pulled a long face and declared that it could last all

---

1 "Monday, actually Tuesday"—Freud confused the days. It should read: "Sunday,
actually Monday, 12:30 A.M."

2 Judith Bernays, eldest child of Eli and Anna Bernays.

night, in which case the child would lose the advantage of being born on Sunday. In the afternoon, however, the pains increased, the main task was soon over, and we had every reason to expect the birth to take place in the evening. But at 5 P.M. the pains ceased, the child didn't advance, and when things hadn't improved by 7:30, Lott finally decided to use the forceps. Martha was all for it, wasn't in the least frightened, and at every free moment joked with her two helpers and her fellow sufferer—that was me—and I am actually so tired I feel as though I had gone through it all myself. Thus at 7:45 P.M. the child arrived. Martha felt very well at once, was given a plate of soup, was terribly pleased when shown the little creature, and in the midst of the physical and moral havoc caused by such an occasion we were both very happy. I have now lived with her for thirteen months and I have never ceased to congratulate myself on having been so bold as to propose to her before I really knew her; ever since then I have treasured the priceless possession I acquired in her, but I have never seen her so magnificent in her simplicity and goodness as on this critical occasion, which after all doesn't permit any pretenses. I am very happy and join the doctor in hoping that everything else will continue as well as all the signs so far promise.

Goodnight; I trust you will soon write again to your little family consisting of Martha, Mathilde and

Sigmund

104 *To* EMMELINE AND MINNA BERNAYS

Vienna, Friday
October 21, 1887

Dear Mama and Minna

I was highly amused by Mama's last letter. It shows so much family pride, and the emphasis on the two families' peculiarities emerges quite innocently at a time when the new little creature seems to display a combination of them both. For I announce to you solemnly: although Mathilde sucks her fingers, she shows a striking resemblance to me—according to general opinion; indeed, several people point to gaps in my face where the little girl has been cut out.* She has already grown much prettier, sometimes I think even

---

* I.e., is the image of her father.—*Tr.*

very pretty. Furthermore, she has inherited from me the determination to get enough to eat, and alas the tendency toward feeding difficulties, too, as I shall explain later. So far two offers for her little hand have been received, one for Ludwig Paneth[1] from his mother, and the other for the fat Karl Kassowitz,[1] admittedly only through the agency of his uncle. The decision, however, is still outstanding, and so is the dowry. I actually owned a gold coin, as a start for the dowry, but let myself be persuaded into giving it to the midwife. As for the wet nurse who arrived the night before last from Roznau[2] we are not very satisfied with her, and we are making it dependent on Kassowitz' decision whether she is to be sent back at once or allowed a few days to improve.

Monday, October 24, 1887

Impossible to finish this letter. Reasons later. Latest news from the lying-in room is as follows: Martha's condition is no longer interesting, no grounds for concern. (Lott has taken the day off.) She is in good health, has an excellent appetite, and enjoys everything enjoyable. She is the object of the most friendly attention from every side. I can't even remember how often Frau Breuer, Frau Hammerschlag, Paneth, *et al.* have called, but it was not until today that she began receiving her lady visitors. Each one brings a bouquet of roses; my two recorders in the hospital sent a lovely *Sempervirens*[3] for the young citizen of the world; Kassowitz listens indefatigably whenever I bore him with nursery tales; Breuer blows in and out like a comet. The story of the wet nurse is as follows: she produced less and less milk; at the same time she devoured outrageous quantities of everything imaginable; finally she got indigestion, felt miserable, and on top of it all the child developed a *green* stool. Only a heartless aunt could laugh at this. Mama will know what it means. Just now the second wet nurse arrived on whom the child has already fastened itself; the first one is going home this evening, enriched by a beautiful memory. Let us hope for our Mathilde, who until now looked more like a Kamille,* that the second nurse will be the last. The child shows increasing

---

[1] Ludwig Paneth and Karl Kassowitz, the eldest sons of Dr. Paneth and Professor Kassowitz.

[2] A small spa in Moravia.

[3] An evergreen plant.

* German for Camilla as well as for camomile.

signs of being definitely influenced by the paternal side. She has a great appetite, and when she is hungry she screams without any self-control. Almost simultaneously with her birth my practice underwent a revolution which couldn't be more radical. The previous six weeks had been the quietest of the whole year; then, when Martha's labor started, I was asked to attend a joint consultation with Chrobak at Frau L.'s on Monday. . . . Yesterday evening there was a consultation with Kassowitz, a few days ago I put Frau Dr. Z. on a fattening cure—in short, activities galore. My consulting room is full of new faces, more than I usually see in two months. I have to admit that none of this has earned me anything as yet, nor will every contact lead to something, but things are picking up as if the birth of a daughter were equal to a certificate of qualification for the medical profession. Even my lectures, which were seriously endangered, have been saved. One of them (on brain anatomy) starts next Monday with four or five students. Martha has already taken a five-florin voucher.

I wish you both the pleasantest weather and the best of health and I send you my fondest regards.

Sigmund

105 *To* JOSEF BREUER

Vienna
May 3, 1889

Dearest Friend and best loved of men!

By now you are of course convinced that I am mortally offended because I still haven't reacted to your letter—but this is far from the case: the night before last there was a card party at Paneth's which lasted till 1 A.M. and yesterday I was so tired that I couldn't write. I must postpone my answer for a quiet evening or, should it suit you better and your need for rest permit, I could replace it altogether with a nocturnal visit.

I have examined myself thoroughly and come to the conclusion that I don't need to change much. Almost everything you say is correct—almost, because one thing even my wife found incorrect, but it doesn't particularly hurt my feelings, and all in all you have taken care to coat the bitter pill with the sweet ingredient of one

word. Fundamentally you are proud of me all the same, which you
wouldn't have been had I behaved differently.

With warm greetings

Your

Sigmund

You owe me one little reparation nevertheless for your letter,
which shows that you don't believe in my star, even if it is only a
miserable little meteor. In our relationship, for instance, you might
stray more consistently from the *Sie* to the *Du*, or something like
that, s.v.p.

106 *To* WILHELM FLIESS

Reichenau[1]

August 1, 1890

Dear Friend

    I am writing to you today to tell you very reluctantly that
I cannot come to Berlin; it is not that I care so much for Berlin
or the Congress, but I am disappointed that I won't be able to see
you there. My decision has been upset not by one big reason but
by a combination of smaller ones, which so easily arise in the life
of a doctor and father of a family. I just cannot manage it from any
point of view—not professionally, since my most important patient
is just going through a kind of nervous crisis . . . and not with the
family because all sorts of things have gone wrong with the children (I now have a daughter and a son). My wife, moreover, who
as a rule has no objections to my taking a short trip, is very much
against my leaving here just now. And so on and so forth. In short,
I just cannot manage it, and as I have looked upon this trip as a
great luxury that I was granting myself, I feel obliged to abandon it.

    Very reluctantly, for I had expected so much from this meeting
with you. Although quite satisfied, even happy if you like, I nevertheless feel very isolated, scientifically stagnant, lazy and resigned.
When I talked to you and realized that you think something of
me, I actually began to think something of myself, and the picture

---

[1] Martha had taken the children to Reichenau, about two hours from Vienna, for
the summer vacation. Freud spent the week ends with them.

of convincing energy which you offered did not fail to make an impression on me. Professionally, too, I probably would have profited from you and possibly from the Berlin atmosphere, as it is years now since I have met anyone who could teach me anything and I am almost entirely immersed in the treatment of neuroses.

Is there no chance of seeing you anywhere except at the Congress in Berlin? Aren't you taking a trip afterwards? Or won't you be returning in the autumn? Please don't lose patience with me for leaving you without an answer and for now refusing your exceedingly kind invitation. Do let me know that there is a chance of seeing you for several days so that I won't lose you as a friend.

With cordial greetings
Yours sincerely
Dr. Sigm. Freud

107 *To* MINNA BERNAYS

Vienna
July 13, 1891

Dear Minna

The consulting hour offers plenty of opportunity for writing letters just now. I consider hanging up my photograph in the waiting room with the inscription: *Enfin seul.* Unfortunately it wouldn't find any admirers there.

I see with great satisfaction from your card from Süllberg[1] and from many other signs that you are very well, and I am delighted and hope moreover that you will make good use of this sojourn.

Here (giving precedence to the weather) it never stops raining. In spite of it I climbed the Rax[2] yesterday and again brought home some edelweiss. I can't send you any as Martha has pressed them all. Next time. The brats are thriving. Martin[3] is becoming quite enchanting, so affectionate and good-natured and rather intelligent; he can speak a fair number of words, repeats a lot, and understands

---

[1] A village near Blankenese, in the vicinity of Hamburg.
[2] A mountain not far from Reichenau.
[3] Jean Martin, Freud's eldest son, born 1889, named after Professor Charcot.

almost everything, except of course scientific and technical terms. Oliver[4] still screams like a maniac but is good-looking and attentive, gains 4½ ounces a week and has some excellent achievements to show. Only the little woman[5] is a nuisance; her face has such a wild look, she is up to all kinds of mischief, says No on principle to any suggestion and considers herself under no obligation to obey. Add to this the ghastly methods of education on the part of the Nanny (whom I am soon going to pension off), combined with Martha's weakness in not daring to rebuke the old woman for her most unwarranted criticisms. But I hope the little creature will survive this influence, too, and get used to girlish ways again. . . .

The "Aphasia"[6] has just come out, as you will see by the enclosed, and has already caused me deep disappointment. Breuer's reception of it was such a strange one; he hardly thanked me for it, was very embarrassed, made only derogatory comments on it, couldn't recollect any of its good points, and in the end tried to soften the blow by saying that it was very well written. I believe his thoughts were miles away. In the midst of all this he inquired whether Dr. P. had arrived, and when the latter did turn up and shrank back at sight of me, I, of course, left at once. The breach between us is widening all the time, and my efforts to patch things up with the dedication have probably had the opposite effect.

On Thursday I am going to escape to Reichenau. Things are really too stupid here.

Kind regards
Your
Sigmund

[4] Freud's second son, born 1891.
[5] Freud's first child, Mathilde, born 1887.
[6] *Zur Auffassung der Aphasien*, Vienna, 1891. (*On Aphasia*, London and New York, 1953.)

108 *To* MARTHA FREUD

Vienna IX, Berggasse 19
Thursday, June 7, 1894

My beloved darling

Yesterday evening was almost unbearably sultry. In the morning I wake at six, find the room light, think to myself that this rare event of waking up so early should be put to some use, knock at Marie's door and order a bath, then lie down again. Half an hour later I wake up again, find the room so dark that I can't see the clock, quickly dismiss the notion that I could have gone blind, dash to the window and see a strip of black sky. A few minutes later comes a roar of thunder, in no time the street is completely white, horses break loose, hailstones of a fantastic size beat against the windows. I rush to the back, there I find the study window already smashed in three places, the writing table flooded, the wings of the window wide open, of course. The terrace looked almost grandiose, the doors had been blown open and the hailstones covered the floor as far as the sideboard. The storm lasted for half an hour. The devastation in the city is terrific. On one side of almost every street (in the Berggasse it's opposite) nearly all the windowpanes are smashed, especially in the upper stories; for long stretches not one pane is intact, as though boys had bombarded them with stones; at street corners and where windows have not been protected by moldings it looks simply ludicrous. A woman who came to see me this morning was quite right when she said that the windows looked like circus hoops after dogs had jumped through them. Other streets have been spared. The trees have suffered most; in our garden there are more leaves on the ground than on the branches, and the poor tree itself is torn and battered; it looks as though it had been thrashed with whips and then chewed by caterpillars. Anything that was a garden must be in an awful state. I am told that the windows—in our case only one—are to be repaired by the landlord. I am anxious to know if you had anything like this in your part of the world; it could have been quite a mess. I hope not, and that it was only local.

No news otherwise. Yesterday I did my writing in the evening and hope to do the same today. Business middling; today it would be more profitable to be a glazier than a doctor.

Fond greetings
Your
Sigmund

109 *To* MARTHA FREUD

Venice, on board, en route to the Lido
Tuesday morning
August 27, 1895

My precious darling

We agreed that you won't get many detailed descriptions. The trance which Venice puts everyone into makes it impossible. We[1] are enormously well and spend all day walking, cruising, gazing, eating, and drinking. Every morning to the Lido, twenty minutes, to bathe in the sea, the most delicious sand underfoot. Yesterday was cool and the sea rather rough; today has started hot. Yesterday we also went up the tower of St. Mark's, strolled from the Rialto through the town, which allows one to see the strangest things, visited a church, Frari, and the Scuola S. Rocco, enjoyed a surfeit of Tintorettos, Titians, and Canovas, went four times to the Café Quadri on the piazza, wrote letters, entered into negotiations about some purchases, and the two days seem like six months. *Zanzare** definitely exist. Needless to say, I am already very anxious to hear your news. The only letter so far is from Minna, *poste restante*. I hope you and all the brats are very well.

Fond greetings
Your
Sigmund

1 Freud's traveling companion was his younger brother Alexander.
* Mosquitoes.—*Tr.*

110 *To* WILHELM FLIESS

Vienna IX, Berggasse 19
April 2, 1896

My dear Wilhelm

. . . If we are both granted a few more years of quiet work, we shall certainly leave behind us something that will justify our existence. It is this knowledge that strengthens me to face all the

worries and cares of daily life. As a young man my only longing
was for philosophical knowledge, and now that I am changing over
from medicine to psychology I am in the process of fulfilling this
wish. I became a therapist against my will; I am convinced that,
given certain conditions in the person and the case, I can definitely
cure hysteria and obsessional neurosis.

So till we meet again. We have honestly earned a few beautiful
days together.

When you say goodbye to your wife and son at Easter, please
greet them on my behalf.

Your
Sigm.

111 *To* WILHELM FLIESS

Vienna IX, Berggasse 19
November 2, 1896

My dear Wilhelm

I find writing so difficult just now that I have taken far too
long to thank you for the moving words in your letter. By one of
those obscure paths behind official consciousness the death of the
old man[1] has affected me profoundly. I valued him highly, understood him very well, and with that combination of deep wisdom and
romantic lightheartedness peculiar to him he had meant a great
deal to me. His life had been over a long time before he died, but
his death seems to have aroused in me memories of all the early
days.

I now feel quite uprooted.

Otherwise I am working on children's paralyses[2] (Pegasus put
to the yoke!), am pleased with my seven cases and especially with
the prospect of talking to you for several hours. I feel isolated, but
I take this for granted. Perhaps I shall tell you a few strange
things in return for your great ideas and discoveries. Business, on

---

[1] Freud's father, Jacob Freud, had died at the age of eighty-one.

[2] "Die Infantile Cerebrallähmung" ("Infantile Cerebral Paralysis"), II Teil, II
Abt., von Nothnagel's *Spezielle Pathologie und Therapie*, 9, Vienna.

which my mood invariably depends, is not so satisfactory this year. . . .

I recently heard the first reaction to my venturing into psychiatry. "Ghastly, awful, old wives' psychiatry" is one quotation. Rieger[3] in Würzburg. I was highly amused. And of all things about paranoia, which has become so clear!

Your book still hasn't appeared. Wernicke[4] recently sent me a patient, a lieutenant in the officers' hospital.

I must tell you of a nice dream[5] I had the night after the funeral. I was in a shop and there on a board I read a notice:

> You are requested
> To close the eyes

I recognized the shop at once as the barber's where I go every day. On the day of the funeral I was kept waiting, and as a result I arrived rather late at the house of mourning. My family was displeased with me at this time because I had arranged for the funeral to be quiet and simple, something they later recognized as quite justified. They also rather resented my being late. So the notice on the board has a double meaning. It means that one should do one's duty toward the dead in two senses: (a) apology, as if I hadn't done my duty and my conduct needed to be overlooked; (b) the duty in the literal sense. Thus the dream is an outlet for that tendency toward self-reproach which death invariably leaves among the survivors. . . .

My warmest greetings to I. F.[6] and R. W. F.;[6] perhaps my wife is already with you.

Your

Sigm.

[3] Dr. Konrad Rieger (1855-1939), Professor of Psychiatry at the University of Würzburg.

[4] Dr. Karl Wernicke (1848-1905), Professor of Neurology at the Universities of Breslau and Vienna.

[5] This dream was used as an example in *The Interpretation of Dreams,* where it is told in great detail, apparently based on notes.

[6] The initials of Frau Fliess and her elder son.

112   *To* JOSEF BREUER

Vienna IX, Berggasse 19
January 7, 1898

Dear Dr. Breuer

I am taking the liberty of answering for my wife, who is not familiar with the two subjects of your letter and of explaining the reasons for the return of 350 florins (via post office savings account).

So far as my debt to you is concerned, there is no doubt that it exists. I have not forgotten it and have always intended to pay it; nor have I assumed that you expected anything else. You once told me that you weren't sure of the sum; according to my memory, which I know is not very reliable, it came to about 2300 florins. On paper I have been solvent for several years, but the profit was always made up of outstanding fees, as is so often the case in our profession, and the constantly recurring difficulty of obtaining the cash for daily living, even if it is backed by securities, has made it impossible for me to start the repayment until now. Only during this past year, the most lucrative for my practice, has my profit in cash been large enough to let me dare reduce it by a certain sum. Originally this sum had been far larger than the amount received by you; but part of it had to go to England in order to take care of a need that had arisen in the family. It was circumstances of this kind which I described in my accompanying letter as the obstacles preventing my rise to "prosperity." You may rest assured that if I have begun to settle my debt this year, and not before, it isn't because of any other change[1] that has occurred in the meantime. In one respect at least we see eye to eye: for neither of us are financial relationships the most important in life, nor do they seem incommensurable with other relationships. That I had to testify to this theory in an active and passive way, as receiver and giver, while you were able to confine yourself to the active role, is something you yourself always used to describe as a matter of luck rather than of merit.

The second concern, that of my treating Fräulein C., has not the

---

[1] An allusion to the change in their relationship caused by scientific disagreement.

slightest connection with the above. If you had ever felt that you owed me anything for this service, you surely wouldn't have waited till I sent you money and then use part of it to settle your debt. The incorrectness of your remarks about my "claims" on Frl. C. also proves that you are trying to confuse the second concern with the first in order to spare me expenses for reasons I cannot judge. I will try to shed some light on the situation. In her exuberance Frl. C. demanded to be treated like any other patient. I was interested in preventing any excessive gratitude from arising; on the other hand I had no intention of depriving the poor girl of her modest means. So we agreed that I would charge her five florins per session "like anyone else," that she would at once pay 150 florins as part of the total fee of 750 florins (150 sessions), and the remaining 600 florins later on, after having inherited from her mother, who is still alive. So you see I cannot be compelled to accept any payment for the treatment of Frl. C. before she has inherited; and as soon as this occurs she herself will no doubt pay off her debt, which she takes very seriously. So I don't think Frl. C. has anything to do with the return of the 350 florins.

May I express the hope that you will accept this sum without raising any further objections.

With sincere thanks
Your
Dr. Freud

113 *To* WILHELM FLIESS

Vienna IX, Berggasse 19
April 14, 1898

My dear Wilhelm

I consider it a good rule for letter writing to leave unmentioned what the recipient already knows, and instead tell him something new. I shall therefore not comment on what I heard: that you had a bad time at Easter; you know this anyway. I would rather tell you about my Easter vacation, which I spent in a grumpy mood but from which I returned refreshed.

We (Alexander and I) left on Friday evening from the South Station and at 10 A.M. on Saturday reached Gorizia, where we wandered about in bright sunshine between whitewashed houses, saw trees covered in white blossom, and were able to eat oranges and crystallized fruit. We kept comparing memories, the view from the fortress reminded us of Florence, the Fortezza itself of S. Pietro in Verona and of the castle in Nuremberg. The first impression that invariably assails one on Italian soil, that of missing meadow and forest, was of course very vivid, as is inevitable on such transitions. The Isonzo is a magnificent river. On the way we passed three ranges of the Julian Alps. On Sunday we got up early so as to catch the local Friulian train which took us close to Aquileia. The former capital is a little dump, but the museum contains an inexhaustible wealth of Roman relics: tombstones, amphorae, medallions of the gods from the amphitheater, statues, bronzes, and jewelry. . . .

At 10 A.M., when the tide happened to be rather low, a little steamer was towed into the canal of Aquileia by a strange-looking motor which had a rope round its body and smoked a pipe while at work. I should love to have brought the steamer back to the children, but since it was the only link between the world and the Spa of Grado, it couldn't be spared. A 2½-hour journey through the most desolate lagoons brought us to Grado, where at last we were able to collect shells and echinites again on the Adriatic coast.

We returned to Aquileia the same afternoon, after a meal on board from our provisions, washed down by some delicious Istrian wine. Several hundred of the most beautiful Friulian girls had gathered in the cathedral for Easter Mass. The splendor of the ancient Romanesque basilica came as a relief amidst the poverty of modern times. On our way back we saw a strip of Roman road laid bare in the middle of a field. A contemporary drunkard lay on the antique paving stones. The same evening we returned to Divača on the Carso, where we spent the night in order to visit the caves on the next and last day, Monday. In the morning we went to Rudolf's Cave, fifteen minutes from the station. The cave is filled with all manner of strange stalactitic formations shaped like pewter-grass, pyramid cakes, tusks seen from below, curtains, corn-cobs, heavy folds of tents, hams and poultry hanging from above. Strangest of all was our guide, pretty pickled but quite safe on his legs, gay and humorous. He was the discoverer of the cave and evidently a decayed genius; he kept talking about his death, his conflict with the clergy, and his conquests in this subterranean world. When he mentioned that he had already entered thirty-six "holes" in the Carso, I recognized him as a neurotic and his exploits of a conquistador as an erotic substitute. A few minutes later he confirmed this by responding to Alexander's question as to how far one can penetrate into the cave with the words: "It's like with a virgin; the further one goes the nicer it is."

The man's ideal is to come to Vienna to get ideas in the museums for the naming of his stalactites. I overtipped the "greatest scoundrel in Divača," as he called himself, by several gulden, so that he could drink himself to death all the faster.

The caves of St. Canzian, which we visited in the afternoon, are an awe-inspiring phenomenon of nature, a subterranean river running through magnificent vaults, waterfalls, stalactites, pitch darkness, slippery paths protected by iron railings. Sheer Tartarus. If Dante saw anything like this he didn't have to strain his imagination for his description of the Inferno. The Ruler of Vienna, Herr Dr. Carl Lueger,[1] was with us in the cave, which after 3½ hours spat us all back into the light.

On Monday morning we started on our return journey. The following day I recognized from the re-emergence of fresh ideas while at work that the apparatus had benefited from the holiday.

Your
Sigm.

[1] The then Lord Mayor of Vienna.

114 *To* WILHELM FLIESS

Aussee
August 20, 1898

My dear Wilhelm

Your lines have brought back to me the pleasures of my vacation. It was really wonderful, the Engadine composed in simple lines out of few elements, a kind of post-Renaissance landscape, Maloja with Italy beyond and an Italian atmosphere which may have been merely the product of our anticipation. For us[1] Leprese was a magic idyll enhanced by the reception given us there and by

[1] The traveling companion was Minna.

the contrast produced by our walk up from Tirano. We walked along this road, which isn't exactly level, in the most ghastly dust storm, and arrived half dead. The air made me gay and bellicose as I have hardly ever been before. The soundness of my sleep was in no way affected by the five-thousand-foot increase in altitude.

The sun didn't worry us in Maloja until the last day. Then it grew hot, even for up there, and we hadn't the courage to go down to Chiavenna—i.e., to the lakes. I think this was sensible, for several days later in Innsbruck we were both overcome by a paralyzing weakness. Since then it has become even hotter, and here in our beautiful Obertressen[2] we lie about in different kinds of chairs from ten in the morning till six in the evening, without venturing a foot outside our little estate.

Little Anna[3] describes, not inappropriately, a small Roman statuette which I bought in Innsbruck, as "an old child."

I seem to be so removed from all intellectual activities that I can hardly follow your fascinating comments on old age, and my mind is mainly occupied with annoyance that so much of the vacation has already passed. My deep regret that you two should have been tied all this time to town is mitigated by the thought that you have had your vacation and that Ida is looking forward to a beautiful compensation.

Yes, I too have glanced through Nansen, whom everyone here is raving about—Martha because Scandinavians (Grandma, our guest at the moment, still speaks Swedish) evidently revive in her a youthful ideal which she hasn't realized in life, and Mathilde because she is transferring her admiration from the Greek heroes, who have hitherto occupied her mind, to the Vikings. Martin, as usual, has reacted with a poem—not a bad one—to the three volumes of adventure.

I shall be able to make some use of Nansen's dreams, which are practically transparent. I know from my own experience that his psychological condition is typical of someone who is daring to do something new which demands confidence, and probably does discover something new by a false path, but finds that it is not so

---

[2] Part of the resort Alt-Aussee in the Salzkammergut, where Freud's family spent many summer vacations.

[3] Anna, Freud's youngest child, born 1895, named after Anna Lichtheim (née Hammerschlag).

much as he imagined. The secure harmony of your nature fortunately protects you from such things. . . .

I send you and your dear wife my warmest greetings. I am not yet reconciled to the distance that separates us while we are at work and that is so rarely spanned during vacations.

Your
Sigm.

115 *To* HEINRICH GOMPERZ

Vienna IX, Berggasse 19
November 15, 1899

Dear Mr. Gomperz

I have only one objection to accepting your proposal, and it is for you to overrule it. If in interpreting your dreams you meet with such difficulties, in other words if you have built up in yourself such strong resistance to a number of psychic impulses, then instructing you in the interpretation of your dreams would amount to embarking on a self-analysis. Once this has begun it is not so easy to bring it to an end, and perhaps you are in the midst of work which should not be disturbed or interrupted. If you can disregard this obstacle and forgive me for the indiscretion with which I should have to explore you and for the unpleasant effects which I shall probably have to arouse in you—in short, if you are willing to apply the philosopher's unrelenting love of truth also to your inner life, then I would be very pleased to play the role of the "other" in this venture. Your opinion that several chapters would have to be added to psychology if our conception of dreams is correct tallies completely with mine. I, needless to say, have already accepted the additional proviso. The prospect of convincing you to any extent of the correctness of my findings is very tempting to me, and your hint that you may perhaps approach the subject scientifically sounds to me almost like a wish fulfillment. For this reason I can easily resist the other temptation, that of using your dreams as material for a complete interpretation. I don't think it would be possible to conceal from the public the material, the thoughts and memories, contained in your dreams. Moreover, I consider you a person subject to hysteria—which of course does not prevent you from being healthy and resilient as well.

No doubt we shall run into difficulties—I don't know of what nature, for someone of your intellectual constitution has so far never put himself at my disposal.

I am far from being fully occupied. I am almost always free at 6 P.M. and would be pleased to see you tomorrow, Thursday, at this hour or to receive your suggestion for another appointment.

With kind regards and sincere thanks for your letter.

Your

Dr. Freud

116 *To* WILHELM FLIESS

Vienna IX, Berggasse 19

June 12, 1900

My dear Wilhelm

We have had family visitors. The day before Whitsun my eldest brother arrived with his youngest son, Sam, who is already over thirty-five, and they stayed till Wednesday evening. His visit was very refreshing, for he is an excellent man, vigorous and mentally alert despite his age (sixty-eight to sixty-nine), and he has always meant a great deal to me. He then left for Berlin, now the family headquarters. . . .

Life at Bellevue[1] is shaping very pleasantly for us all. The evenings and mornings are lovely; the scent of acacia and syringa has succeeded that of lilac and laburnum; the wild roses are in bloom and all this, as even I can see, seems to be happening overnight.

Do you think that one day there will be a marble tablet on this house, saying:

> In this House on July 24, 1895
> the Secret of Dreams was revealed to
> Dr. Sigmund Freud

[1] A house on the slopes of the Wienerwald, where the family spent the summer.

So far the chances seem rather slight. Yet when I read the recent psychological books (Mach's *Analysis of Emotion*, second edition; Kroell's *Structure of the Psyche*, and so on) and see what they have to say about dreams, I can't help feeling as pleased as the dwarf in the fairy tale[2] because "the Princess doesn't know."

I haven't had any new patients—or rather there is one taking the place of another who arrived in May and dropped out, so I am now back where I was before. But this latest is an interesting case, a thirteen-year-old girl whom I am expected to cure at top speed and who for once shows on the surface what I usually have to excavate from beneath superimposed layers. I don't need to tell you that it is the same old story. We shall talk about the child in August provided she is not prematurely taken away from me. For in August I shall see you for certain unless I am disappointed about the 1500 kronen which I expect on July 1. Or rather, I shall come to Berlin in any case and . . . then try to get some fresh air and new vigor for 1901 in the mountains or in Italy. Low spirits yield as little as economizing does.

I had heard about Conrad's[3] accident and its fortunate outcome. Now I am entitled to get news of you and your family again.

My heartiest greetings to you and to them.

Your
Sigm.

[2] An allusion to "Rumpelstiltskin," fairy tale by the Brothers Grimm.
[3] Fliess's younger son.

117 *To* ELISE GOMPERZ

Vienna IX, Berggasse 19
November 25, 1901

Your Highness

We weren't warned seriously enough at school against the "inclined[*] plane." The speed that one can acquire on it in a short time is incredible. This, for instance, is the fourth letter I have written on my own behalf in the past twenty-four hours, since becoming a "climber."

[*] *Schief*—which also means "crooked."—*Tr.*

But I have to write this letter in order to say what I had no chance of putting into words during yesterday's visit, owing to the sudden interruption. Namely, my warm thanks. An effort like yours, and the manner in which you suggested it to me some weeks ago, is neither customary nor something that can be taken for granted. May I never be in the position to return it! This is the best expression of gratitude I can think of.

I wrote at once to Nothnagel and Krafft-Ebing[1] asking them to renew my application [for the title of Professor]; I have also informed Exner[2] of the outcome of your effort (in writing, so as to take up as little as possible of his precious time). He knew of my plan to accept your assistance and encouraged me to do so.

As for the further "directive" which you mentioned as I was taking my leave, I think it would be a good idea if you jogged the old fox's[3] memory next time you see him. Or would you rather give him to understand that I cannot be unknown to him? When I presented myself to him 4½ years ago, he professed to have heard "excellent things" about me. Once my latest work[4] has gained popularity, "forgetting" will no longer be a valid subterfuge.

I wish you a quick recovery. In my household three people are ill.

Cordially and gratefully yours
Dr. Freud

[1] Dr. Richard von Krafft-Ebing (1840-1902), Professor of Psychiatry at the University of Vienna.

[2] See note 3 to letter of June 8, 1885.

[3] The then Minister of Education.

[4] "Vergessen von Eigennamen," not published at the time the letter was written, appeared in *Psychopathologie des Alltagslebens,* Berlin, 1904. ("On Forgetting and Suppressing," in *The Psychopathology of Everyday Life,* Standard Ed., 6.)

118 *To* ELISE GOMPERZ

Vienna IX, Berggasse 19
December 8, 1901

Dear Protectrix

Here are further developments: My three letters to Exner, Nothnagel, and Krafft-Ebing have called forth three answers, of which I enclose two *in natura*; the third, from Exner, I am quoting in part.

It is the most official,* the unfriendliest and unpleasantest of the three.

Exner writes: "In accordance with my promise, I talked about you to the Minister, who also told me of Frau *Hofrat* Gomperz' intercession on your behalf. Unfortunately I did not gain the impression that all these efforts are likely to meet with success in the immediate future, during the next few weeks or months. As I already pointed out during your visit, you will have to be prepared for [a wait of] six months or a year, and even for these dates I can offer no guarantee. So let us hope."

Nothnagel and Krafft-Ebing both wrote only after they had done what I had asked them to do. So the application has now been renewed. His Excellency can no longer plead ignorance.

So now—i.e., during the first weeks following the renewal of the application—I shall ask you to repeat your intercession at a suitable occasion and, if you are offered even half a promise, to emphasize the element of *time* which, according to the astronomer Seni[1] and my friend Fliess, is the most important element in all human concerns. Having already waited 4½ years I ought, as someone sufficiently convinced of the omnipotence of an Excellency, to be spared any further waiting. In the event of a protracted delay it is after all not impossible that my sudden demise could put an end to the intention of honoring me (which would be for good, since we don't live in China where they have something like a posthumous elevation to rank). Indeed, it is perhaps even more likely that by postponement till summer or autumn the Minister himself will reach that fatal region in which his most amiable intentions would be as immaterial to the person concerned as his most hostile ones; a possibility, by the way, which can be suggested to a Minister only in the most subtle manner.

But if the Minister, as one might conclude from Exner's diplomatic hints, really doesn't want to act on my behalf for some definite reason, isn't one allowed to ask what this reason is? If it is the content and trend of my work, in which, according to Exner, certain circles find so much to "laugh at," then I at least know what it is and can consider whether to alter this trend or to renounce the title. The way to improvement should not be barred to a lost soul.

* *Hofrätlichste*—which is a coined word derived from *Hofrat*, meaning "court councilor."—*Tr.*
[1] Giovanni Baptista Seni (1600-1656), Italian astrologer.

Or, before I arrive at this decision, would your husband, *Hofrat*
Gomperz, who is not a stranger to the scientific world, like to
point out to His Excellency what little argumentative power lies
in the laughter of circles unknown to me, and how from time immemorial mankind has always been wont to laugh at an idea to
which one day it is obliged partly or entirely to submit? Up to
now I have heard that my activities only annoy people, and this I
find quite understandable and as a first reaction just as it should be.

So these are my further requests, dear Providence. As soon as I
can assume you would welcome a visit I will do myself the honor
of coming to see you. I trust you have recovered long ago and ask
you to give my regards to the *Hofrat*.

Most sincerely yours
Dr. Freud

119 *To* WILHELM FLIESS

Vienna IX, Berggasse 19
March 11, 1902

My dear Wilhelm

It just goes to show what an "Excellency" can do! It can even
enable me to hear your familiar voice again in a letter. But since
you associate the news with such wonderful things as recognition,
outstanding achievements, etc., my habitual compulsion to be honest, which has so often done me harm, obliges me to tell you how
it all came about.

It was all my own doing. On my return from Rome my zest for
life and work had somewhat increased and that for martyrdom
somewhat diminished. I found my practice considerably reduced,
I withdrew my last publication from the printer because in you I
had recently lost my last remaining audience.[1] I came to the conclusion that I could easily spend the greater part of my life waiting
for recognition and that in the meantime none of my fellow men
would lift a finger on my behalf. And I so much wanted to see Rome

---

[1] Johann Nestroy (1801-1862), Viennese comedian and author of popular comedies
and farces, on looking through a peephole of the stage curtain before a benefit performance and seeing only two people in the orchestra, is said to have exclaimed:
"I know one 'audience'; he has a free ticket. Whether the other audience also has
a free ticket I don't know."

again, help my patients, and keep my children happy. So I decided
to break with my strict principles and take some practical steps
as other human beings do. One has to seek one's salvation somewhere, and I chose the title of Professor to be my savior. For four
long years I hadn't uttered a word to further it; now I decided to
call on my old teacher, Exner. He was as unpleasant as possible,
almost rude, refused to disclose any reasons for my having been
ignored, and generally assumed the airs of the high official. Only
after I had provoked him with a few sarcastic remarks about the
goings-on of those in high places did he drop some obscure hints
about personal influences which were prejudicing His Excellency
against me, and he advised me to seek some counterinfluence. I was
in the position to tell him that I would approach my old friend and
former patient, the wife of *Hofrat* Gomperz. This seemed to impress
him. Frau Elise was most amiable and took an immediate interest.
She called on the Minister in question and in answer to her request
was greeted by a face of astonishment: four years? "And who may
that be?"—The old fox gave the impression he had never heard of
me. In any case it turned out that I had to be proposed all over
again. I wrote to Nothnagel and Krafft-Ebing (who was about to
retire), asking them to renew their former proposal. Both men were
charming. A few days later Nothnagel wrote: "We have handed in
the proposal." But the Minister consistently avoided Gomperz and
it looked as though the whole thing would fall through once more.

Then another power appeared on the scene; one of my female
patients . . . had heard about the case and began taking matters
into her own hands. She refused to rest until she had made the
Minister's acquaintance at a party, managed to ingratiate herself
with him and via a mutual lady friend made him promise to confer
a professorship on the doctor who had cured her. But being well
enough informed to know that a first promise from him was as good
as none, she approached him personally, and I believe that if a certain picture by Böcklin[2] had been in her possession and not in that
of her aunt, I would have been made a professor three months
earlier. As it is, His Excellency will have to be satisfied with a
modern painting for the gallery which he intends to open, not of
course for his private use. Thus at last, when dining at her house,

2 Arnold Böcklin (1827-1901), Swiss artist. His picture "A Castle in Ruins" would
have been welcomed by the Modern Gallery in Vienna, which was founded at that
time.

the Minister graciously informed my patient that the documents had already been sent to the Emperor and that she would be the first to be told when the appointment had gone through.

Then one day she arrived radiant for her session, brandishing a pneumatic[3] letter from the Minister. So it had gone through! The *Wiener Zeitung* hasn't published it yet, but the news has spread like wildfire from the Ministry. "The public enthusiasm is immense." Congratulations and bouquets are already pouring in as though the role of sexuality had suddenly been realized by His Majesty, the importance of dreams confirmed by the Council of Ministers, and the necessity of treating hysteria by psychoanalytic therapy accepted in Parliament by a two-thirds majority!

I have obviously become respectable again; admirers who had recently fought shy of me now greet me from afar in the street.

I myself would any day exchange five congratulations for one decent case of steady treatment. I have learned that this Old World is ruled by authority as the New One is by the dollar. Having made my first bow to authority, I now live in the hope of reward. If the effect it has on a wider circle is as great as that on the closer, I have every reason to hope.

In the whole story there is one person who has proved himself to be a real ass and one whom you don't sufficiently appreciate in your letter: that is me. If I had taken these steps three years ago, I would have been appointed three years earlier, and spared myself a great deal. Other people are just as clever without having to go to Rome first. So this was the glorious event to which I owe, among other things, your kind letter. Please keep the contents of this one to yourself. With thanks and cordial greetings

Your
Sigm.

[3] Local letters were propelled by compressed air through tubes for faster delivery.

120 *To* THE FAMILY

Waidbruck (South Tyrol)
April 20, 1905

My dear ones

It works, Alex has cleaned my pen and until the roast and salad arrive I have time to describe the events of the day. The white Traminer, by the way, is excellent, as R. will find out for himself in a few years.

Well, we went up to Dreikirchen in beautifully warm weather, although the sky was misty so that one couldn't see where the many snow-white mountains ended and the clouds began. You must imagine it like this: from Waidbruck, directly opposite the Trostburg (one more castle belonging to a Kannitverstan[1] of Wolkenstein), you climb a hill about as high as the Leopoldsberg;[2] the road isn't much better, either; it winds up in steep hairpin bends to a height of about 1350 feet, and is so narrow that only a light little carriage can go up it. According to Herr Vonmetz[3] the drive is more of a task than a pleasure; what's more, the carriage can't hold more than two people, a mother and child, and this expedition costs three florins fifty—which in our case would come to four times seven kronen. This time of course it didn't cost anything because we went up on foot. It was pretty warm, vines grow on lovely terraces on the mountain, which proves that it is a warm climate. The first climb of the season is not easy, but the new shoes proved a great success, I feel so at home in them I might have been born in them. So long as the people in the Führichgasse sell shoes like this, they may so far as I am concerned stick their tongues out at their customers before and after the deal.—So we went on climbing and stopped after about forty-five minutes in the shade of a few particularly beautiful forget-me-nots, until we felt rested. After another forty-five minutes we came to a village, St. Barbian, and here the comparison with the Leopoldsberg comes to an end. Then we continued up a road paved with stones, quite steep and slanting, through a beautiful forest which had the one drawback of never coming to an end. Here we were over three thousand feet up, and as a result transported back into early spring; below, there had been all kinds of flowers, of which I enclose a few samples, but higher up there was only heather, and in the end crocus, white and purple, the first harbingers of spring, like their cousins, the autumn crocus, the last of the season. Finally we did emerge from the forest, and the view that confronted us was just like a map— the whole Eisack valley from Klausen to Waidbruck; above it, all the mountains unknown to me as far as the Schlern; the Grödner

---

[1] See note 4 to letter of Jan. 28, 1884.
[2] Hill near Vienna.
[3] Innkeeper at Waidbruck.

valley opposite looking like a slight crack, and then at last we came to a large wooden house with not a soul inside—which shows that all information must be obtained in the valley. The walls of this house must once have been wooden boxes, for how else could DO NOT OVERTURN have been written on them? It can't have been a warning to the wind. There was a delightful solitude—mountain, forest, flowers, water, castles, monastery, and not one human being. On our way back it began to rain, but gently. After this, supper tasted very good. Tomorrow we are off to St. Ulrich and Wolkenstein.

Fondest greetings

Papa

121   *To* ALEXANDER FREUD

Rapallo

September 17, 1905

Dear Alex

If I haven't written to you before, there is no need to feel offended. Even for those in Aussee I haven't produced anything but picture postcards, and this letter—the first since my departure—is already being written in anticipation of the return to the life of ordinary mortals. One just doesn't find time for anything; the heavenly sun and the divine sea—Apollo and Poseidon—are enemies of all mental activity. I realize that the only thing that has hitherto kept us going has been that remaining sense of obligation to identify—Baedeker in hand—new regions, museums, palaces, ruins; since this obligation does not exist here I am simply drowning in a life of ease.

Now for the sea. The beach is fine mud in which one can walk out for a quarter of an hour, head above water, and a little further on there are wonderful rocks (like those we saw in Capri) with pools in which one can sit, sloping carpets of rocks on which one rolls like a monster in one of Böcklin's pictures, completely alone and losing all account of time. Since I am all by myself I am not going in for serious fishing. But on the first day I managed to tickle an octopus, which refused to leave the water, then I got myself pricked by sea urchins, stung by jellyfish, tried in vain to detach some sea anemones, and to catch some of those wily crabs

whose legs here too are torn out by Italian urchins of the beach.
The morning is taken up with bathing, which I find so much to my
taste that next year I seriously consider making Viareggio (which
I can't see this time) my sole aim.

It is impossible to talk of the land unless one is a poet or quotes
others. Everything grows here as in Sorrento except that there is
a regular surfeit of palms, and one may well wonder about the
merciful "retribution."[1] Isola Madre[2]—admittedly belonging to another setting—was the gem of my lake trip. Here by the sea one is
disturbed only by the regret that, ill advised as one was when
choosing one's profession, one didn't start dealing in pigs like that
Mr. Brown or enter the world a Spinola[3] or some other nobleman of
Genoa. Envy often spoils the enjoyment of gardens and country
houses.

You know Genoa, of course; it is impressive, solid, almost defiant,
clean, prosperous; what strikes one is the general use of the German language in hotels and shops; Martha would have found here
the peace of mind she missed in Venice. There are certainly more
German shop signs in Genoa than in Trieste or Prague. The only
thing occupying the public mind here is the *terramoto* [earth-
quake] in Calabria. Austria, as we know, is no longer mentioned
in the papers.

I shall be back at the end of the week, my royalties spent. I shall
soon be seeing the last olive tree, orange, magnolia, etc.

Cordial greetings to you in your northern exile.

Your

Sigm.

[1] Allusion to a quotation from Goethe's "Elective Affinities": "Nobody walks
without retribution under palm trees."

[2] Island in Lake Maggiore.

[3] Ambrosio Spinola (1569-1630), Spanish general, born in Genoa.

122 *To* KARL KRAUS

Vienna IX, Berggasse 19
January 12, 1906

Dear Sir

That I find my name repeatedly mentioned in the *Fackel*[1]
is presumably caused by the fact that your aims and opinions par-

[1] Viennese polemical journal edited by Karl Kraus.

tially coincide with mine. On the basis of this impersonal relationship, I am taking the liberty of drawing your attention to an incident whose further developments will probably create a considerable stir among the contributors to your journal.

Dr. Fliess of Berlin has induced R. Pfenning to publish a pamphlet[2] attacking O. Weininger[3] and H. Swoboda[4] in which both young authors are accused of the most flagrant plagiarism and abused in the cruelest fashion. The cogency of this fabrication may be judged by the fact that I myself, a friend of Fliess for many years, am accused of being the one who imparted to Weininger and Swoboda the information (gained from my contact with Fliess) which has served as a basis for their unauthorized publications.

I trust it is not necessary for me to defend myself in detail against such absurd slander. Dr. Swoboda, who is alive, will see to his own defense, which he will find easy. As for the deceased Weininger, I assume that his friends will intercede on his behalf and that you, sir, as on former occasions, will place your journal at their disposal for this purpose. Now it should be mentioned here that I personally do not share the high esteem for Weininger as expressed in the *Fackel*. But in this case I feel obliged to side with Weininger's friends and to warn them of the danger suggested by the very repulsiveness of the indictment's character. For the assertion divulged in this pamphlet and substantiated by the unauthorized publication of my private letters[5]—that Weininger did not discover the idea of bisexuality on his own but in a roundabout way, leading back to Fliess via Swoboda and myself—is actually based on truth. The undoubtedly brilliant young man cannot be spared the reproach of having failed to divulge the source of this idea and, instead, of passing it off as his own inspiration. The extent to which this event mitigates the harsh opinion expressed in the unlawfully published private letter to Fliess can

[2] A. R. Pfenning, *Wilhelm Fliess und seine Nachentdecker: Otto Weininger und H. Swoboda* (*Wilhelm Fliess and his Echoers: Otto Weininger and H. Swoboda*), Berlin, 1906.

[3] Otto Weininger (1880-1903), author of *Geschlecht und Charakter* (*Sex and Character*), Vienna, 1903.

[4] Hermann Swoboda (1873-   ), author of *Perioden des menschlichen Organismus* (*Periods of the Human Organism*), Vienna, 1904.

[5] See letter of Aug. 15, 1924.

be decided only after a thorough investigation of Dr. Swoboda. I myself did not know Weininger before he wrote his book.

I am writing this letter in order to enlist your help in preventing the friends of the deceased from taking up an untenable position in his defense. There remain enough respects in which his name can be successfully defended against the Fliess-Pfenning slanders.

I hope, sir, that you will regard this letter as nothing but a token of my respect and an assumption of your interest in a cultural matter. What concerns us here is the defense against the presumptuousness of a brutal personality and the banning of a petty personal ambition from the temple of science. Please consider these lines as a private communication, and rest assured that I shall be at your disposal at any time with comments suitable for publication.

Respectfully yours
Dr. Freud

123 *To* ARTHUR SCHNITZLER

Vienna IX, Berggasse 19
May 8, 1906

Dear Dr. Schnitzler

For many years I have been conscious of the far-reaching conformity existing between your opinions and mine on many psychological and erotic problems; and recently I even found the courage expressly to emphasize this conformity ("Fragment of an Analysis of a Case of Hysteria," 1905). I have often asked myself in astonishment how you came by this or that piece of secret knowledge which I had acquired by a painstaking investigation of the subject, and I finally came to the point of envying the author whom hitherto I had admired.

Now you may imagine how pleased and elated I felt on reading that you too have derived inspiration from my writings. I am almost sorry to think that I had to reach the age of fifty before hearing something so flattering.

Yours in admiration
Dr. Freud

124 *To* C. G. JUNG

Vienna IX, Berggasse 19
May 26, 1907

Dear Colleague

Many thinks for your praise of the "Gradiva"!¹ You won't believe how few people are capable of doing just that; it is actually the first time I have heard a friendly word about it (no, I mustn't be unjust to your cousin Riklin).² This time I knew my little work deserved praise; it was written during sunny days and I derived great pleasure from doing it. Of course it doesn't contain anything new for us, but it allows us to enjoy our wealth [of insight]. I certainly don't expect it to open the eyes of the relentless opposition; I long ago gave up listening in that direction, and because I have so little hope of converting the experts I have also, as you have correctly detected, shown only a halfhearted interest in your galvanometric experiments, for which you now have punished me. A confession such as yours, by the way, is more valuable to me than the approval of a whole medical congress, not least because it promises the approval of future congresses.

If you are interested in the fate of the "Gradiva" I will keep you informed. So far only one review has appeared, in a Viennese daily, laudatory but about as lacking in understanding and feeling as one would expect from your dementia patients. This kind of journalist who is evidently incapable of grasping the passionate emphasis on abstract values doesn't shrink from writing: The mathematicians tell us that 2 x 2 frequently equals 4, or: We are assured that 2 x 2 usually doesn't equal 5.

What Jensen³ himself has to say? He wrote very warmly about it. In his first letter he expressed his pleasure, etc., and declared that the analysis agreed in all important points with the purpose of the story. Of course he wasn't referring to our theory, since as an old gentleman he seems altogether incapable of entering into

---

¹ "Der Wahn und die Träume in W. Jensen's 'Gradiva,'" Vienna. ("Delusions and Dreams in Jensen's 'Gradiva,'" Standard Ed., 9.)

² Franz Riklin, Swiss psychoanalyst.

³ Wilhelm Jensen (1837-1911), German author.

any other but his own poetic intentions. He suggested the agreement could be ascribed to poetic intuition, and partly perhaps to his early medical studies. Then in a second letter I became indiscreet and asked for information about the subjective share of the work, where the material came from, where his own person is hidden, etc. I now learned from him that the antique relief actually exists, that he possesses a reproduction of it, but has never seen the original. It was he himself who invented the story of the relief representing a woman from Pompeii; it was he too who loved to daydream in the noon heat of Pompeii, where he once experienced an almost visionary state. Otherwise he knows nothing about the origin of the subject matter; the beginning suddenly occurred to him while he was working on something else, whereupon he dropped everything, started to write it down, never hesitated, found all the material as though it had been waiting for him, and finished it in one spurt. All of which suggests that a continued analysis would lead through his childhood to his own most intimate erotic experiences. So the whole thing is once again an egocentric phantasy.

In conclusion let me express the hope that one day you too will stumble on something you consider suitable for laymen, and that you will present it to my "Collection"[4] rather than to the *Zukunft*. . . .[5]

Thank you very much for the two missiles from the enemy camp. I am not tempted to keep them for more than a few days, until I can read them without affect. It is after all nothing but emotionally-tinged drivel. First they write as though we had never published a dream analysis, a case history, or the interpretation of parapraxes; then if the evidence is forcibly brought to their notice, they say: "Yes, but that's no proof, that's arbitrary."—Just try to produce a proof to someone who is set against it! There is nothing to be done with logic; one can say of it what Gottfried von Strassburg[6] said, I think rather irreverently, about the divine ordeal:

> "That [even] the good Christian
> Is as wavering as the sleeve of a coat in the wind."

[4] *Schriften zur angewandten Seelenkunde*, Deuticke, Vienna. *Bull. Amer. Psychoan. Assn.*, 8 (1952), 214; Standard Ed., 9.

[5] A Berlin periodical edited by Maximilian Harden.

[6] German medieval poet, *circa* A.D. 1200.

But just let five to ten years go by and analysis *aliquis*,[7] which is inconclusive at the moment, will have become conclusive without anything about it having changed. The only thing to do is to go ahead and work, not waste too much energy on refutations, and let the fertility of our ideas have its effect on the sterility of those we are fighting against. Envy, incidentally, is what shows in every line of Isserlin's[8] work. And certain things are really too stupid; everything testifies to ignorance.

And yet, don't worry, everything will turn out all right. You will live to see it, even if I don't. We aren't the first who have had to wait for the world to understand their language. I always think we have more secret followers than we know; I am convinced that you won't stand alone at the Congress[9] in Amsterdam. Every time we are laughed at anew I am more than ever convinced that we are in the possession of something great. In the obituary you are to write for me one day, don't forget to bear witness that all the opposition has not once succeeded in diverting me from my purpose.

I trust that your chief[10] will soon recover and that you will then be less burdened by work. I miss your letters so much when you don't write for a long time.

Yours very cordially
Freud

[7] Reference to *Psychopathology of Everyday Life,* Chapter II.

[8] Dr. Max Isserlin, Professor of Psychiatry.

[9] International Congress for Psychiatry and Neurology, Amsterdam, 1907.

[10] Dr. Eugen Bleuler (1857-1939), Professor of Psychiatry at the University of Zürich, Director of the Cantonal Sanatorium Burghölzli, psychoanalyst.

125 *To* C. G. JUNG

Hotel Wolkenstein
in St. Christina, Gröden[1]
August 18, 1907

Dear Colleague

So the impoverishment of my personality brought about by the interruption of our correspondence has fortunately come to an end. Lazy myself and rambling about the world with my family,

[1] Resort in the Dolomites.

I realize that you are back at work and your letters will remind me of what has become the most interesting topic to us both. Don't despair; probably what you wrote was only a figure of speech. It is immaterial whether or not one is understood by official representatives at the moment. Among the masses who anonymously hide behind them, there are enough people who *want* to understand and who, as I have often seen happen, will suddenly step forth. After all, one works primarily for the annals of history, and there your lecture in Amsterdam will stand out as a landmark. What you call the hysterical side of your character, the desire to create an impression and to influence people—the very qualities that make you a teacher and a pioneer—will come into its own, even if you haven't made any concessions to fashionable opinions. And if in addition you succeed to an even greater extent in imparting your personal leaven into the fermenting brew of my ideas, then there will no longer be any difference between your cause and mine.

I am not well enough to risk the prospective September journey to Sicily, where the sirocco is said to prevail at that time, so I don't know where I shall be spending those weeks. I shall stay here till the end of August, taking mountain trips and picking edelweiss; I won't be back in Vienna till the end of September. On the whole it would be better for you to write to my Vienna address, as the summer mail in the mountains is very unreliable. My little pocket notebook doesn't show a single entry for the past four weeks, which goes to show how thoroughly I have emptied all my intellectual channels. But I shall remain very grateful if you jog my memory a little.

Germany probably won't take any notice of psychoanalysis until some bigwig has solemnly recognized it. Perhaps the quickest way would be to attract the interest of Kaiser Wilhelm, who is known to understand everything. Have you any connections that reach that far? I haven't. Perhaps Harden, editor of the *Zukunft* [*Future*], can sniff the future psychiatry in your work? You can see that at the moment I am very much in the mood for joking. I trust the enforced vacation from work has brought you all the rest which I hope to find here by deliberately avoiding it.

Yours ever
Dr. Freud

126 *To* C. G. JUNG

Hotel Annenheim und Seehof
on the Ossiacher Lake (Carinthia), Annenheim
September 2, 1907

Dear Colleague

Here I sit thinking of you in Amsterdam, shortly before or just after your inflammable lecture, busily defending psychoanalysis, and it strikes me as almost an act of cowardice that I should meanwhile be looking for mushrooms in the woods or bathing in a peaceful Carinthian lake instead of myself representing my cause or at least standing by your side. For my peace of mind I tell myself that it is better for psychoanalysis, furthermore that you will be spared a part of the opposition that would be awaiting me, that nothing but useless repetitions would be heard if I were to say the same things all over again, and that you are more suitable as a propagandist, for I have invariably found that something in my personality, my words and ideas strike people as alien, whereas to you all hearts are open. If you, a healthy person, consider yourself a hysterical type, then I must claim for myself the class "obsessive," each member of which lives in a world shut off from the rest.

Whether you have been or will be lucky or unlucky I don't know; but I would like to be with you at just this time to enjoy the sensation of being no longer isolated, and to tell you, in case you need some encouragement, of my many years of honorable but painful isolation which started after I had had my first glimpse into the new world; of my closest friends' lack of interest and understanding; of the anxious periods when I myself believed I had been mistaken and wondered how I could still turn to advantage a bungled life for the sake of my family; of the gradually growing conviction that kept clinging to the interpretation of dreams as to a rock in a stormy sea, and of the calm assurance that finally took possession of me and bade me wait till a voice from the unknown answered mine. That voice was yours; I also realize now that Bleuler can be traced back to you. Let me thank you for this, and don't lose confidence that you will live to see and enjoy victory.

Luckily I don't as yet have to claim your sympathy for my suffering condition. I am celebrating my entrance into the climacteric age with a dyspepsia (following influenza) which proved rather stubborn but which, during these beautiful weeks of tranquility, has receded except for some very slight reminders.

I decided long ago to come to Zürich. But I think of it as a Christmas or Easter trip, right out of the middle of work, stimulated and full of problems, not as I am now when all channels are emptied almost as in sleep. I also feel the need to spend several hours talking to you.

> With warm greetings (and best wishes!)
> Your
> Dr. Freud

127 *To* C. G. JUNG

Rome
September 19, 1907

Dear Colleague

On arrival here I found your letter describing the further developments of the Congress. It didn't depress me and I realize with satisfaction that it didn't depress you, either. For you, I think this experience will have good results, or at least those I like best. As for me, it has increased my respect for psychoanalysis. I was on the point of saying to myself: "What? Already on the way to recognition after barely ten years? That can't be sound." Now I can believe again that it is. But you will see that the conditions for the tactics you have been using up to now do not exist. People just don't want to be enlightened. That is why for the present they can't understand the simplest thing. Once they are ready for it, you will see they are capable of understanding the most complicated ideas. Until then there is nothing to be done but to go on working and to argue as little as possible. After all, the only thing one can say is either: You're an idiot! or: You're a fraud! And one is quite rightly prohibited from expressing such opinions. Besides, we know they are poor devils who are partly afraid of giving offense, thus endangering their careers, and partly shackled by the fear of their own repressions. We have got to wait until they die out or gradually dwindle into a minority. Any young, fresh mind that turns up is bound to be on our side.

Unfortunately I can't quote by heart the beautiful verses from C. F. Meyer's[1] *Huttens Letzte Tage* which end like this:

> The small bell here chimes out with so much mirth:
> A little Protestant has come to earth.

. . . You quite rightly emphasize the absolute sterility of our opponents who are condemned to exhaust themselves with one outburst of abuse or with identical repetitions, while we can go ahead and work, and so can anyone who joins us. The old man who astonished you certainly won't be the only one; in the year to come we shall hear of unexpected disciples and you will win others in your flourishing school.

Now for my *ceterum censeo:* Let us found our journal.[2] People will grumble, buy and read it. One day in retrospect the years of struggle will strike you as the most beautiful. But please don't make so much of me. I am too human for that role. Your desire to possess my photograph makes me express a similar request, which is probably easier to fulfil. I have not willingly sat for a photograph for fifteen years, because I am so vain I can't quite face the physical deterioration. Two years ago I had to be photographed for the Exhibition of Hygiene (by order), but I detest the picture so much that I won't lift a finger to let you have it. At about the same time my boys took a picture of me which is quite natural and far better. If you like, I will try to find you a copy in Vienna. The best and to me most flattering is the medallion C. M. Schwerdtner[3] made for my fiftieth birthday. As soon as I hear that you would like it, I will have it sent.

I am quite alone here in Rome, engrossed in all kinds of daydreams, and don't intend coming home till the end of the month. My address is Hotel Milano. I buried science very deep at the beginning of the holidays and am now trying to come to and to produce something out of myself. This incomparable city is the perfect place for just that. Even though my main work may have been accomplished, I nevertheless would like to keep up with you and the younger people as long as possible. . . .

With best greetings and in anticipation of your answer

Yours sincerely

Dr. Freud

---

[1] Conrad Ferdinand Meyer (1825-1895), Swiss author.
[2] *Internationale Zeitschrift für Psychoanalyse.*
[3] Carl Maria Schwerdtner (1874-1916), Austrian sculptor.

128 *To* THE FAMILY

Rome
September 21, 1907

My dear ones

Don't be surprised by my writing so seldom; I have already warned you that in Rome one is continually oppressed by self-imposed tasks and one doesn't get around to anything. I have had a wonderful day again; the Villa Borghese is a huge park with castle and museum which until recently belonged to a Roman prince, but is now the property of the city and generally accessible, for the good prince made some unfortunate speculations and was obliged to sell everything for three million lire. Very cheap, by the way; the museum contains what must be the loveliest of all Titians, called "Sacred and Profane Love," for which alone the Americans would have paid that sum. You are sure to know the picture; the title doesn't make any sense; what the painting actually means is not known; that it is very beautiful is quite enough.

One couldn't wish for a lovelier park than this, except that instead of green lawns you have to imagine barren ground; at least that's how it is now. It is filled with the most noble trees—umbrella pines, cypresses, all sorts of unknown things, between them spacious playgrounds filled with innumerable children, stone tables and benches on which the *petite bourgeoisie* eat their meals, a lake with an island on which there stands an Aesculapian temple, all sorts of other artificial ruins and reproductions of temples—in other words a Schönbrunn[1] condescending to be a Prater. It is also like Schönbrunn in that strange animals have homes here—gazelles, pheasants; I noticed a monkey whose life was being made a hell by some street urchins. Peacocks strut freely about, taking their chicks (which look rather insignificant) for a walk. All kinds of things are forbidden in this park of course, but not more than is necessary, and I am under the impression that everyone makes himself comfortable and no one is very law-abiding. Here and there in what used to be the prince's private garden are also some samples of genuine antiquity, a beautiful sarcophagus, a column, a broken statue. So one is always reminded of being in Rome.

[1] Formal park around the palace of Schönbrunn on the outskirts of Vienna.

On one avenue stands a statue of Victor Hugo presented by the French to further the brotherhood of nations. He looks like Verdi, Joachim,[2] *et al.* It was this statue which so annoyed Kaiser Wilhelm that out of envy he had Eberlein make the statue of Goethe which he had set up in the same garden. It is quite clever but nothing wonderful. Goethe looks too youthful; after all, he was over forty when he first came to Rome; he stands on a shaft of a column, or rather on a chapiter, and the pedestal is surrounded by three groups: Mignon with the harpist (who is perhaps the best); Mignon herself has an empty face; Faust reading from a book while Mephisto looks over his shoulder; Faust himself quite good, the devil rather grotesque, a Jewish face with a cock's comb and horns, and a third group which I cannot interpret, perhaps Iphigenia and Orestes, but if so, it's hard to recognize.

In the museum itself there are not only classical but also modern works of sculpture, the Princess Pauline Borghese, a sister of Napoleon, as Venus, by Canova, famous groups by Bernini and others. All the classical sculpture is restored, which makes it difficult to form an opinion. Recent discoveries are nowadays treated more carefully, which makes me look forward to the museum in the Thermae of Diocletian, which I intend to visit tomorrow.

The most difficult thing in Rome, where nothing is easy, is shopping. Until now I have been very modest, but today a beginning has been made with several marble bowls. They will be a treat for the Wednesday gentlemen.[3] Your Aunt and Mama would certainly find all kinds of other things to buy. For once this marble is genuine and not painted, and therefore somewhat more expensive. I am still waiting to hear what you want.

I hope the vaccination hasn't been so unpleasant for you as it was for your uncle. Fond greetings to you, to Mama and everyone else at home.

Your

Papa

Heavenly weather, maximum 80° at siesta time; full moon, clear sky.

---

[2] Joseph Joachim (1831-1907), Hungarian violinist who, like Victor Hugo and Verdi, wore a beard.

[3] Members of the "Wednesday Society" of psychoanalysts, who met every Wednesday in Freud's apartment.

129 *To* MARTHA FREUD

Rome
September 21, 1907

E quindi uscimmo a riveder le stelle.[1]

Who knows this? Until nightfall I was with the dead in a Roman columbarium, in Christian and Jewish catacombs. It is cold, dark and not very pleasant down there. In the Jewish ones the inscriptions are Greek, the candelabrum—I think it's called Menorah —can be seen on many tablets. The (female) guide—I was the only visitor—forgot to bring the key of the exit, so we had either to go all the way back or stay down below. I chose the former.

Affectionately
Papa

[1] Last line of Dante's "Inferno": "And thence we came forth to see the stars again."

130 *To* THE FAMILY

Rome
September 22, 1907

My dear ones

On the Piazza Colonna behind which I am staying, as you know, several thousand people congregate every night. The evening air is really delicious; in Rome wind is hardly known. Behind the column is a stand for a military band which plays there every night, and on the roof of a house at the other end of the piazza there is a screen on which a *società Italiana* projects lantern slides (*fotoreclami*). They are actually advertisements, but to beguile the public these are interspersed with pictures of landscapes, Negroes of the Congo, glacier ascents, and so on. But since these wouldn't be enough, the boredom is interrupted by short cinematographic performances for the sake of which the old children (your father included) suffer quietly the advertisements and monotonous photographs. They are stingy with these tidbits, however, so I have had to look at the same thing over and over again. When I turn to go I detect a certain tension in the crowd, which makes me look again, and sure enough a new performance has begun, and so I stay on. Until 9 P.M. I usually remain spellbound; then I begin to feel too lonely in the crowd, so I return to my room to write to you all after having ordered a bottle of fresh water. The others who promenade in couples or *undici, dodici*[1] stay on as long as the music and lantern slides last.

In one corner of the piazza another of these awful advertisements keeps flashing on and off. I think it is called Fermentine. When I was in Genoa two years ago with your aunt it was called Tot; it was some kind of stomach medicine and really unbearable. Fermentine, on the other hand, doesn't seem to disturb the people. In so far as their companions make it possible, they stand in such a way that they can listen to what is being said behind them while seeing what is going on in front, thus getting their full share. Of course there are lots of small children among them, of whom many women would say that they ought to have been in bed long ago. Foreigners and natives mix in the most natural way. The clients of the restaurant behind the column and of the confectioner's on one side of the piazza enjoy themselves, too; there are wicker chairs to be had near the music, and the townspeople like sitting on the stone balustrade round the monument. I am not sure at the moment whether I haven't forgotten a fountain on the piazza, the latter is so big. Through the middle of it runs the Corso Umberto (of which it is in fact an enlargement) with its carriages and an electric *tranvia,* but they don't do any harm, for a Roman never moves out of a vehicle's way and the drivers don't seem to be aware of their right to run people over. When the music stops everyone claps loudly, even those who haven't listened. From time to time terrible yells are heard in the otherwise quiet and rather distinguished crowd; this noise is caused by a number of newspaper boys who, breathless like the herald of Marathon, hurl themselves onto the piazza with the evening editions, in the mistaken idea that with the news they are putting an end to an almost unbearable tension. When they have an accident to offer, with dead or wounded, they really feel masters of the situation. I know these newspapers and buy two of them every day for five *centesimi*

[1] Song from an operetta (*Eleven, twelve* . . .), popular at this time.

apiece; they are cheap, but I must say there is never anything in them that could possibly interest an intelligent foreigner. Occasionally there is something like a commotion, all the boys rush this way and that, but one doesn't have to be afraid that something has happened; they soon come back again. The women in this crowd are very beautiful (foreigners excepted); the women of Rome, strangely enough, are beautiful even when they are ugly, and not many of them are that.

I can hear the music plainly from my room; but of course I cannot see the pictures. Just now the crowd is clapping again.

Fond greetings
Your
Papa

131 *To* THE FAMILY

Rome
September 23, 1907

Dear children

Since describing the Piazza Colonna to you I have had a good look at it—one really ought to do so before—and I must correct a few things. There is indeed a lovely fountain in it, and people sit on its rim, but trolleys don't pass by, only horse-drawn buses. This shows how difficult it is to observe correctly. Today, by the way, was the first time the band didn't play.

Fond greetings.
Papa

132 *To* THE FAMILY

Rome
September 24, 1907

My dear ones

If I had stayed to the end, it would have been 1 A.M. But this I must explain to you, otherwise you won't understand it.

Because my evenings are so dreary I decided long ago to pay a visit to the theater, but waited for something I knew. This evening

Carmen was being played in the Theater Quirino. It began at
9 p.m., the hour I usually go up to my room. Now you understand
the beginning. Although I arrived only ten minutes before the
curtain, my checkroom ticket was No. 1; for 2.60 lire I got a seat
in the second row of the stalls, right in the middle, in other words
very grand; the theater filled up slowly, but it never got very full.
In this theater the front orchestra merges into the cheaper section
(*platea*), and above it are three rows of boxes. The center of the
third row is left open for a balcony. Not very elegant, nor was the
audience; either the distinguished Romans are still out of town or
they don't frequent this theater. But the others are probably not
much more elegant. The stage curtain showed all kinds of advertisements: sewing machines, underwear, furniture by installments,
as well as a Roman Dr. Stern offering his services to stutterers and
guaranteeing brilliant success. The rest of the orchestra, on a level
with the front orchestra and with nothing between them, was almost deserted when I arrived. Only one conscientious elderly
gentleman was trying out his violin and evidently hadn't had the
time to tune and clean it at home. One by one others appeared, the
younger men hitching up their trousers before sitting down; then
they produced instruments which they cleaned, put together, and
on which they proceeded to make the most appalling sounds. I
suspected some of them to have brought with them little whistles
off the street which could be used for the pretext of creating an
atmosphere. Gradually some thirty musicians, with varying amounts
of hair, arrived; one of them was a girl, at all events young and
dark, if not pretty, who sat down beside an enormous gilded harp
straight out of a picture book. This harpist's sister (to judge by her
looks) sat in the otherwise empty row in front of me, and it seemed
that the lady artist was on excellent terms with all the younger
members fo the orchestra. Only one of the musicians looked like a
rather boozy Bavarian who had strayed to Italy; otherwise they
had good Roman heads. Each of them had over his music rest a
kind of tin mug covered by a piece of green baize. I was hoping
for some electric light to shine forth from behind this weird apparatus. Then a man appeared dragging with him a heavy load of
books on each of which was written *Carmen;* these he distributed
among the musicians, evidently without any fear or favor. At last
came the conductor, tall and elegant, not quite slender enough,
looking like a confidence-inspiring headwaiter. He remained on his

platform, his back to the audience, for quite some time without moving a muscle. He clearly needed this rest, it was the calm before the storm. Then it suddenly broke out, the lights actually went on, all the apparent loafers in the orchestra began working away like mad, even the lady plucked at her harp and laughed only in the intervals. The set with the *Manufacturas di Tabagos* was quite in character; needless to say, the soldiers and street urchins (some of whom I thought I knew by sight) found it easier to look Spanish than ours do. I received my first shock when there appeared on the stage a lady whom I took to be Carmen, until I remembered that she must be the good Micaela. For she was very long and thin, her teeth long and bare, with a carrot-colored wig. In short, she looked a little like those English ladies sent abroad to scare the public, and her presence kept tormenting me until I realized her striking resemblance to poor Tina Urbitsek. The second, more enduring shock came soon after, for when the bell had stopped ringing there appeared the ardently awaited girls from the cigar factory, each with a cigarette in her mouth, but so exquisitely ugly, as if they had been provided by an infirmary, an almshouse, or a domestic agency. That they smoked was the only human thing about them, but this they shared with the audience. When I arrived the conscientious violinist had been smoking a Virginia, in the audience people smoked without restraint, and in the intervals even I—for whom smoking in the theater has always been made difficult by an ineradicable education —joined in.

At last Carmen made her appearance. I greeted her as an old acquaintance; you too would have recognized her. It was Katy Reich with her round face, her little snub nose and colossal proportions. Or rather, which also tallied better with her age, she was her mother, old Frau Gerstl, whom Mama will remember well. The resemblance was so great that I kept looking around for Heinz; but he didn't appear. As a matter of fact, Frau Reich's mother actually was an opera singer. It is strange that very fat people usually have little snub noses; so it seems that a longish nose can spare one considerable discomfort. Carmen, by the way, had a very powerful voice and acted with great vivacity but often rather overdid it. The colossal lady did not seem to be quite the right object for tenderness, and later when she began to dance, in order to detain José, it wasn't exactly graceful. "What a treat for the eye, if elephants could fly." But one could believe her wickedness; she made the most malicious faces. In Vienna they use clothes to show the contrast between the wicked Carmen and the virtuous Micaela, Carmen being dressed in red and Micaela in blue. Here they use size rather than color to express the contrast. There is no doubt that the wicked one was more substantial than the good one. Don José was a trifle elderly and again too fat; I have still not seen a slender José, although this is what he should be. He seemed rather downcast and took everything very seriously but he had a fine voice. His acting, too, was exaggerated; in the beginning he took little notice of Carmen, as though she were a fly buzzing round his newspaper, although she almost shook him by the hair. In the third act he grew furious. Escamillo cut a brilliant figure; unfortunately he mumbled so that one recognized his couplet only when the chorus repeated it. That they didn't restrain themselves when it came to the "March of the Toreadors" you can well imagine. The conductor dished it out in all directions, and I had an earful. The great melodies actually came off very well, but everything was just a bit coarsened, noisy; also some jokes that were added struck me as new. The officer and the chief smuggler, by the way, were excellent, you felt their joy in dressing up and they were a pleasure to watch.

In the long intervals the audience leaves the auditorium, walks about in the corridors and smokes; if one has forgotten to bring cigars, they can be bought in the theater. Because I like the music which accompanies the card game so much, I stayed on for the third act. What I noticed in it was that Carmen lifted poor Micaela's sash from behind and joined with the others in making fun of her dress. Such mean tricks would not be tolerated in Vienna. I left after this act; it was midnight; each act plus the interval lasted an hour. After all, I could imagine the end: Don José had already grown very furious and twice flung Carmen to the ground. How delicious the night air felt after this! With the sureness of a Roman I took the shortest way home.

What a pity one can't live here always! These brief visits leave one with an unappeased longing and a feeling of insufficiency all the way round.

I greet you all warmly and will see you soon after the arrival of this last long letter.

Papa

133 *To* MARTHA FREUD

Rome
September 24, 1907

Very pleased with your card. I can imagine that the little cabinet came as a surprise. It is my present for you from my journey. A small mirror frame ought to have arrived, too. I am getting tired of traveling, plan to leave here on Thursday evening and to arrive on Saturday morning, depending on a few small purchases for which I am still negotiating. Just imagine my joy when, after being alone so long, I saw today in the Vatican a dear familiar face! The recognition was one-sided, however, for it was the "Gradiva,"[1] high up on a wall. The weather is getting more and more glorious, the city more and more marvelous. Yesterday I handed in my linen to be washed.

Warmest greetings
Your
Sigmund

I'll start work again on Monday, the thirtieth.

[1] Antique relief in the Museum of the Vatican (see letter of May 26, 1907).

134 *To* KARL ABRAHAM

Vienna IX, Berggasse 19
October 8, 1907

Dear Colleague

My first reaction to your letter was one of regret, which I soon overcame. To a youthful man like yourself it doesn't do any harm to be forcibly turned loose into the world, *"au grand air,"* and the fact that as a Jew it will be more difficult for you may, as with all of us, have the effect of stimulating your productivity. It goes without saying that my sympathy and best wishes accompany you on the new road,[1] and if it is at all possible I shall try to do more. If I were still on intimate terms with Dr. W. Fliess of Berlin,

[1] Dr. Abraham planned to start work as a psychoanalyst in Berlin.

the path would be paved for you, but unfortunately this channel is now completely closed. When consulted by patients from Germany during the past year I was continually in the position of having to express regret that I had no representative there to whom I could recommend them. In the event of such cases returning this year, I shall know what to do. Should my reputation in Germany increase, you will certainly be able to profit by it, and if I may describe you as my pupil and disciple—you don't strike me as man who would be ashamed of this—then I can take active steps on your behalf. On the other hand you yourself know the extent of the antagonism I still have to deal with in Germany. I trust you won't even try to find favor with your new colleagues who are like colleagues everywhere, if anything rather more brutal, and that instead you will address yourself directly to the public. When the fight against hypnosis was being waged most violently in Berlin, a highly unsympathetic hypnotizer by the name of Grossmann quickly built up for himself a large practice. One would imagine that with the help of psychonalysis you should succeed even better.

You hint in your letter that you have something to show me; I hope you know that I shall put myself at your disposal to the best of my ability. Isn't it quite easy to travel from Zürich to Berlin via Vienna?

I will try to answer your last letter as soon as you are settled enough to discuss scientific matters again.

With my most heartfelt good wishes
Yours sincerely
Dr. Freud

135 *To* THE ANTIQUARY HINTERBERGER

(Undated [1907])

You ask me to name "ten good books"[1] and refuse to add a word of explanation, thus leaving to me not only the choice of the books but also the interpretation of your request. Accustomed to paying attention to small details, I have to take my clue from the very words in which you clothe your mysterious demand. You don't say "the ten greatest works" (of world literature), in which

---

[1] This request had been put to a number of people.

case, like so many other people, I should feel obliged to mention:
Homer, the tragedies of Sophocles, Goethe's *Faust*, Shakespeare's
Hamlet, Macbeth, etc. Neither do you say "the ten most significant
books," in which case I should have to mention such scientific
achievements as those of Copernicus, the old physician Johann
Weier on witchcraft, Darwin's *Descent of Man*, and so on. You
don't even ask for "favorite books," among which I should not have
forgotten Milton's *Paradise Lost* and Heine's *Lazarus*. So I assume
that a special emphasis falls on the word "good," and that by it
you mean books with which one is on the kind of terms one is with
"good" friends, books to which one owes some part of one's knowledge of life and philosophy, books one has enjoyed and likes to
recommend to others, but in relation to which the element of shy
reverence, the conviction of one's own smallness versus their greatness, does not play a special part. So I shall name for you ten
such "good" books that come to mind without much reflection:

> MULTATULI: letters and works
> KIPLING: Jungle Book
> ANATOLE FRANCE: Sur la Pierre Blanche
> ZOLA: Fécondité
> MEREZHKOVSKI: Leonardo da Vinci
> G. KELLER: People of Seldwyla
> C. F. MEYER: Hutten's Last Days
> MACAULAY: Essays
> GOMPERZ: Greek Thinkers
> MARK TWAIN: Sketches

I don't know what you intend to do with this list. It appears
rather odd even to myself, and I really shouldn't let it out of my
hands without some comment. The problem why I have chosen just
these and not other equally "good" books I won't even attempt to
solve; I will merely try to throw some light on the relationship
between the author and his work. Not in every case is the relationship so close as in that between Kipling and his *Jungle Book*.
Frequently I could just as well have chosen another work by the
same author, for instance Zola's *Dr. Pascal*. The man who has
given us one good book has very often presented us with several
good books. In the case of Multatuli I found myself incapable of
giving precedence to the *Love Letters* over the *Private Letters*,
and vice versa, and for this reason I have written: "letters and

works." Books of poetry of purely aesthetic value have been excluded from this list probably because the request "good" books did not seem to aim at this category. In the case of C. F. Meyer's *Hutten,* for instance, I place its "goodness" far above its beauty, its "edifying" qualities above its aesthetic ones.

With your invitation to name "ten good books" you have touched on something of which there is inevitably a great deal to be said. So I am concluding this letter in order to avoid becoming too effusive.

Yours sincerely
Dr. Freud

136 *To* C. G. JUNG

Vienna IX, Berggasse 19
February 18, 1908

Dear Friend

Don't be alarmed; after this, I promise you a long silence. This is merely a postscript to emphasize yesterday's suggestion to offer Bleuler the presidency[1] in Salzburg. You will be doing me a great favor if you would pass on to him this wish of mine as a personal request. I consider it quite fitting, even more dignified, if he rather than I were to take the chair. It might even look strange if I as the outlawed knight were to preside over the private parliament which has been called in defense of my rights against king and country. On the other hand it would be an honor for me and would also create a deeper impression abroad if he [Bleuler], as the oldest and most important of my followers, were to lead the movement on my behalf.

My Viennese colleagues too would behave better under his presidency; in short, all would be plain sailing if he accepted. I hope that you agree with me and that you will use your influence with him.

I have decided not to write about anything else, so I send you my kind regards and thanks for your efforts.

Your
Freud

[1] Of the First Psychoanalytical Congress in Salzburg.

137 *To* MATHILDE FREUD

Vienna IX, Berggasse 19
March 19, 1908

My dear Mathilde

It is the first time that you have asked me for help and you don't make it difficult for me this time, for it is easy to see that you are very much overrating your trouble and that you are drawing conclusions which according to my knowledge and information are quite out of place. I am not going to offer you any illusions, neither now nor at any other time; I consider them harmful and I know that the suspicion that they are illusions is enough to spoil the pleasure in them. But none is needed, anyhow. Meran[1] is intended to strengthen you physically, for which it is very suitable; of course it cannot cure the local complaint; this has to take care of itself for the time being. It will probably go on giving you pain for several months (as a matter of fact your last attack sounds as though it may have been caused by a floating kidney), but it is harmless in itself; it is bound to shrink more and more and finally leave you altogether. Women often contract such things after childbirth, and they disappear without causing any trouble in later life. By the time the question of marriage arises in your life, you will be completely free of it. You know that I have always intended to keep you at home until you are at least twenty-four, until you are strong enough for the duties of marriage and possibly of bearing children, and until the weakness, which those three serious illnesses in your early life left behind, have been repaired. In social and material circumstances like ours, girls quite rightly do not marry during their early youth; otherwise their married life would be over too soon. You know that your mother was twenty-five when she married.

I think you probably associate the present minor complaint with an old worry about which I should very much like to talk to you for once. I have guessed for a long time that in spite of all your common sense you fret because you think you are not good-looking enough and therefore might not attract a man. I

---

[1] Mathilde had been sent to Meran in South Tyrol to recuperate; she stayed with the family of Dr. Raab.

have watched this with a smile, first of all because you seem quite attractive enough to me, and secondly because I know that in reality it is no longer physical beauty which decides the fate of a girl, but the impression of her whole personality. Your mirror will inform you that there is nothing common or repellent in your features, and your memory will confirm the fact that you have managed to inspire respect and sympathy in any circle of human beings. And as a result I have felt perfectly reassured about your future so far as it depends on you, and you have every reason to feel the same. That you are my daughter shouldn't do you any harm, either. I know that finding a respected name and a warm atmosphere in her home was decisive in my choice of a wife, and there are certain to be others who think as I did when I was young.

The more intelligent among young men are sure to know what to look for in a wife—gentleness, cheerfulness, and the talent to make their life easier and more beautiful. I would be terribly sorry if your despondency were to make you change your direction, but let us hope that it is only a passing phase in a situation which many things have combined to produce. You have inherited your physical disposition from two of your aunts, both of whom you resemble more than your mother. I would rather you took after Aunt Minna than Aunt Rosa.

You have, my poor child, seen death[2] break into the family for the first time, or heard about it, and perhaps shuddered at the idea that for none of us can life be made any safer. This is something all we old people know, which is why life for us has such a special value. We refuse to allow the inevitable end to interfere with our happy activities. So you, who are still so young, may as well confess that you really have no reason to be downhearted. I am very pleased to hear that the sun in Meran is doing you good. We would have pulled a very long face if you had returned as you left. You had better stay on as long as the Raabs are there and are willing to put up with you, let us hope until well into the month of May.

I greet you warmly and hope to hear from you again soon.

Your loving
Father

[2] Dr. Heinrich Graf, husband of Freud's sister Rosa, had died.

138 *To* MARTHA FREUD

Vienna IX, Berggasse 19
April 29, 1908

My beloved Old Dear

I am very pleased that you found Mama so much better than you expected; this gives your journey quite a different character.

The Congress[1] was a great success and I think it has left a good impression on all those who took part in it. For me it has meant a great deal of work, but also some relaxation. The whole enterprise will probably not be without results. While we were having dinner on Monday I happened to glance behind me into the hall and saw a back that struck me as familiar, unlikely though the meeting seemed. I rushed to inspect the back from the front and sure enough it was Emanuel,[2] who was revenging himself in this way for the surprise in Wiesbaden. We met again, the following morning when everyone had left, and in the evening he took me to the station. The hours between we spent talking and drinking beer—mostly he—went up to the Fortress,[3] to Hellbrunn,[4] and so on. For seventy-five he is amazingly vigorous, but for the first time he showed definite signs of having aged; the last bout of influenza has taken it out of him. He was on his way to Berlin, was very interested to hear about Mama, and was as aghast about Rosa[5] as we all are.

There were more patients today at consultation than I could deal with. Tomorrow an Englishman and an American[6] are coming to dinner.

Fond greetings
Your
Sigmund

---

[1] See letter of Feb. 18, 1908.
[2] Freud's half brother from England.
[3] Hohensalzburg.
[4] Castle and park near Salzburg.
[5] Allusion to Rosa's excessive grief at the death of her husband.
[6] The first visits of Ernest Jones (London) and A. A. Brill (New York).

139 *To* STEFAN ZWEIG

Vienna IX, Berggasse 19
May 3, 1908

Dear Mr. Zweig

I was absent from Vienna during the beginning of last week and on my return found so much work to catch up on that I had to postpone thanking you for your kind gift.[1] I know that you are a poet from reading the *Frühen Kränze* [*Early Wreaths*][2] and the sounds of the fine powerful verses which come flowing toward me whenever I open the book promise me an hour of pleasure which I shall very soon try to steal from the relentless work. I can guess the connection, and not that you are charitable enough to kill off the man who, according to the ancient poets, returned from Troy unscathed.

Accept my best thanks once more
Yours very sincerely
Freud

[1] *Tersites*, by Stefan Zweig, Inselverlag, Leipzig, 1907.
[2] *Poems*, by Stefan Zweig, Inselverlag, Leipzig, 1906.

140 *To* ANNA FREUD

Vienna IX, Berggasse 19
July 7, 1908

My dear Anna

Indeed, if I weren't already very impatient to get to the Dietfeldhof,[1] your letter would make me so. I also liked it very much when I was there for the first time in April, when white snow still lay between the clumps of yellow primroses. Strawberries and mushrooms[2] are a very welcome prospect; we are bound to dis-

[1] The house in Berchtesgaden in the Bavarian Alps, where the family spent the summer of 1908.
[2] The gathering of wild strawberries and mushrooms was one of Freud's favorite pastimes when on vacation.

cover some lovely walks in no time. Perhaps we can rent the
Aschauer pond for ourselves alone, so that there will be room
for us all to bathe.

Your brother Martin is very proud of having passed his exam[3]
so well and will soon appear in your midst with a new traveling
bag and velour hat. Lampl[4] will be coming with him, but they
will soon start off on their great holiday trip with which they
intend to celebrate their independence. As for us, we will read and
write and wander about in the woods; if only the Almighty doesn't
spoil the summer for us by making it rain all the time, it can be
very beautiful.

My greetings to all your brothers and sisters; before you have
reread this letter several times, we will be there, too—the morning
of July 16.

Love from your
Papa

[3] Martin had passed his matriculation.
[4] Hans Lampl, school friend of Martin's. He later became an intimate friend of
all the brothers and sisters. See List of Addressees.

141 *To* SANDOR FERENCZI

Dietfeldhof, Berchtesgaden
August 4, 1908

Dear Colleague

I have just taken a room for you in the Hotel Bellevue, not
too far from us; it is not possible to get anything nearer on account
of the isolated position of our house. I may say that we are looking forward to your arrival, although the pleasure I anticipated
has been thwarted by a change of plans. For on September 1 I have
to go straight to England, where my brother's family is expecting
me, and as a result I cannot claim your company as I had intended
for a tour of the Dutch towns during the preceding week.

I am inundated with things to be written, find everything more
difficult and slower than I had expected, and hope that you will
occasionally pull me out of this condition. I haven't benefited much
from the vacation so far. My boys will be pleased if you are a moun-
taineer and would feel like joining them on a tour they are planning

to the Hochkönig. Our house will always be open to you, but I want you to feel independent and I hope you won't be surprised if I carry on with my obligations.

Farewell for the present.

Yours very sincerely

Freud

142 *To* THE FAMILY

Salò[1]

September 25, 1908

My dear ones

Everything comes to an end. You owe this letter to the reckless purchase a few days ago of a blue stamp and to today's rain. I myself am going to follow very soon, but I am quite contented with the postponement, for after the superabundance of this summer's experiences I am in need of a few days' comfort without any activity. For this, Salò and the hotel are just right. It is very comfortable without oppressing one with elegance. You know the landscape, so far as one can get to know anything from passing through. It becomes infinitely more beautiful when one spends some time in it. The day before yesterday we took a trip to S. Vigilio by motorboat (private); the spot is among the most beautiful on the Lake Garda, thus as beautiful as anything anywhere. It is a place for solitude, and of course unsuitable for a family. In passing by we had a chance to study the island, with the usual feeling of shameless envy at not being Prince Borghese. Yesterday and today we confined ourselves to taking walks, which makes one feel very well for no particular reason. Today it is raining gently but steadily. Minna is resting in her room, I am thinking of eating a pomegranate (ten *centesimi*), of having a smoke, and of playing the new patience. In my old age I seem to be developing a lot of talent for the enjoyment of life. In reality, of course, it is only the calm before the storm.

Tomorrow, Sunday, we plan to start the complicated journey home, leaving by boat at 11:20 A.M. and arriving in Bolzano at 6:45 P.M. As there is some doubt about the weather and no doubt about its being Sunday, it seems the best way of spending the

---

[1] Resort on Lake Garda.

time. I shall still have half a day in Bolzano to do some shopping. Tuesday morning in Vienna, as announced by telegram. Then once again we will share together the cares and tasks of this year; putting things in order and finishing my corrections will quickly fill the remainder of this miserable remnant of September.

Looking forward to seeing you again soon

Affectionately

Papa

143 *To* SANDOR FERENCZI

Vienna IX, Berggasse 19

January 18, 1909

Dear Colleague

Be prepared to laugh at me. Today I glanced at the first 14 sheets of our *Jahrbuch*[1] and was enchanted. Jung has done a splendid job. A real revenge for Amsterdam![2] It is something to be proud of.

And now an awful thing has occurred to me: that you are not represented in a volume which cannot be repeated so soon in this excellent and creditable fashion. Then I realized that it was *my* fault because, from some mistaken discretion or God knows why, I had prevented your excellent paper about the transference from being published in the second half-volume. I told myself that you could easily write another essay for the second half-volume, but this was followed by the thought that you could hardly produce anything of similar significance within a year. In short, I feel compelled to ask you to look upon my objection of that time as canceled, and to reconsider whether you would not rather withdraw the paper from Brodman's journal and offer it to the *Jahrbuch*. This seems to me at the moment *much the better* thing to do. So far as Jung is concerned you can by all means lay the odium of inconsistency on me. As for Brodman, you don't have to worry about him, just give him any reason. So have a good laugh at me and please take the advice I am giving you now.

Yours very sincerely

Freud

---

[1] *Jahrbuch für Psychoanalytische und Psychopathologische Forschungen.*
[2] See letter of May 26, 1907.

144 *To* OSKAR PFISTER

Vienna IX, Berggasse 19
January 18, 1909

Dear Sir

I cannot content myself with thanking you for sending me your essay "Delusion and Student Suicide"[1]; I must also tell you how pleased I am that our psychiatric researches have aroused the interest of a clergyman who has free access to the souls of so many young and healthy* individuals. We often reproach our psychoanalysis half jokingly but actually quite seriously for requiring a state of normality for its application, and for finding a barrier in the organized [organic?] abnormalities of mental life, with the result that psychoanalysis meets the optimum of favorable conditions where its practice is not needed—i.e., among the healthy. Now it seems to me that this very optimum exists in the field in which you work.

Our mutual friend, C. G. Jung, has often mentioned your name to me; I am glad that I am now in a position to associate a more definite idea with that name, and I hope that you will do me the honor of showing me your future essays.

Yours gratefully
Freud

[1] Pfister's essay published in *Schweizer Blätter für Schulgesundheitspflege,* 1909.
* *vollwertig.*

145 *To* C. G. JUNG

Vienna IX, Berggasse 19
January 25, 1909

Dear Friend

I know that for everyone who has digested his first successes in psychoanalysis there follows a difficult and bitter period during which he curses it and its founder. After a while, however, this feeling subsides and he arrives at a *modus vivendi*. These are the realities! *C'est la guerre.* Perhaps my essay on technique[1] (which I am trying hard to finish) may redress the worst, but certainly not very much. Still, it is only by struggling with difficulties that one learns, and I am not at all sorry that Bleuler has deprived you of your teaching job. You are destined to be a teacher anyway; sooner or later you will have more than enough of it, whereas into psychoanalytical practice one has to be forced. It is good to have no choice.

"He does his best who cannot do otherwise" is more or less what C. F. Meyer makes his man on the Ufenau say. To salve my conscience I often tell myself: Above all don't try to cure, just learn and earn some money! These are the most useful conscious aims.

Meanwhile I have received from Pfister an intelligent letter bursting with ideas. Me and the *Protestant Monthly,* just imagine! But I don't mind. In some respects the clerical analysts work under better conditions; moreover, they have nothing to do with money. Teachers really should be familiar with all our ideas, if only for the sake of the healthy child. Which is why I greet the news of your course for teachers with a cheerful *Prosit!*

I acknowledge my slip of the pen with a smile. Good intentions are of no avail in the struggle against these devilish little tricks of the demon, so one just has to take them in one's stride.

With best regards, also for the now complete family.[2]

Your
Freud

[1] Presumably an essay not published until 1912: "Ratschläge für den Arzt bei der Psychoanalytischen Behandlung." ("Recommendations to Physicians Practicing Psychoanalysis," Standard Ed., 12.)

[2] Allusion to a couple with son and daughter.

146 *To* OSKAR PFISTER

Vienna IX, Berggasse 19
May 10, 1909

Dear Pastor Pfister

The Matterhorn is now covering the unanswered letters on my desk. I happily accept that little part of Switzerland in the symbolic sense you suggest, as homage from the only country where I feel a man of property in the knowledge that the hearts and minds of good men are well disposed towards me. I don't intend to defend myself personally. I have purposely set myself up only as an example, never as a model, far less as one to be venerated.

It is easy enough to endow the Matterhorn with another and less exalted meaning. The scale of 1:50,000 may be roughly the proportion in which fate fulfils our wishes, and in which we ourselves carry out our good intentions. Incidentally, it has struck me how difficult it is for us to imagine numbers; I find it extremely hard to believe that I would have to pile a mere 50,000 little squares on top of one another in order to reach the summit of a giant mountain; I should have thought it would have taken more than a million.

I would like to mention even a third significance that the Matterhorn holds for me. It reminds me of a remarkable man who came to see me one day, a true servant of God, whose very existence struck me as highly unlikely: I mean in the sense of someone who feels the need to be of some spiritual benefit to everyone he meets. So you have also done me good. After your exhortations I asked myself why I don't feel really happy, and I soon found an answer. I renounced the unattainable desire to become rich in an honest fashion, decided after the loss of one patient not to replace him, and have been well and happy ever since and realize you were right. Since then, moreover, I have remained faithful to this decision three times. But for your visit and your influence I should never have been able to accomplish this; my own father complex— as Jung would say—i.e., the need to correct my father, would not have permitted it.

Your remarks about transference and compensation I shall give the consideration which they deserve. I think you are right; it is the condition for a lasting success. There is, above all, a certain type of woman who refuses to be compensated by ideas, and demands either some kind of tangible happiness in life or tries to cling to the transference. They are those of whom the poet says they have a feeling only for "the logic of soups-and-dumpling arguments."[1]

[1] Allusion to a quotation from Heine's *Zeitgedichte* (*Topical Poems*), No. 24.

Now accept my warm thanks. Do go on writing valiantly and
let me hear more and more about your struggle and successes.

> With kindest regards
> Your
> Freud

147 *To* MARTHA FREUD

Palermo
September 15, 1910

Beloved Old Dear

Have just fetched my mail, Martin's praiseworthy communication and your letter, so I know that at this moment you are
preparing to leave Berlin and will be back home tomorrow morning while we[1] are on our way to Girgenti, the last stop before
Syracuse.

Palermo was an incredible feast, something in which one really
shouldn't indulge alone. I solemnly promise that whatever drudgery
may be in store for me during the coming year I shall remember
that I have already had my share [of the feast] and consumed it.
Such a wealth of color, such views, such fragrant smells, and such
a sensation of well-being I have never experienced all at once. Now
this is over, will be locked up and opened again only for others.
But perhaps even more beautiful things are awaiting me in Syracuse.

I am really awfully sorry that you cannot all be with me. But in
order to enjoy all this in a group of seven or nine, or even of only
three—in short, of *undici, dodici, tredici*[2]—I shouldn't have become
a psychiatrist and an alleged founder of a new school in psychology,
but a manufacturer of something useful such as toilet paper,
matches, shoe buttons. To change professions now it is much too
late, and so I continue—selfishly but on principle regretfully—to
enjoy everything on my own.

Only shopping creates the greatest difficulties. There is absolutely nothing here which cannot be found elsewhere and which

[1] The traveling companion was Ferenczi—see List of Addressees and letter of
Oct. 2, 1910.
[2] See note to letter of Sept. 22, 1907.

tempts one to bring it home—carriages and mules are of course out of the question, and flowers don't keep—and I really must ask you to consider as canceled all promises made in this respect, and to accept cash as a substitute when I am back in Vienna. Only Robert[3] will receive the sulphur he asked for—which goes to show that he uttered a practical, geographically appropriate wish.

As it is far easier to take an interest in familiar rather than unfamiliar things, I inform you that I have rendered the silk suit wearable by the purchase of a new portfolio. The heat, pure sun with a lovely breeze, has bothered me so little that I have so far felt no need for it. Here one understands the lines:

> How delightful 'tis to wander
> By the breath of evening fann'd. . . .[4]

or however it goes.

In order to give you not too rosy a picture and not to arouse the envy of the gods, I must mention that today for the first time we read rumors about the cholera in Naples. We intend to check up on this news and make our plans accordingly. In Sicily so far no trace has been found, thus less than in Vienna or Budapest. As a matter of fact in Vienna with its bad weather there is little cause for fear. Caution in the eating of fruit can never do any harm.

And now in conclusion and as a farewell to Palermo, fondest greetings to all who are reunited at home.

From

Papa

[3] Mathilde's husband Robert Hollitscher (1875-1959).
[4] First lines of the letter-duet from Mozart's *Marriage of Figaro*.

148 *To* SANDOR FERENCZI

Vienna IX, Berggasse 19
October 2, 1910

Dear Friend

Your letter reminded me that I am the same person who picked papyrus in Syracuse, had a scuffle with the railway staff in Naples, and bought antiques in Rome. The identity has been re-established. It is strange how easily one gives in to the tendency to isolate parts of one's personality.

You will believe me when I say I look back upon your company during the journey with nothing but warm and affectionate feelings, although I frequently felt sorry for you because of your disappointment, and in some respects I would have liked you to be different. You were disappointed because you probably expected to swim in constant intellectual stimulation, whereas I hate nothing more than striking up attitudes and out of contrariness frequently let myself go. As a result I was probably most of the time a quite ordinary elderly gentleman, and you in astonishment kept measuring the distance between me and your phantasy ideal. On the other hand I often wished that you would pull yourself out of the infantile role and place yourself beside me as a companion on an equal footing, something you were unable to do, and in a practical respect I would like you to have carried out your share of the responsibility, orientation in space and time, a little more reliably. But you were inhibited and dreamy. So much for my educational efforts.

The most beautiful and most interesting patient at my first consultation has promised to come to you for a thorough treatment. I don't have to recommend Frau Dr. T. to you.

Warm greetings for today to you and Frau Gisela.[1] Soon your idyll will be over, too.

Your faithful
Freud

---

[1] Ferenczi's future wife.

149 *To* ELSE VOIGTLÄNDER

Vienna IX, Berggasse 19
October 1, 1911

Dear Dr. Voigtländer

Many thanks for your communication. I was very interested to hear of the attitude taken by a philosophically trained mind toward my "wild" discoveries; and I was very pleasantly surprised to see that—in spite of being unfamiliar with the empirical foundations of the theories—you look upon so many things as useful and significant.

Some objections I am not in a position to appreciate fully, for I lack the understanding for the content of certain philosophical terms. On one point, however, I think I have understood you completely, and this is the one I wish to answer.

You suggest that I overestimate the importance of accidental influences on character formation and in contrast you stress the importance of constitutional factors, of disposition, which selects from among the experiences and allows them to become significant. Everything you say on this subject is excellent, except that its polemical application seems to be based on a misunderstanding. For, with a minor modification, we say the same.

We find in psychoanalysis that we are dealing not with *one* disposition but with an infinite number of dispositions which are developed and fixed by accidental fate. The disposition is so to speak polymorphous. We also believe that this is again a case in which scientifically thinking people distort a cooperation into an antithesis. The question as to which is of greater significance, constitution or experience, which of the two elements decides character, can in my opinion only be answered by saying that δαίμων καὶ τύχη[1] [fate and chance] and not one *or* the other are decisive. Why should there be an antithesis, since constitution after all is nothing but the sediment of experiences from a long line of ancestors; and why should the *individual* experience not be granted a share alongside the experience of ancestors?

Now, it seems that in single cases all these possibilities of variation are realized in such a way that in each individual sometimes this, sometimes that part of the inherited disposition becomes so dominating that it chooses some experiences and rejects others, whereas on the other hand accidental influences work here and there so powerfully that they arouse and fix this or that part of the originally dormant disposition.

If in our analytical work we concentrate more on the accidental influences than on the constitutional factors, we do so for two reasons. First, because the former have been overlooked and now have to be proved, whereas the latter are only too readily admitted; second, because on the basis of our experience we know something

---

[1] Titles of the first two stanzas of Goethe's poem "Urworte. Orphisch."

about the former, while about the latter we know as little as—non-analysts. This predilection for the accidental, however, by no means signifies a denial of the constitutional. We are inclined more toward overdetermination and less toward antithesis than other observers.

We are also of the opinion that by appreciating the importance of the accidental we have taken the right road toward the understanding of constitution. It is the correct line of procedure. What remains inexplicable after a study of the accidental may be put down to constitution.

The exposition in *Three Essays on the Theory of Sexuality*[2] and in the essay on Leonardo[3] (which tries to prove an especially crass example of the impact of accidental family constellations) are based throughout on this point of view.

I shall be very pleased to learn that you have not turned your interest away from psychoanalysis.

Yours very sincerely
Freud

[2] *Drei Abhandlungen zur Sexualtheorie,* Vienna, 1905. (*Three Essays on the Theory of Sexuality,* Standard Ed., 7, 125.)

[3] "Eine Kindheitserinnerung des Leonardo da Vinci," Vienna, 1910. ("Leonardo da Vinci and a Memory of his Childhood," Standard Ed., 11.)

150 *To* KARL ABRAHAM

Vienna IX, Berggasse 19
January 2, 1912

Dear Friend

Having filled up the holidays with the writing of 2½ papers (which I don't like) and all kinds of personal discontents, I am finally getting around to wishing you, your wife and children all the prosperity in the New Year which you deserve and which would give me such pleasure, too. Common interests and personal sympathies have linked us together so intimately that we have no need to doubt the sincerity of each other's good wishes.

I am aware that your position in Berlin is far from easy and I continually admire your serene spirits and tenacious optimism. The chronicle of our enterprise may not always be a happy one, but most chronicles are probably much the same; it will produce quite an interesting chapter in history, nevertheless. . . .

For myself I don't expect much; we are passing through a gloomy time, and recognition will no doubt come only to the next generation. But we have the incomparable pleasure of gaining the first insights. My work on the psychology of religion[1] is proceeding very slowly, and I would prefer to drop it altogether. I have to write a kind of introduction for the new journal *Imago*,[2] something on the psychology of savage tribes. Reik's[3] essay is too long for the collection; I have just learned today from the author (who is one of our members) that it is to appear as a book.

Farewell and don't write too rarely to

> Your faithful friend
> Freud

[1] *Totem und Tabu*, Vienna, 1912/13. (*Totem and Taboo*, London, 1950; New York, 1952; Standard Ed., 13,1.)

[2] *Zeitschrift für die Anwendung der Psychoanalyse auf die Geisteswissenschaften*, founded in 1912.

[3] Theodor Reik, Ph.D. (born 1888), psychoanalyst in New York (see letter of Mar. 8, 1925).

151 *To* LUDWIG BINSWANGER

> Vienna IX, Berggasse 19
> April 14, 1912

Dear Dr. Binswanger

An old man like me who shouldn't complain (and has decided not to complain) if his life comes to an end in a few years, feels especially aggrieved when one of his flourishing young friends, one of those who is meant to continue his own life, informs him that his life is in danger. On thinking it over, however, I gradually collected myself and remembered that in spite of the existing suspicions all the odds are in your favor and that you have only been reminded a little more conspicuously of the precariousness in which we all live and which we are so ready to forget.

But *you* will not forget it just now and life, as you say, will hold a special and enhanced charm for you. As for the rest, we will hope for the best, something our present-day knowledge permits us to do without self-deception. I shall of course keep the secret, as you have asked me to, proud of the privilege you have granted me. But it goes without saying that I would like to see you, as soon as this is possible without disturbing you. Perhaps at Whitsun? I trust you will let me know if this would suit you.

I am pleased to hear that the plan for your paper preoccupies you more than ever, and I will now answer your questions. . . .

I have concerned myself very little with the individual great men whom you mention. Nor have I taken much interest in the whole species. It has always seemed to me that ruthlessness and arrogant self-confidence constitute the indispensable condition for what, when it succeeds, strikes us as greatness; and I also believe that one ought to differentiate between greatness of achievement and greatness of personality.

Dr. P. of Bolzano is an excellent man and dear friend. The woman whom you mention is his stepdaughter, of a rather different ilk. I know both her and her mother and have always found that with all their kindness and charm these women lack something like a moral backbone. They are as though in a permanent state of erotic intoxication. But it is very possible that the grim experience of her marriage has brought out a more serious side in the young woman, and I would be very pleased to hear that under your guidance she had turned into a useful person.

Imago has appeared; I hope that my second essay (on taboo)[1] in the third number will be more interesting than the first.

And now I send you and your dear nurse my warmest wishes for your recovery and for the maintenance of your excellent, manly, and courageous spirits.

Your faithful
Freud

[1] See footnote 1 to letter of Jan. 2, 1912.

152 *To* MAX HALBERSTADT

Vienna IX, Berggasse 19
July 7, 1912

Dear Sir

My little Sophie,[1] whom we had sent on a few weeks' vacation to Hamburg, returned two days ago, gay, radiant, determined, and broke the surprising news that she is engaged to you. We realize that this fact makes us, so to speak, superfluous and that

[1] Freud's second daughter Sophie (1893-1920) had become engaged to Max Halberstadt, a photographer in Hamburg.

there is nothing left for us to do but go through the formality of bestowing our blessing. Since our wishes have always been for our daughters to feel free to follow their hearts in the choice of a husband, as our eldest daughter has actually done, we have every reason to be satisfied with this event. But we are parents after all, burdened with all the delusions that accompany this condition, feel obliged to assert our importance, and we would therefore like to set eyes on the energetic young man whose determination has infected our child before we declare a solemn Yes and Amen.

It benefits your plans that you are not entirely unknown to us. True, I brought back from my visit to your studio only a fleeting though most pleasant impression, but the two mothers—my wife and my sister-in-law—know you, your mother, your family and, according to them, have always considered you a member of the intimate circle of our relations. So from this side too all hearts are disposed in your favor. I have good reason to believe that our daughter, hardly out of her teens, will find a safe home at the side of a serious, loving, clear-headed, and intelligent husband.

Should this be the case then it is our duty as parents to welcome her choice. Now, I feel the urge to get to know you better soon, and it is only a few practical considerations that compel me to temper my impatience. Sophie, who already completely identifies herself with your interests (which seems so strange), tells me that you can get away in the immediate future only with great difficulty. We ourselves are dissolving our household during this coming week; my wife will be very occupied, and I myself am working more than nine hours a day. It therefore seems impractical to invite you to Vienna at this time. After the fifteenth, however, we will both be in Karlsbad where we intend to stay about four weeks, and during this time we would like very much to have you with us, although the attractive members of the family, the young, will not be on exhibition. We shall then not only have the task of consolidating our acquaintance, we shall also have to exchange a few words about the financial foundations of your marriage ceremony and the technical questions of time and place. It is most certainly our wish too that after this visit we shall address you by a more affectionate name.

Meanwhile my cordial greetings and best wishes

Your

Freud

153 *To* SOPHIE FREUD

Karlsbad
July 20, 1912

My dear Sophie

Another father in my situation would write that he cannot understand how a telegram saying "Mama Papa Max congratulate you" could possibly be construed in any other sense than "We congratulate you on your engagement, greet you as a bride"—and he wouldn't be able to understand how such a greeting could possibly produce dissatisfaction. I, however, can imagine that you are somewhat bothered by your conscience for having ignored us so completely when you decided to get engaged, and this at least is to your credit. The degree of your remorse may be judged by the fact that you even succeeded in upsetting your aunt,[1] normally so imperturbable.

Anyhow, everything is all right and Max, although still somewhat shy, has been very nice and amiable. Since then you will have learned of the further arrangements from the second telegram or from my letter to your aunt. And now I must tell you that I played a little trick on him as he went away. He had paid behind our backs his bill for the wretched attic room, which was the only thing we could get for him, and so I showed him a little knitted purse which I carry with me for foreign money and pretended that it was an old piece of needlework of yours, and asked him to keep it. This purse, however, contained the 6.80 kreutzer, which he had paid to Frau Schubert. Now you can explain the whole thing to him and disown the needlework.

I greet you fondly and wish you a good rest until we meet in Bolzano.

Papa

[1] Martha Freud's sister Minna Bernays.

154 *To* MAX HALBERSTADT

Karlsbad
July 24, 1912

My dear Son-in-law

You are right in assuming that our common interest in making our little Sophie happy will soon bring us close. But apart from this I hope you may soon discover that we will both serve as quite useful parent substitutes, and that we will grow fond of you for your own sake. At our first meeting we were all quite naturally somewhat inhibited. But when we meet again on the Lago di Carezza[1] we shall have outgrown this stage. We cannot as yet fix the exact date; it has been suggested from Lovrana[2] that you shouldn't come to Bolzano, so as to avoid having to make the acquaintance of your new relatives when you are tired after the journey, feel grumpy, or are suffering from migraine. This would postpone everything by one day. But you can fight this out for yourself and make all necessary arrangements with the others on the Adriatic.

Sophie has continued her tactics toward us for some little while. Nothing we have written to her has seemed detailed or loving enough for her. I trust that by now she is pacified and is helping you to count the days. It is very strange to watch one's little daughter suddenly turn into a loving woman.

In the meantime we have been busy spreading the news privately. As a result at least two people have divulged their intention of cunningly having their portraits taken by you in order to make your acquaintance. I am not allowed to tell you who they are, so be on your guard!

I hope you have handed on our greetings to your mother and our other relatives, and for myself express the wish that you are in good health so that we can make a show of our son-in-law and son in front of the others.

With warm affection
Your old future father-in-law
Freud

[1] In the Dolomites.
[2] Resort on the Adriatic where Sophie stayed with her aunt Minna.

155 *To* MAX HALBERSTADT

Karlsbad
July 27, 1912

Dear Max

There is no doubt that we don't yet know you properly. Who would have thought that you are such an industrious correspondent! We had been told the opposite. In this respect you are perfectly suited for a protracted engagement. In others admittedly less. I thought we had decided that the engagement should be published simultaneously in Vienna and Hamburg on the twenty-eighth. Now you haven't been able to wait and Vienna has to limp behind.

There is no reason to admire us for the four (actually 4½) years. There was no merit in that; we just couldn't help it and had nothing to show for the engagement except plenty of impecunious relatives. I was not yet the possessor of five high distinctions as you are, but had to produce everything from scratch; admittedly I was only twenty-five and when I married not younger than you are today. You are quite right in not wanting to follow our example. I have really got along very well with my wife; I am thankful to her above all for her many noble qualities, for the children who have turned out so well, and for the fact that she has neither been very abnormal nor very often ill. I hope that you will be equally fortunate in your marriage, and that the little shrew will make a good wife.

Today your aunt announced her visit to us; we have arranged in writing a first meeting tomorrow morning (after appeasing our early appetite). The engagement correspondence has followed right on the heels of the birthday letters. On July 26[1] next year you will join in the celebration as Mama's latest acquisition.

The two issues of the newspaper have arrived. I spell my name Sigm. without an *e*, but there is no doubt about the identity.

Cordial greetings, also from my wife.

Your new and old
father-in-law
Freud

[1] Martha Freud's birthday.

156 *To* MARTHA FREUD

Rome
September 20, 1912

My beloved Old Dear

Just received with great satisfaction your first letter from Vienna, and am very pleased by its news, especially the change in Mathilde's condition, although I wasn't really worried about her. Do try and persuade her to pass some of her temperature onto the air outside.

By the same mail I received an offer from an English publisher to have *The Interpretation of Dreams* translated (this is already the third offer I have had to refuse)[1] and the announcement of an urgent case, a woman from Cracow. There is no doubt that we shall again have enough to live on so long as I can work.

Rome was certainly the best choice for me. I enjoy it more than ever, probably because this hotel is so beautifully situated. My plan for old age is made: not Cottage,[2] but Rome. You and Minna will like it, too.

I can't give you a better idea of the beauty of the weather, the sun, wind and fresh air, than by reminding you of S. Cristoforo.[3] For comfort this hotel can be compared with that in Klobenstein.[4] Ferenczi has not left me; in fact, since I am all right again he has become a stimulating and understanding companion. Affectionate he always is.

My trouble is not entirely gone, it recurs every two days, but it is infinitely better than it was. Last night after dinner we even went to the theater to see a new patriotic musical comedy. This was a bit too much for me; perhaps it was also the coffee in the entr'acte which didn't agree with me. But now—just before lunch—I am all right again. I must confess I have never taken such care of myself and lived so idly, giving in to every wish and whim. Today I even found and bought a gardenia, the scent of which has put me in the best of moods. Minna knows the flower; it is even nobler than the camellia.

So I hope to return in a refreshed and productive condition; I really have no talent for being seedy. But don't be surprised if I follow Ernst's example and dismiss the idea of curtailing my trip; "I will stay"[5]—so long as the money lasts.

Fond greetings to all our young and old.

Your

Sigm.

---

[1] Freud had already authorised A. A. Brill to undertake the translation.

[2] The name of a suburban district of Vienna.

[3] In South Tyrol.

[4] Near Bolzano.

[5] Allusion to an utterance of Freud's youngest son Ernst (born 1892), often quoted in the family circle. Ernst refused to leave Lovrana at the end of the summer vacation in 1894.

157   *To* MARTHA FREUD

Rome
September 25, 1912

My dearest
Just received the letter announcing the return of our young
gentlemen and the requests of two patients. Frau Dr. M. is looking
for a reliable nursemaid in Vienna. Perhaps you can help her.

Yesterday Ferenczi went to Naples; he is returning on Friday
morning and will then accompany me as far as Udine. I am enjoying a delicious, somewhat melancholy solitude, walk a lot in the
heavenly weather, on the Palatine among the ruins, in the Villa
Borghese, an enormous park but quite Roman, and every day I pay
a visit to Moses in S. Pietro in Vincoli, on whom I may perhaps
write a few words.[1] Feeling completely well and sleeping soundly,
I am at last looking forward to coming home and to work.

Today I made a few little purchases, I hope to everyone's satisfaction. In this respect Rome is even more dangerous than Munich.
I sport a gardenia every day and act the rich man who lives only
for his whims. Life will be serious soon enough. In any case it was
wonderful while it lasted.

I hope to find you all in good health and send you my best
greetings.

Papa

[1] "Der Moses von Michelangelo," Vienna, 1914. ("The Moses of Michelangelo,"
Standard Ed., 13, 211.)

158   *To* ANNA FREUD

Vienna IX, Berggasse 19
November 28, 1912

My dear Anna
I haven't been able to write to you before because since
your departure, which already seems ages ago, life here has been
hectic; and the Sunday in Munich[1]—with the night journeys before

[1] Freud had attended a meeting of psychoanalysts in Munich.

and after, conversations in between lasting from 9 A.M. till 11:40
P.M.—wasn't exactly a rest cure. In any case I know the ladies of
the house keep you regularly informed about anything worth
hearing. Today's letter, however, is meant as a birthday congratulation. As you know, I am always premature on such happy occasions. Today I gave Heller[2] an order which I trust will arrive in
time and be what you want. Your monthly allowance will reach
you in your new home via the post office savings bank.

I have no doubt at all that you will put on more weight and feel
better once you have grown accustomed to idleness and the sunshine. You might as well abandon the knitting until after the
wedding;[3] it probably isn't very good for your back. Otherwise
keep well and enjoy all that is offered you by the winter in Meran
and the care of your sister-in-law,[4] Frau Marie.

I don't think I shall be going away for Christmas; as a matter of
fact I am expecting Dr. Abraham to visit me here. As you know,
I am no longer master of my free time, a condition, however, that
I quite enjoy.

I don't think you have seen my room since it was refurnished,
or have you? It has turned out very well. Before you return for
good we will do your room too; writing table and carpet are in
any case assured.

I send you fond greetings, wish you everything good for your
seventeenth (hard to believe that I too was once so young!) and
please give my kind regards to Frau Marie and Edith.[5]

Your

Papa

[2] Hugo Heller, Viennese bookseller and publisher.
[3] Of Sophie Freud and Max Halberstadt.
[4] Marie Rischawy, sister-in-law of Mathilde Hollitscher (née Freud).
[5] Edith Rischawy, Frau Marie's daughter.

159 *To* ANNA FREUD

Vienna IX, Berggasse 19
December 13, 1912

My dear little Anna

I hear you are already worrying again about your immediate
future. So it seems that putting on 3¼ pounds still hasn't changed

294

you much. I now want to set your mind at rest by reminding you
that the original plan was to send you to Italy for eight months in
the hope that you would return straight and plump and at the
same time quite worldly and sensible. Actually we hadn't dared to
hope that a few weeks in Meran would achieve this transformation,
and so had already prepared ourselves at your departure for not
seeing you at the wedding or so soon afterwards in Vienna. I think
you must now slowly accustom yourself to this terrible prospect.
The ceremony can be performed quite well without you, for that
matter also without guests, parties, etc., which you don't care for
anyhow. Your plans for school can easily wait till you have learned
to take your duties less seriously. They won't run away from you.
It can only do you good to be a little happy-go-lucky and enjoy having such lovely sun in the middle of winter.

So now, if you are reassured that your stay in Meran won't be interrupted in the immediate future, I can tell you that we all enjoy
your letters very much but that we also won't be worried if you
feel too lazy to write every day. The time of toil and trouble will
come for you too, but you are still quite young.

Give my kind regards to Frau Marie and Edith and feel as well
and happy as

Your Father
wants you to be.

160 *To* C. G. JUNG

Vienna IX, Berggasse 19
December 22, 1912[1]

Dear Dr. Jung
The reason why the Viennese local branch rejected the
change of title was mainly that the announcements, proofs, etc.,
had already been printed—i.e., sent out—and consideration for the
publisher made the alteration inadvisable. It was actually too late
for these changes. I hope that the whole business, being of little
importance, won't cause any trouble. *Therapeutic* was hardly a
good substitute; the pedagogues will discover that the new period-

---

[1] This letter, apparently not sent, was found among Freud's papers.

ical[2] is going to exclude their collaboration as little as did the old one.—

I regret that my reference to your slip of the pen has irritated you so much, and I feel that your reaction is out of proportion to the occasion. Your reproach that I abuse psychoanalysis for the purpose of keeping my pupils in infantile dependency and that therefore I myself am responsible for their infantile behavior toward me, as well as everything else that you base on this assumption, I won't judge, because all judgment concerning oneself is so difficult and doesn't carry conviction. I will just furnish you with some factual material for the basis of your theory and leave it to you to revise it. In Vienna, for instance, I am accustomed to hearing the opposite reproach—i.e., that I concern myself too little with the analysis of my "pupils." In actual fact, Stekel,[3] for example, in all the ten years since he ceased to be treated by me, has not heard another word from me relating to the analysis of his own person, and in the case of Adler[4] I have avoided it even more carefully. Whatever analytical comment I made about these two men was uttered to others and chiefly at a time when they were no longer in contact with me. So I don't know why you feel so sure in assuming the opposite.

With cordial greetings

Your

Freud

[2] See note 2 to letter of Sept. 19, 1907.
[3] See List of Addressees and letter of Jan. 13, 1924.
[4] Alfred Adler (1870-1937), follower of Freud, who later founded his own school.

1913

———————————————

1924

161 *To* JAMES J. PUTNAM

Vienna IX, Berggasse 19
January 1, 1913

Dear Colleague

I take pleasure in embarking on the activities of the New Year with a letter to you. My warmest thanks for declaring your intention to contribute to our periodical. I am equally grateful because your words give me reason to believe that you have not lost faith in me despite the many personal attacks to which I am exposed at the moment and which will probably continue for some time. You will also believe me when I say that I am not very upset by these attacks because I understand too well the psychological necessity for such occurrences. I am thinking not so much of Stekel, the loss of whom was actually a gain, but chiefly of Jung, whom I greatly overestimated and for whom I felt considerable personal affection.

Scientific differences are after all inevitable in the development of a new science and even errors, as I have experienced in myself, have many aspects that are beneficial. But that such deviations and reforms of a theoretical character have to go hand in hand with so much wounding of justified personal feelings does little credit to human nature.

That I look upon Jung's new views as "regressive" errors goes without saying, but this won't necessarily be convincing to others. In cases of this kind everyone should consult his own experiences and the impression made on him by the arguments. On me all this has the effect of a *déjà-vu*. I have already experienced in the resistance of the nonanalyst what is now repeating itself in the resistance of the half-analyst.

We are looking forward to your forthcoming papers with great

299

interest. Please don't let our philosophical ignorance interfere with your production or publication. Although we are people not competent in this field because we work in others, we listen to you piously nevertheless, and we will be followed by other and freer analysts for whom your ideas may prove very fruitful.

With cordial good wishes for your well-being and that of your family in the New Year

Your faithful
Freud

162 *To* MAX EITINGON

Vienna IX, Berggasse 19
January 7, 1913

Dear Dr. Eitingon

If it is not too late I want to thank both you and your fiancée, of whom I approve simply because you have chosen her, for your New Year wishes. I can use them very well this year, for all the evil spirits have been let loose against me, but I have been familiar with them for years and am not greatly afraid of them. Needless to say, it is the intention behind the wishes that matters to me, and in this respect I know I am safe with you. You were the first emissary to reach the lonely man, and if I should ever be deserted again you will surely be among the last to remain with me. . . .

I continue to work undeterred for the applause of the few who want to understand me. If you hear that I have grown old, weak, and neurotic, don't believe it, and do convince yourself again in person (with your fiancée or wife) of the contrary.

With best wishes for 1913
Your faithful
Freud

163 *To* SANDOR FERENCZI

Vienna IX, Berggasse 19
July 9, 1913

Dear Friend

Is it possible? Am I already congratulating you on your fortieth birthday? Your nostalgic letter moved me very much, first because it reminded me of my own fortieth birthday, since when I have changed my skin several times, which, as we know, occurs every seven years. At that time (1896) I had reached the peak of loneliness, had lost all my old friends and hadn't acquired any new ones; no one paid any attention to me, and the only thing that kept me going was a bit of defiance and the beginning of *The Interpretation of Dreams*. But when I look at you, I can't help thinking of you as luckier in many respects, even if the congratulations haven't become any louder. You stand there firmly established, the road ahead is clear, you are highly respected by an unusually select circle of friends whose leader you are destined to become. One possession[1] of which I felt sure at your age you haven't yet acquired, and I know that one misses most keenly and appreciates most highly what one hasn't been able to attain.

On your fortieth birthday I think I may drop my reserve and confess to you that the only reason why I didn't advise you strongly against D. was that I was afraid that according to the pattern of neurotic behavior, you would insist all the more. What are you planning to do now? For each of us fate assumes the form of one (or several) women, and your fate bears several unusually precious features.

I think you know that this year I have put off the Emdens,[2] charming company as they are, in order to spend several weeks in Marienbad free of analysis. My closest companion will be my little daughter,[3] who is developing very well at the moment (you will long ago have guessed the subjective condition for the "Choice of the Three Caskets."[4] I hope to be fit enough in San Martino[5] to enjoy your company better than last year when everything and everyone were too much for me. Abraham is to visit us at the end of August and will probably accompany us to Munich.[6] Your hope of my being able to tell you something new by then I cannot as yet support. My good ideas actually occur in seven-year cycles: in 1891 I started working on aphasia; 1898/9, the interpretation of dreams; 1904/5, wit and its relation to the unconscious; 1911/12, totem

[1] That of a wife.

[2] The Dutch psychoanalyst Dr. Jan van Emden (1868-1950) and his wife.

[3] Anna.

[4] "Das Motiv der Kästchenwahl," Vienna, 1913. ("The Theme of the Three Caskets," Standard Ed., 12.)

[5] In the Dolomites.

[6] The Fourth Psychoanalytical Congress was held in Munich.

and taboo; thus I am probably in the waning stage and won't be able to count on anything of importance before 1918/19 (provided the chain doesn't break before). . . .

Let us carry on our work in calm self-confidence. That assurance that the children will be provided for,[7] which for a Jewish father is a matter of life and death, I expected to get from Jung; I am glad now that you and our friends[8] will give me this.

With hearty cheers for the next two-thirds of your individual existence.

In sincere friendship
Freud

[7] Here Freud is thinking not of his own children, but of the future of the psychoanalytical movement, the product of his mind.

[8] A group of leading analysts, the Committee: Abraham, Eitingon, Ferenczi, Jones, Rank, and Sachs. (See letter of Nov. 11, 1928.)

164 *To* KARL ABRAHAM

Eden Hotel, Rome
September 21, 1913

Dear Friend

Many thanks for your friendly words and good news, especially for your promise to serve our periodical, which we will now have to run entirely on our own means.

I have quickly recovered my spirits and zest for work in the incomparably beautiful Rome, and in the free hours between visits to museums, churches, and the Campagna I have managed to write an introduction to the book about totem and taboo, an extension of the lecture given at the Congress; and a draft of an essay on narcissism;[1] in addition I have corrected the proofs of my propaganda article for *Scientia*.[2] My sister-in-law, who warmly returns your and your wife's greetings, sees to it that the actual task of exploring Rome is kept within moderate bounds. She herself has taken all the inevitable exertions unexpectedly well in her

[1] "Zur Einführung des Narzissmus," Vienna, 1914. ("On Narcissism: an Introduction," Standard Ed., 14.)

[2] "Das Interesse an der Psychoanalyse," published in German in *Scientia* (Italian scientific periodical, published in Bologna), 1913. ("The Claims of Psychoanalysis to Scientific Interest," Standard Ed. 13, 165.)

stride and it is pleasant to watch her feeling more at home and growing more enthusiastic about Rome every day.

Yesterday I received a belated letter of admiration and reassurance from Mäder[3] with an added: "Here I stand. I can do no other"[4] (an appropriate remark for someone taking a risk, but hardly for someone withdrawing from it). He will receive a cool, not very exhaustive answer.

In a week from now, alas, the whole Roman décor will be laid aside and replaced by a more sober and familiar one.

With warm greetings and in the hope that I shall continue to have good news of you and your family

Your faithful
Freud

[3] Alphonse Mäder, Swiss psychoanalyst.
[4] Quoting Martin Luther.

165 *To* ELISE GOMPERZ

Vienna IX, Berggasse 19
November 12, 1913

Dear Frau *Hofrat*

Please accept my sincere thanks for your efforts to let me have the paper[1] I wanted and which in the meantime I have received as a present from your son. It would actually have been sufficient if you had pointed out to me where this short essay was reprinted.

The little notebook containing the handwriting of your unforgotten husband reminded me of that time lying so far behind us, when I, young and timid, was allowed for the first time to exchange a few words with one of the great men in the realm of thought. It was soon after this that I heard from him the first remarks about the role played by dreams in the psychic life of primitive men—something that has preoccupied me so intensively ever since.

As ever, respectfully yours
Freud

[1] Theodor Gomperz's paper on Plato. When still a student Freud had translated into German one volume of John Stuart Mill's works, edited by Gomperz. It contained "Enfranchisement of Women," "Plato," "The Labor Movement," "Socialism." The book was published in December, 1880. (See note 5 to letter of Nov. 15, 1883.)

166 *To* STANLEY HALL

Vienna IX, Berggasse 19
November 23, 1913

Dear Mr. Hall

The few years that have elapsed since my visit to your house and your university[1] have not lessened my gratitude for your hospitality. A letter from you represents for me the most welcome reminder of that significant time.

I am glad that your interest in our work has remained undiminished. You probably realize even from afar how much everything is in a state of ferment and transition, and yet the trend toward progress has on the whole been maintained, despite all deviations from the path. I will continue making it my business to send you everything that has been published on our side.

That it is just the question of sexual symbolism to which you take exception does not worry me. You will surely have observed that psychoanalysis creates few new concepts in this field, rather it takes up long-established ideas, makes use of them, and supports them with a great deal of evidence. Possible exaggerations will wear off, but most of it I believe will stand the test of time.

The only unfavorable developments within the psychoanalytical movement concern personal relationships. Jung, with whom I shared my visit to you at that time, is no longer my friend, and our collaboration is approaching complete dissolution. Such changes are regrettable but inevitable.

In the hope that you and Mrs. Hall are enjoying continued good health, I beg to remain

gratefully and sincerely yours
Freud

---

[1] Clark University, Worcester, Mass., where Freud had given a course of lectures in 1909, and was made a Doctor *honoris causa.* These lectures were published in German under the title: "Über Psychoanalyse," Vienna, 1910. ("Five Lectures on Psychoanalysis," Standard Ed., 11.)

167 *To* HERBERT AND LOE JONES

(Undated [1914])[1]

My dear Friends

These wretched times, this war, which impoverishes us as much in spiritual as in material goods, have prevented me from thanking you earlier for the clever and practical fashion in which you returned my little daughter[2] and for all the friendship that lies behind it. She is very well, but I suspect she sometimes pines for the country of our enemies.

I was very pleased to hear via an obvious channel that all is well with you and that you are about to move into a new home, which I would like to bless with the most heartfelt good wishes for you both. Till we meet again, God alone knows when!

Your faithful

(*Unsigned*)

[1] This letter is without date or signature, having been sent during the first weeks of World War I by way of a neutral country.

[2] The outbreak of war had found Anna in England where she was visiting friends. She returned home with the Austrian ambassador to England.

168 *To* HERMANN STRUCK

Vienna IX, Berggasse 19
November 7, 1914

Dear Sir

I have noticed with admiration how seriously you take the work on my portrait. Probably this is the way you always work.

I am looking forward very much to your promised comments on "Moses" [of Michelangelo]. Since they haven't yet arrived, I don't want to go on postponing the return of the proof impressions and the writing of this letter. I hasten to tell you, even before hearing your comment, that I am well aware of the cardinal weakness of this work of mine. It lies in the attempt to assess the artist in a rational way as though he were a scholar or technician, whereas he is actually a being of a special kind, exalted, autocratic, villainous, and at times rather incomprehensible.

A little book on Leonardo da Vinci[1] which I have written is not likely to be to your taste. It takes for granted that the reader is not shocked by homosexual topics and that he is fairly familiar with the devious ways of psychoanalysis. As a matter of fact it is also partly fiction. I wouldn't like you to judge the trustworthiness of our other discoveries by this example. I am sending you at the same time under separate cover a trifle[2] which comes at least close to another art.

And now I will start on the critical comments (incorporating those of my wife and others) which you asked me for. They are made possible only by your remark that we may rest assured of our incompetence. Otherwise I wouldn't dare to offer any criticism. I myself am without doubt the most incompetent of all. But if you wish to hear my opinion, I hope you will accept the following remarks in good grace. The etching strikes me as a charming idealization. This is how I should like to look, and I may even be on the way there, but it seems to me I have got stuck halfway. Everything that is shaggy and angular about me you have made smooth and rounded. In my opinion an element of unlikeness has been introduced by something unimportant, by your treatment of my hair. You have put my parting on one side, whereas, according to the lithograph, I wear it on the other. Furthermore, my hairline runs across the temple in a rather concave curve. By rounding it off you have greatly improved upon it. Very likely this correction was intentional. In a word, I feel the etching to be a great honor. Each time I look at it I like it better.

My relations with the lithograph are less friendly. Whatever is Jewish about the head has my full agreement, but something else has struck me as alien. I have come to the conclusion that it is the exaggerated opening of the mouth, the stretching forward of the beard, and the prominence of its outer contour. In trying to discover where these features could come from, I remembered the beautiful, malicious orchid, the *Orchibestia karlsbadiensis,* which we shared. This would produce a hybrid phenomenon (as it is called in *The Interpretation of Dreams*) of Jew and orchid! Now I have come to the end of my remarks and wish to recommend myself once more to your mercy. I will be very grateful to

[1] See note 3 to letter of Oct. 1, 1911.
[2] Probably Freud's essay on Jensen's *Gradiva.* (See footnote 1 to letter of May 26, 1907.)

receive the completed prints—for the time being a rather ineffective gratitude, I'm afraid.

Perhaps we would have made better use of each other's company in Karlsbad had we guessed its end would be so sudden and its repetition so unlikely.

My wife asks me to send you her best greetings to which I add mine.

Yours very sincerely
Freud

169 *To* JAMES J. PUTNAM

Vienna IX, Berggasse 19
July 8, 1915

Dear Friend

Your book *Human Motives* has arrived at last, long after it was announced. I have not yet finished reading it, but I have read what are for me the most relevant sections on religion and psychoanalysis and yield to the urge to write to you about them.

You will surely not demand praise and recognition from me. It pleases me to think that it will make an impression on your compatriots and break down the deeply rooted resistance of many of them.

On page 20 I found the passage I must consider as most applicable to myself: "To accustom ourselves to the study of immaturity and childhood before . . . undesirable limitation of our vision, . . ." etc.

This I recognize as my case. I am certainly incapable of judging the other aspect of the subject. I must have needed this one-sidedness in order to see what remains hidden from others. That is the justification of my defensive reaction. The one-sidedness did have its own usefulness, after all.

On the other hand, that the arguments for the reality of our ideals failed to make an impression on me is less important. I cannot see the connection between our ideas of perfection having a psychical reality and the belief that they have a material existence. This will not surprise you. You know, after all, how little one can expect from arguments. I should add that I stand in no awe whatever of the Almighty. If we were ever to meet I should have more reproaches to make to Him than He could to me. I would ask Him why He hadn't endowed me with a better intellectual equipment, and He couldn't complain that I have failed to make the best use of my so-called freedom. (By the way, I know that every individual represents a chunk of life energy, but I don't see what energy has to do with freedom—i.e., not being conditioned by circumstances.)

I think I ought to tell you that I have always been dissatisfied with my intellectual endowment and that I know precisely in what respects, but that I consider myself a very moral human being who can subscribe to Th. Vischer's excellent maxim: "What is moral is self-evident." I believe that when it comes to a sense of justice and consideration for others, to the dislike of making others suffer or taking advantage of them, I can measure myself with the best people I have known. I have never done anything mean or malicious, nor have I felt any temptation to do so, with the result that I am not in the least proud of it. I am taking the notion of morality in its social, not its sexual, sense. Sexual morality as defined by society, in its most extreme form that of America, strikes me as very contemptible. I stand for an infinitely freer sexual life, although I myself have made very little use of such freedom. Only so far as I considered myself entitled to.

The emphasis placed on moral laws in public life often makes me feel uncomfortable. What I have seen of religious-ethical conversions has not been very attractive. . . .

On one point, however, I see that I can agree with you. When I ask myself why I have always aspired to behave honorably, to spare others and to be kind wherever possible, and why I didn't cease doing so when I realized that in this way one comes to harm and becomes an anvil because other people are brutal and unreliable, then indeed I have no answer. Sensible this certainly was not. In my youth I didn't feel any special ethical aspirations, nor does the conclusion that I am better than others give me any recognizable satisfaction! You are perhaps the first person to whom I have boasted in this fashion. So one could cite just my case as a proof of your assertion that such an urge toward the ideal forms a considerable part of our inheritance. If only more of this precious inheritance could be found in other human beings! I secretly believe that if one had the means of studying the sublimation of instincts as thoroughly as their repression, one might find quite natural psychological explanations which would render your humanitarian assumption unnecessary. But as I have said before, I know nothing about this. Why I—and incidentally my six adult children as well—have to be thoroughly decent human beings is quite incomprehensible to me. Which leads to another reflection: if the knowledge of the human soul is still so incomplete that my poor mental faculties have managed to produce such ample discoveries, it is evidently premature to declare oneself for or against such assumptions as yours.

Permit me to correct yet another minor error which is quite irrelevant to world history: I was never Breuer's assistant, never saw his famous first case, and only learned of it years later from Breuer's report. This historical error seems to be the only one that I have detected in your work. Everything else you say about psychoanalysis I can subscribe to without sacrificing anything. For the time being psychoanalysis is compatible with various *Weltanschauungen*. But has it yet spoken its last word? For my part I have never been concerned with any comprehensive synthesis, but invariably with certainty alone. And it is worth sacrificing everything to the latter.

With cordial greetings and best wishes for lasting health and zest for work. I myself am using the break in my work at this time to finish off a book containing a collection of twelve psychological essays.[1]

Yours very sincerely
Freud

[1] "Triebe und Triebschicksale," Vienna, 1915. ("Instincts and their Vicissitudes," Standard Ed., 14.)

170 *To* LOU ANDREAS-SALOMÉ

Rudolfshof, Karlsbad
July 30, 1915

My dear Frau Lou

I am writing from an idyll which we, my wife and I, have defiantly and stubbornly created for ourselves but which is continually being interrupted by the demands of the times. About a week ago our eldest son wrote that a bullet had gone through his cap and another had grazed his arm, neither of which, however, had interfered with his activities. And today the other warrior[1] announces that he has received his marching orders for tomorrow, also northwards. My little daughter, whom you may remember and who is staying with her eighty-year-old grandmother[2] in Ischl,[3] has written to us in concern: "How am I going to take the place of six children all by myself next year?" Since we don't dare to look into the future, we just live for the day and try to get out of it what it is willing to yield.

Your letters are now a doubly precious reward for my dispatches. I say now, for I am almost entirely alone; of all collaborators I see only Ferenczi resisting the military influence and sticking to the group. But since he too is tied to his garrison in Pápa,[4] I often feel as alone as during the first ten years when I was surrounded by a desert, but I was younger then and still endowed with an infinite capacity for holding out. Fruit of the present time will probably take the form of a book consisting of twelve essays beginning with one on instincts and their vicissitudes. But I seem to remember that I have already told you about this. The book is finished except for the necessary revision caused by the arranging and fitting in of the individual essays.

Every time I read one of your discerning letters I am surprised by your talent for going beyond what has been said, for completing it and making it converge at some distant point. Needless to say, I don't go along all the way. I so rarely feel the need for synthesis. The unity of this world seems to me something self-understood, something unworthy of emphasis. What interests me is the separation and breaking up into its component parts what would otherwise flow together into a primeval pulp. Even the assurance most clearly expressed in Grabbe's[5] *Hannibal* that "We will not fall out of this world" doesn't seem sufficient substitute for the surrender of the boundaries of the ego, which can be painful enough. In short, I am evidently an analyst and believe that synthesis offers no obstacles once analysis has been achieved.

From the same point of view I also object to your justification of

---

[1] Ernst, who, like Martin, fought on the Russian front.
[2] Freud's mother.
[3] A spa in the Salzkammergut.
[4] Small town in Hungary.
[5] Christian Dietrich Grabbe (1801-1836), German dramatist.

the "desire to kill," if it is supposed to be one. One must not under-estimate the role played by unpleasure and its potential function as a barrier.

Your letter also contains a precious promise. I would very much like to read "Anal and Sexual,"[6] and if our periodicals can hold out I will see that it gets printed. But how and where should the manu-script be sent? I shall be staying here till about August 10; anything sent to Vienna, where my house will be open, is bound to reach me. It has been pointed out to me that all written material sent through the mail is subject to strict censorship. I hope, nevertheless, that it will reach me.

I wonder when all of us dispersed members of an unpolitical community will meet again and whether it will turn out that politics has corrupted us. I cannot be an optimist and I believe I differ from the pessimists only in so far as wicked, stupid, sense-less things don't upset me because I have accepted them from the beginning as part of what the world is composed of. My friend Putnam maintained in a recent book which is based on psycho-analysis that perfection has not only a psychic but also a material reality. That man can't be helped, he must become a pessimist!

I hope things will continue to go well for you in these difficult times and that you will remember me even if I don't have anything to send you.

Yours very sincerely
Freud

[6] Lou Andreas-Salomé's "Anal and Sexual" was published in *Imago*, 1915/16.

171 *To* EDUARD HITSCHMANN

Vienna IX, Berggasse 19
May 7, 1916

Dear Doctor

Only a funeral oration at the Central Cemetery is normally as beautiful and affectionate as the speech you did not deliver. I realized that you are able to express things well—you have often proved this in publications—but on this occasion I am downright moved, probably because I myself am the subject. No doubt I have meant to be and to do everything you say about me, but will it be possible in soberer hours to maintain that I succeeded? I don't know, but what I do know is that in order to live one needs a few people who believe it.

So please accept my warmest thanks for your words of appreciation and devotion, which go far beyond compensating me for my usual disappointment in human beings. I am not bitter and know that I have no reason to be; I am grateful for all the good things that come my way. I hope you will remain at my side in your endeavor to further our science. And let us continue the friendly interest in each other's personal fate.

With warm greetings to yourself and your dear wife
Your faithful
Freud

172 *To* LOU ANDREAS-SALOMÉ

Vienna IX, Berggasse 19
May 25, 1916

My dear Frau Lou

I cannot believe there is any danger of your misunderstanding any of our assertions; if so it must be our, in this case my, fault. After all, you are an "understander" *par excellence;* and in addition you invariably understand more and improve upon what is put before you. I am always especially impressed when I read your comment on one of my papers. I know that in writing I have to blind myself artificially in order to focus all the light on one dark spot, renouncing cohesion, harmony, edifying effects and everything which you call the symbolic element, frightened as I am by the experience that any such claim, such expectation, carries within it the danger of distorting the truth, even though it may embellish it. Then you come along and add what is missing, build upon it, putting what has been isolated back into its proper context. I cannot always follow you, for my eyes, adapted as they are to the dark, probably can't stand strong light or an extensive range of vision. But I haven't become so much of a mole as to be incapable of enjoying the suggestion of something brighter and more comprehensive, or even to deny its existence.

Your card, nevertheless, did bring me one minor disappointment.

I was under the impression that the writing of your essay was finished and wouldn't keep us waiting much longer. I implore you not to put it off and not to give me precedence in time. My book containing twelve essays of this kind cannot be published before the end of the war, and who knows how long after this ardently longed-for date? Spans of life are unpredictable and I would so much like to be able to have read your contribution before it is too late. But should you be referring to my lectures,[1] they contain absolutely nothing that could tell you anything new.

Today I received the first galleys of your "Anal and Sexual."

With many hearty greetings

Your

Freud

[1] *Vorlesungen zur Einführung in die Psychoanalyse,* Vienna 1916/17. (*Introductory Lectures on Psychoanalysis,* revised ed., London, 1929; *A General Introduction to Psychoanalysis,* New York, 1935. Standard Ed., 15-16.)

173 *To* LOU ANDREAS-SALOMÉ

Hotel Bristol, Salzburg

July 27, 1916

My dear Frau Lou

I hasten to tell you that in view of the difficulties of this summer we have settled here in the hotel for a prolonged stay. The stimulations and comforts of the beautiful town, the guarantee of the postal connections and the food supply, have caused us to renounce the otherwise so greatly needed contact with the country. At the moment we are looking forward to the furlough of our son Ernst, in whom you observed a resemblance to R. M. Rilke.[1] The latter, whom I should like to congratulate on having regained his freedom[2] as a poet, has made it quite clear to us in Vienna that "no lasting alliance can be forged with him."[3] Cordial as he was on his first visit, it was impossible to persuade him to pay a second one.

As soon as I get back I will dispatch the books you ask for (Holt's

[1] Rainer Marie Rilke (1875-1926).
[2] Rilke had just been released from military service.
[3] Quotation from Schiller's "Das Lied von der Glocke."

The Freudian Wish and Putnam in the *Psychoanalytical Review*), unless by that time (end of September) you let me know you have read them already or—that you will make the journey instead.

I was of course very pleased by your friendly reception of Part I of the lectures. I expect, however, that in Part II, the one on the dream, you will soon find that I have not succeeded in maintaining this kind of preparatory, indirect approach which educates rather than lectures the reader. While here I have started to draft the series of lectures on the theory of neuroses and have already completed the first lecture. I would have felt safe throughout my life if my production had remained predictable at all times and in all conditions. Unfortunately this has never been the case. There have always been days in between when everything has refused to function and when I have been in danger of losing all ability to work and to struggle, owing to certain minor fluctuations in mood and physical health. A most unsuitable constitution for a man who is no artist and doesn't aim at being one.

Among all my books the *Everyday Life*[4] is making the best career for itself. At the moment I am preparing the fifth impression and in front of me lies a Dutch edition published by J. Stärcke,[5] which looks very presentable.

My curiosity continues to be aroused by a small work on psychoanalysis which has been announced to me from a certain quarter.

With cordial wishes for your well-being in these "trying"[*] times

Yours sincerely
Freud

[4] See footnote 4 to letter of Nov. 25, 1901.
[5] Dr. Johann Stärcke, Dutch psychoanalyst.
[*] Written in English.

174 *To* JOSEF POPPER-LYNKEUS

Hotel Bristol, Salzburg
August 4, 1916

My dear Sir

I received from you on the first of the month a consignment with which I felt very pleased and honored. I waited to see whether it would be followed by an explanatory letter, but since this did not arrive—one can't trust the mail nowadays—I don't want to postpone thanking you any longer and asking how you came upon the little book by Straus.[1]

As a reader of the *Vossische Zeitung* I was already familiar with your beautiful, envy-arousing appreciation of your deceased friend, Mach.[2] Owing to my narrower point of view I have unfortunately never been able to appreciate his work and had to consider his way of treating psychic phenomena as unpsychological. The physicist and psychology don't go together very well. I remember my surprise years ago when I discovered that you (as the only one) recognized that the distortion of dreams is the result of censorship ("Dreaming like Waking" in Lynkeus).[3] The all but hundred-year-old dissertation by Dr. Heinrich Straus is indeed very remarkable. It contains several things with which a former friend of mine, W. Fliess of Berlin, used to be very much preoccupied. By his own observation the latter revived many of the assertions about the rhythm of vital phenomena, and added the considerable discovery that two such rhythms exist, a masculine one of 23 days and a feminine one of 28, and even after that friendship came to an end I retained some faith in this idea. Furthermore I was pleased to find in the little pamphlet remarks that touch my own sphere of interest—for instance, that human beings try to resume the embryonic position in sleep, the comparison between hibernation and the migration of birds for which the phylogenetic derivation can easily be produced.

With thanks and best wishes for your good health

Yours very sincerely

Sigm. Freud

[1] An early eighteenth-century thesis on dreams.

[2] Ernst Mach (1838-1916), German philosopher and physicist.

[3] From "Phantasien eines Realisten" ("Daydreams of a Realist"), published in 1899 under the pen name Lynkeus.

175 *To* SOPHIE AND MAX HALBERSTADT

Vienna IX, Berggasse 19
September 18, 1916

My dear Children

Good, it shall be as you wish. Of course I am enormously pleased about the promised visit and for this very reason don't wish to exercise any undue influence. Perhaps Mama takes the food problem too seriously. I am sure it will work out all right; we will do all we can and Sophie will no doubt realize that at other times we would receive her differently.

You mustn't take your insecurities too seriously. This is the way things are nowadays and some people's fate at the present moment is even more checkered. You are young; for you it is just an episode.

I began consultations today, without much enthusiasm. It will be at least a week before I really get back into work. A little work is only upsetting.

Ernstl's[1] next installment can still be expected before the end of the year from the royalties in German marks on the fifth edition of the *Everyday Life*.

Let us keep cheerful and not take things too much to heart. Hindenburg has just said that our prospects are good.

With fond greetings to all three of you
Papa

[1] Elder son of Max and Sophie Halberstadt.

176 *To* GEORG GRODDECK

Vienna IX, Berggasse 19
June 5, 1917

Dear Colleague

It is a long time since I received a letter which has pleased and interested me so much; it also tempts me to replace in answering it the normal politeness due to a stranger with analytical candor.

I will do my best: I note that you urge me to confirm to you officially that you are not a psychoanalyst, that you don't belong to the flock of disciples, but that you may be allowed to consider yourself as something apart and independent. I evidently would be doing you a great favor by rejecting you and relegating you to the place where Adler, Jung and others stand. But this I cannot do; I must lay claim to you, must insist that you are an analyst of the first order who has grasped the essence of the matter once for all. The man who has recognized that transference and resistance are the hubs of treatment belongs irrevocably to the "Wild Hunt."

Whether he gives the "UCS"[1] the name of "Id" as well makes no difference. Let me show you that the notion of the UCS requires *no extension* to cover your experiences with organic diseases. In my essay on the UCS which you mention you will find an inconspicuous note: "An additional important prerogative of the UCS will be mentioned in another context." I will divulge to you what this note refers to: the assertion that the UCS exerts on somatic processes an influence of far greater plastic power than the conscious act ever can. My friend Ferenczi, who is familiar with this idea, has a paper on pathoneurosis[2] waiting to be printed in the *Internationale Zeitschrift;* it comes very close to your disclosures. The same point of view, moreover, has caused him to make for me a biological experiment to show how a consistent continuation of Lamarck's[3] theory of evolution coincides with the final outcome of psychoanalytical thinking. Your new observations harmonize so well with the reasoning of this work that we would be only too glad if we could refer to your already published paper when we are ready to go into print.

While I should very much like to welcome your collaboration with open arms, there is one thing that bothers me: that you have evidently succeeded so little in conquering that banal ambition which hankers after originality and priority. If you feel assured of the independence of your discoveries, why should you want to claim originality? And besides, can you be so sure on this point? After all, you must be ten to fifteen, possibly twenty years younger than I (1856). Is it not possible that you absorbed the leading ideas of psychoanalysis in a cryptomnemonic manner? Similar to the manner in which I was able to explain my own originality? Anyhow, what is the good of struggling for priority against an older generation?

I especially regret this point in your communication because experience has shown that a man with unbridled ambition is bound at some time to break away and, to the loss of science and his own development, become a crank.

I very much liked the samples of your observations which you offer and hope that even after severe critical sifting many of them will hold their own. Even though the whole field is not new to us,

---

[1] "Das Unbewusste," Vienna, 1915. ("The Unconscious," Standard Ed., 14.)

[2] A special form of neurosis connected with organic disorders. Paper published under the title "Von Krankheits -oder Pathoneurosen," Vienna, 1916.

[3] J. B. Antoine de Lamarck (1744-1829), French naturalist.

examples such as that of your blind man have so far never been given. And now for my second objection: why do you plunge from your excellent vantage point into mysticism, cancel the difference between psychological and physical phenomena, and commit yourself to philosophical theories that are not called for? Your experiences, after all, don't reach beyond the realization that the psychological factors play an unexpectedly important role also in the origin of organic diseases. But are these psychological factors *alone* responsible for the diseases, and do they call the difference between the psychic and the physical into question? To me it seems just as arbitrary to endow the whole of nature with a psyche as radically to deny that it has one at all. Let us grant to nature her infinite variety which rises from the inanimate to the organically animated, from the just physically alive to the spiritual. No doubt the UCS is the right mediator between the physical and mental, perhaps it is the long-sought-for "missing link." But just because we have recognized this at last, is that any reason for refusing to see anything else?

I am afraid that you are a philosopher as well and have the monistic tendency to disparage all the beautiful differences of nature in favor of tempting unity. But does this help to eliminate the differences?

I don't have to say that I should be very pleased to receive an answer from you! I am very anxious to know how you will greet this letter, which may sound far more unfriendly than is the intention behind it.

Yours faithfully
Freud

177 *To* LOU ANDREAS-SALOMÉ

Csorbató[1]
July 13, 1917

Dear Frau Lou

I have to disappoint you. I am not going to say "yes" or "no," nor shall I deal out question marks, but I shall do what I have always done with your comments: enjoy them and let them have their effect on me. It is quite evident that you anticipate and

---

[1] Resort in Hungary.

complement me each time, how you try in a visionary way to complete my fragments, build them into a structure. I am under the impression that this is true to a special degree since I began employing the concept of the narcissistic libido. Without this, I feel you too might have slipped away from me to the system-builders, to Jung, or rather to Adler. But through the ego-libido you have observed how I work, step by step, without the inner need for completion, continually under the pressure of the problems immediately on hand and taking infinite pains not to be diverted from the path. It seems that in this way I have gained your confidence.

If I should be in the position to continue building up this theory, you may perhaps recognize with satisfaction several new things as having been anticipated or even announced by yourself. But in spite of advancing age I am not in a hurry.—

I am sitting here in the Tatra, shivering. If there is anything like a cold paradise, this is it, but in paradise it must be warm, even rather hot, and the wind must blow warm, not as a cold storm which tries to carry off one's notepaper while one writes. And if one were going to write in paradise, certainly not in a *Lodenmantel*. The world blockade and the promise of sufficient food have driven us here. On the whole [we] are really quite comfortable. *Csorbató* means "Lake Csorba." The summits of the high Tatra gaze down on me threateningly while I dare to denounce weather and climate. Have I already told you that I have been suggested for the Nobel Prize? I don't think I shall live to see it, even if the postponement of its distribution should come to an end.

> With cordial greetings
> Your
> Freud

178 *To* MARIA MONTESSORI

> Vienna, IX, Berggasse 19
> December 20, 1917

My dear Frau Montessori

It gave me great pleasure to receive a letter from you. Since I have been preoccupied for years with the study of the child's psyche, I am in deep sympathy with your humanitarian and understanding endeavors, and my daughter, who is an analytical pedagogue, considers herself one of your disciples.

I would be very pleased to sign my name beside yours on the appeal for the foundation of a little institute as planned by Frau Schaxel.[1] The resistance my name may arouse among the public will have to be conquered by the brilliance that radiates from yours.

Yours very sincerely
Freud

[1] Now Mrs. Willy Hoffer, psychoanalyst in London.

179 *To* ALEXANDER FREUD

Vienna, IX, Berggasse 19
April 5, 1918

Dear Alex

Today I had a conversation with Dolfi[1] which causes me to make a suggestion about solving mother's difficulties. As you know, she is quite incapable of getting accustomed to the new value of money; she refuses to leave the financial arrangements to Dolfi, and torments her terribly every time money has to be spent. But Dolfi hasn't much energy left.

So I suggest that Dolfi, unknown to mother, be given a sum enabling her to keep the household going without our old lady realizing the expenses. The five hundred kronen which I sent Dolfi in March, in addition to the usual sum, have already been spent; I now suggest that we each contribute a thousand.

In any case we have to find summer lodgings for them not far from the city, where we can pay all the expenses so that they won't have to count every penny. You realize as well as I do that a summer will come when we won't have the chance of repeating the assistance.

Today I gave the bank orders to send you the quarterly contribution.

I greet you affectionately and await your reply.

Your
Sigm.

[1] Freud's sister Dolfi, who had remained unmarried and lived with her mother.

180 ANONYMOUS

Vienna, IX, Berggasse 19
April 28, 1918

Dear crazy young friend!

I assure you I understand you completely, to me you can write anything that comes into your head, and I am not so easily shocked. But don't you realize that you have shocked your patriarch so often that he no longer has any confidence in your perseverance, and is obliged to bother you with the command not to write to him so often? Why did you do that at the time? I know of course why, and even now you don't have to suppress such attempts at "extortion."

The dear Lord has endowed you with one of his most precious gifts, a delicious, fantastic humor. So why not put it in harness and let it pull your cart through the filthy present till we come once more to a cleaner street? You are young and you know one has to make an effort. After all, whether you learn to take the bit between your teeth now or later in your life makes no difference.

To your friendly questions I answer that we are all still alive, although naturally not very gay. I am told there are enough cigars till 1919 (end or beginning?). Food, they say, will return each year with the harvest, in the distribution of which you have a hand. It is very kind of you to express yourself in such affectionate terms. I realize, of course, that they belong to the very father of whom you complain in the same letter. I hope that once you are an ensign the whole situation will become more tolerable. So bear up.

With cordial greetings
Your father substitute
Freud

The address on your letter must have been inspired by the language-purifying devil. In civilian life it is called "professor and physician."

181 *To* ANTON AND ROZSI von FREUND

Budapest
August 9, 1918

My dear Hosts

We[1] arrived here late yesterday evening after a good journey, very dear friends took us by surprise on the ship, and others gave us a warm welcome on landing. From a first impression I can state that this invitation has once again been a friendly fraud. That something described as "primitive" turns out to be highly organized and fully sublimated would not be accepted in science. Of course we are so modest and adaptable that we will grow accustomed to these conditions, too.

I have simplified the writing table considerably. The picture of our hostess and little girl remains standing in the middle, on the right a bowl of fountain pens, on the left another filled with cigars. Whether science will greatly benefit from this display is still uncertain. The only thing I was able to bring as a gift for our hosts was a poor reproduction of a drawing which gives a faithful picture of how I felt before leaving Vienna.

You may be interested to hear that Heller has become very pliable and easygoing and is preparing a second printing of the lectures.

With cordial greetings to young and old

Your
Freud

---

[1] Freud and his daughter Anna.

182 *To* ANTON von FREUND

Lomnicz[1]
September 17, 1918

Dear Doctor

My stay here is coming to an end. In a few days I shall be able to speak to you and thank you for all your efforts, also

---

[1] Small resort in the Tatra Mountains in Hungary.

in connection with the Congress.[2] Just when this will be I am
not yet certain. If you receive no further news—you realize that
at the moment the telegraph system isn't functioning at all—we
leave here on Friday morning, the twentieth, so as to get to
Budapest that evening. Ferenczi has been asked to get us a room
on the Danube.

As regards the Congress, I still have two requests or reminders.
First, that the official element shouldn't be emphasized; let us leave
banquets, speeches, solemn addresses and so on, all rather mean-
ingless in themselves, for other occasions. At least I would like it
to be known that I am taking no part in all this, won't say a word
in public, won't pull a solemn face, etc. In short, I am going to remain passive. The only thing I can add, if the Congress rises to such
heights, is to suggest a more dignified and generally more interesting topic for a lecture than that anounced in the program. I don't
know as yet just what, for I am enjoying too much comfort here to
have any ideas.

Furthermore, I would like to ask you to bridle your hospitality
toward the participants and to reduce their lodgings as well as
the entertainment to a fairly modest scale. I suggest this in your
interest, for I don't want you to appear in the role of the financial
backer, a role which you quite rightly don't wish to enact. On the
other hand the Association[3] itself has no means of its own at the
moment. The warmth of the reception will nevertheless be quite
obvious to all participants.

With cordial greetings to you, your family and guests

Your

Freud

[2] The Fifth International Psychoanalytical Congress in Budapest, Sept., 1918.
[3] The International Psychoanalytical Association.

183 *To* SANDOR FERENCZI

Vienna IX, Berggasse 19
September 30, 1918

Dear Friend

On the day of my return and on the threshold of a new
working year, I cannot refrain from thanking you for all the proofs

of warm friendship during these past days, and from congratulating you on the splendid success of the Congress in which you had such a large share, as well as on your promotion.[1] Do you remember the prophetic words I uttered before the first Congress in Salzburg, when I said that we expected great things from you?

I am reveling in satisfaction, my heart is light since I know that my problem child, my life's work, is protected by your interest and that of others, and its future taken care of. I shall be able to see better times approaching, even if I do so from a distance.

Now I hope you will strengthen your friendship with the man[2] whom Providence has sent us at the right moment, and with Rank,[3] who cannot be replaced, and let me enjoy the satisfaction of watching how well the younger generation carries out what limited energy and old age can no longer achieve.

Budapest offered little opportunity for an exchange of ideas. There were too many claims made upon one, which is a justification for this belated display of emotion.

On returning here I found a letter from Karger[4] requesting the sixth printing of the *Everyday Life*.

Will you please forward the enclosed lines to Sachs,[5] whose new address I don't know.

With cordial greetings
Your
Freud

---

[1] To the presidency of the Association.

[2] Anton von Freund, who by a considerable donation made possible the founding of the Internationale Psychoanalytische Verlag (publishing house).

[3] Dr. Otto Rank (see List of Addressees and letter of Aug. 25, 1924).

[4] Publisher in Berlin.

[5] Hanns Sachs, LL.D. (1881-1947), psychoanalyst. Author of *Freud: Master and Friend*, Cambridge, Mass., 1944.

184 *To* MAX EITINGON

Vienna IX, Berggasse 19
December 2, 1919

Dear Dr. Eitingon

Your kindness and affection have given me a stormy day. In the morning I received your letter announcing that three thou-

sand Swedish kronor had been put at my disposal, half of which,
converted into Austrian kronen, would be remitted by the bank.
As I was busy with four analyses in the morning, I had no time to
think about it, and read the letter out loud at luncheon during
which, apart from my wife, three sons and our young daughter
(whom you know) were present. It had a strange effect: the three
boys seemed satisfied, but the two women were up in arms and my
daughter declared—evidently she can't stand the demolition of her
father complex—that as a punishment (!) she wouldn't go to Berlin for Christmas. In fact, tomorrow morning I am going to send
you a telegram asking you to cancel the remittance to Vienna, if
possible. For what is the good of it? We don't lack useless kronen.
I possess more than 100,000 of them and every day earn 900-1000
more. If on the other hand I were in need of foreign currency,
which is so difficult to get here, I would have approached you
anyhow, as I have already done about the marks. It certainly
gives me a sense of security to know that you hold a store of
Swedish kronor with the object of helping me out if I should have
to turn to you. But once it has been converted into Austrian
kronen, I wouldn't know what to do with it, and it wouldn't be
easy for us to change it back. It is from this point of view that my
telegram will be composed.

You really are the most reckless member of my family. You
have already lent me 2000 marks, another 1000 to my sister-in-law,
who also can repay you only in kronen; you have taken it upon
yourself to finance Ernst until he gets his salary in pounds sterling,
you send us quantities of material for life (food) and work (cigars)
which we would not like to convert, and even with all this you still
aren't satisfied. Your last precious gift, by the way, met with an
accident—about which Mathilde didn't want me to tell you. Her
trunk arrived one day after her, denuded of anything edible or
smokable; everything else remained intact, a selection which suggests a well-organized gang of railwaymen. Two beautiful saus-
ages, for which I remain obliged to Liebermann[1] and Harnik,[2]
happened to be in the suitcase.

At the moment we are preoccupied with our forthcoming wedding,[3] you probably not less with yours, although it is further off.

[1] Dr. Hans Liebermann (1883-1931), German psychoanalyst.
[2] J. Harnik, Hungarian psychoanalyst.
[3] Wedding of Martin Freud to Esti Drucker.

Ernst is packing up his books to tear himself definitely away from home, which at each visit he so thoroughly enlivens. Oli is depressed because he has learned from his trip to the representative of the Dutch Government in Graz that the decision about his application cannot be expected for another two months, and that his chances are somewhat vague. In the meantime he will probably go to Berlin and Hamburg. Freund's condition is bad, he has a temperature, is kept under morphia, has himself set December 12 as his limit. The situation isn't clear; the sarcoma metastases are probably spreading.

I am working very slowly at instinct- and mass-psychology. I miss Rank very much. The Psychoanalytical Society has admitted Dr. Schilder,[4] assistant at the psychiatric clinic; Hattingberg[5] and Schmideberg[6] are on the enrollment list.

The foundation of your Polyclinic[7] will make Berlin the center again. Abraham is fighting to get the next Congress held there.

I greet you and your dear wife in old friendship but with new motives.

Your
Freud

[4] Dr. Paul Schilder (1886-1940), psychoanalyst, Professor of Psychiatry at the University of Vienna.

[5] Hans von Hattingberg, LL.D., German psychoanalyst.

[6] Walter Schmideberg, psychoanalyst.

[7] The first clinic for psychoanalytic treatment was founded by Dr. Eitingon.

185 *To* AMALIE FREUD

Vienna IX, Berggasse 19
January 26, 1920

Dear Mother

I have some sad news for you today. Yesterday morning our dear lovely Sophie died from galloping influenza and pneumonia. We learned of it at noon from a telephone conversation with Minna in Reichenhall. Oli and Ernst have left Berlin to be with Max. Robert and Mathilde are leaving on the twenty-ninth to try and assist the poor bereaved man. Martha is too upset; one couldn't let her undertake the journey, and in any case she wouldn't have found Sophie alive.

She is the first of our children we have to outlive. What Max will do, what will happen to the children, we of course don't know as yet.

I hope you will take it calmly; tragedy after all has to be accepted. But to mourn this splendid, vital girl who was so happy with her husband and children is of course permissible.

I greet you fondly.

Your

Sigm.

186 *To* OSKAR PFISTER

Vienna IX, Berggasse 19
January 27, 1920

Dear Pfister

You have sent us a charming young lad, who introduced himself, moreover, with the traditional gifts. His voice sounded so much like yours on the telephone that for quite a while I couldn't believe he belongs to the second generation; and when he arrived for luncheon on Sunday he didn't look in the least unfamiliar; he also behaved so naturally that it was a delight to have him.

And yet we haven't seen him again since Sunday! That afternoon we received the news that our sweet Sophie in Hamburg had been snatched away by influenzal pneumonia, snatched away in the midst of glowing health, from a full and active life as a competent mother and loving wife, all in four or five days, as though she had never existed. Although we had been worried about her for a couple of days, we had nevertheless been hopeful; it is so difficult to judge from a distance. And this distance must remain distance; we were not able to travel at once, as we had intended, after the first alarming news; there was no train, not even for an emergency. The undisguised brutality of our time is weighing heavily upon us. Tomorrow she is to be cremated, our poor Sunday child! Our daughter Mathilde and her husband are leaving for Hamburg the day after tomorrow, thanks to an unexpected connection with an Entente train; at least our son-in-law was not alone; two of our sons who were in Berlin are already with him, and our friend Eitingon has gone with them.

Sophie leaves two sons, one of six, the other thirteen months, and an inconsolable husband who will have to pay dearly for the happiness of these seven years. The happiness existed exclusively within them; outwardly there was war, conscription, wounds, the depletion of their resources, but they had remained courageous and gay.

I work as much as I can, and am thankful for the diversion. The loss of a child seems to be a serious, narcissistic injury; what is known as mourning will probably follow only later.

But as soon as the condolences have been coped with, I will ask Pfister junior to come and see us. After all, it is not the boy's fault. He has already sent us a card of condolence.

Feeling sure of your sympathy, with cordial greetings
Your
Freud

P.S. Please thank Oberholzer[1] and apologize to him for my silence with this sad news.

[1] Dr. Emil Oberholzer, Swiss psychoanalyst.

187 *To* SANDOR FERENCZI

Vienna IX, Berggasse 19
February 4, 1920

Dear Friend

Please don't worry about me. Apart from feeling rather more tired I am the same. The death, painful as it is, does not affect my attitude toward life. For years I was prepared for the loss of our sons; now it is our daughter; as a confirmed unbeliever I have no one to accuse and realize that there is no place where I could lodge a complaint. "The unvaried, still returning hour of duty"[1] and "the dear lovely habit of living"[2] will do their bit toward letting everything go on as before. Deep down I sense a bitter, irreparable narcissistic injury. My wife and Annerl are profoundly affected in a more human way.

This week we are expecting Oli who, by the way, seems to have found a job. He, Ernst, and Eitingon were present at the cremation.

[1] Quotation from Schiller's *Piccolomini.*
[2] Quotation from Goethe's *Egmont.*

Yesterday we talked to a visitor from Hamburg who was also there. It seems to have been a very dignified and impressive ceremony.

Mathilde and Robert have been in Hamburg since Saturday. We know that Max is not going to separate from the children, and that the widow of his brother killed in the war is coming to spend a few weeks with him.

I clasp your hand in friendship and would like you to thank your wife warmly for her affectionate words.

Your
Freud

188 *To* GEORG GRODDECK

Vienna IX, Berggasse 19
February 8, 1920

Dear Colleague

I will have the novel[1] returned to you by our publishing house during the next few days. But you are mistaken, I liked it very much. By parts of it I was highly amused. The passages emulating old English humorists come off very well. In one respect it strikes me that it resembles the immortal prototype of every humorous novel, *Don Quixote*. The hero grows in the author's hands into something more serious than seemed originally intended. Your talent for plastic description, which is most unusual, I admired especially in the scenes laid in the train.

On the other hand, I agree with you that the book may not be to everyone's taste. Such an abundance of intelligent, candid, and high-spirited thoughts can't be easily digested. And yet you should try to have it published. Certainly worse productions have appeared in the name of psychoanalysis.

We shall always welcome your contributions to *Imago*. True, we haven't any paper at the moment, but we are trying to get some.

With a colleague's respect
Yours sincerely
Freud

---

[1] *Der Seelensucher* (*The Soul-Seeker*), Inselverlag, Leipzig, 1921.

189 *To* ROZSI VON FREUND

Vienna IX, Berggasse 19
May 14, 1920

Dearest Rozsi

I have been wondering for a long time how I should answer your question; I was under the impression that you already knew what you are asking—namely that Toni[1] was well aware of his fate, which he endured like a hero; but like a true human, Homeric hero he was able from time to time to give free vent to his grief about his lot. Only a good, great, and natural human being could behave like this.

You can imagine how frequently conversations with Katá[2] bring back all the memories of him. You will also know that in the meantime I have had further reason to train myself in the resignation of the survivor.

Little as I can do for you and your children at the moment, I hope you won't forget that we remain closely allied through shared experience.

Warmest wishes
from your
Freud

[1] Anton von Freund.
[2] Anton von Freund's sister Katá Levy (see List of Addressees and letter of June 11, 1923).

190 *To* OSKAR PFISTER

Vienna IX, Berggasse 19
June 21, 1920

Dear Pfister

I started reading your little book[1] about expressionism with as much interest as aversion, and finished it in one sitting. In the

[1] *Der psychologische und biologische Untergrund expressionistischer Bilder*, Berne, 1920. (*Expressionism in Art, Its Psychological and Biological Basis*, London, 1922; New York, 1923.)

end I liked it very much, not so much the part that is sheer analysis and which of course cannot overcome the difficulties of an interpretation for the nonanalysts, but what you make of and associate with the subject. It keeps saying to me: What a decent, benevolent man, incapable of any injustice, this Pfister really is, someone with whom you cannot compare yourself, and how pleasant that you can give your approval to everything which he discovers in his own way. For I think you ought to know that in actual life I am terribly intolerant of cranks, that I see only the harmful side of them and that so far as these "artists" are concerned I am almost one of those whom at the outset you castigate as philistines and lowbrows. And in the end you explain quite clearly and exhaustively why these people lack the right to claim the name of artist.

So let me thank you very warmly for this new enrichment to my psychoanalytical hoard.

> Your devoted
> Freud

P.S. What is your nice son doing?

191 *To* STEFAN ZWEIG

> Vienna IX, Berggasse 19
> October 19, 1920

Dear Mr. Zweig

Having found some peace here at last, I realize that I owe you thanks for the excellent book[1] which I found waiting for me and which I have read in spite of the rush of the first two weeks. I have read it with exceptional pleasure, otherwise there would hardly be any point in writing to you about it. The perfection of empathy combined with the mastery of linguistic expression left me with a feeling of rare satisfaction. What interests me especially are the accumulations and the increasing intensity with which your language keeps groping closer to the most intimate nature of the subject.

It is like the accumulation of symbols in a dream which allow what is concealed to come glimmering through with ever-increasing clarity.

---

[1] *Drei Meister: Balzac, Dickens, Dostoevski,* Inselverlag, Leipzig, 1920. (*Three Masters,* New York, 1930.)

If I were allowed to measure your exposition with a particularly severe yardstick, I would say that in the treatment of Balzac and Dickens you have succeeded completely. Perhaps this was not too difficult, for they are comparatively straightforward types. But with the confounded Russian it was bound to be less successful. Here one feels the gaps and unsolved riddles. Please allow me in this connection to produce some material as it comes to my layman's mind. It is also possible that here the psychopathologist, whose property Dostoevski must inevitably remain, has some advantage.

I don't think you should have confined D. to his alleged epilepsy. It is very unlikely that he was an epileptic. Epilepsy is an organic brain disease independent of the psychic constitution and as a rule associated with the deterioration and retrogression of the mental performance. Only one single occurrence of this disease in a man of high intelligence has been recorded, and this concerns an intellectual giant of whose emotional life little is known (Helmholtz).[2] All other great men who are said to have been epileptics are straight cases of hysteria. (That phantast Lombroso[3] didn't yet know how to make a differential diagnosis.) This difference, however, is no medical pedantry, but something very fundamental. Hysteria springs from the psychic constitution itself and is an expression of the same organic basic power which produces the genius of an artist. But it is also a symptom of an especially strong and unresolved conflict which rages between these basic tendencies and which later splits the psychic life into two camps. I feel that the whole case of D. could have been built on his hysteria.

Overwhelmingly great as the factor of constitution may be in a hysteria like D.'s, it is nevertheless interesting that the other factor to which our theory attaches importance can also be demonstrated in this case. Somewhere in a biography of D. I was shown a passage which traced back the later affliction of the man to the boy's having been punished by his father under very serious circumstances—I vaguely remember the word "tragic," am I right? Out of "discretion," of course, the author didn't say what it was all about. You will find it easier than I to trace this passage. It was this childhood scene—and I don't have to elaborate on that to the

[2] Hermann von Helmholtz (1821-1894), German physician, physiologist and physicist.

[3] Cesare Lombroso (1836-1909), Italian psychiatrist, author of *Genius and Madness*.

author of *First Experiences*—which gave to the later scene before
the execution the traumatic power to repeat itself as an attack,
and D.'s whole life is dominated by his twofold attitude to the
father-czar-authority, by voluptuous masochistic submission on the
one hand, and by outraged rebellion against it on the other.
Masochism includes a sense of guilt which surges toward "re-
demption."

What you, avoiding the technical word, call "dualism," we call
"ambivalence." This feeling of ambivalence is a legacy from the
psychic life of primitive races; with the Russian people, however,
it is far better preserved and has remained more accessible to
consciousness than elsewhere, as I was able to point out only a few
years ago in the detailed case history of a typical Russian. This
strong tendency toward ambivalence combined with the childhood
trauma may have partly determined the unusual violence of D.'s
case of hysteria. Even Russians who are not neurotic are also very
noticeably ambivalent, as are the characters in almost all of D.'s
novels.

Nearly all the peculiarities of his production, hardly one of which
has eluded you, can be traced back to his—for us abnormal, for the
Russian fairly common—psychic constitution, or more correctly, sexual constitution, which could be illustrated very well in detail.
Above all, everything that is alien and tormenting. He cannot be understood without psychoanalysis—i.e., he isn't in need of it because
he illustrates it himself in every character and every sentence. The
fact that *The Brothers Karamazov* deals with D.'s most personal
problem, that of parricide, and bases it on the analytical theory of the
equivalence of the actual deed and the evil intention would be but
one example. The strangeness of his sexual love, which is either
blind lust or sublimated compassion, the doubts of his heroes as to
whether they love or hate, whom they love, when they love, etc.,
also shows clearly from what soil his psychology has sprung.

With you I do not have to fear the misunderstanding that this
emphasis on the so-called pathological is intended to belittle or
explain away the splendor of D.'s creative power. I conclude this
already too long letter not because the material is exhausted, but at
the suggestion of the long-suffering sheet of paper.

With repeated thanks and cordial greetings
Your
Freud

192 *To* HEREWARD CARRINGTON

Bad Gastein
July 24, 1921

Dear Mr. Carrington

I am not one of those who dismiss *a priori* the study of so-called occult psychic phenomena as unscientific, discreditable or even as dangerous. If I were at the beginning rather than at the end of a scientific career, as I am today, I might possibly choose just this field of research, in spite of all difficulties.

I must ask you nevertheless to refrain from mentioning my name in connection with your venture, and this for several reasons.

First, because in the sphere of the occult I am a complete layman and newcomer and have no right to claim so much as a hint of authority in this connection.

Second, because I have good reason to be interested in sharply demarcating psychoanalysis (which has nothing occult about it) from this as yet unexplored sphere of knowledge, and in not offering any occasion for misunderstanding in this respect.

Finally, because I cannot rid myself of certain skeptical materialistic prejudices which I would bring with me into the study of the occult. Thus I am utterly incapable of considering the "survival of the personality" after death even as a scientific possibility, and I feel no different about the "idroplasma."

In consequence I believe it is better if I continue to confine myself to psychoanalysis.

Yours sincerely
Freud

193 *To* OSKAR RIE

Bad Gastein
August 4, 1921

Dear Oskar

I received your letter when the thermometer registered 100° F., and am answering it at 48°. Hardly has the wish for a change been granted when one regrets its fulfillment. Isn't life often like this?

The unexpected heat wave in Gastein has also brought me a week of great fatigue, which was conquered, however, before the fall in temperature. The baths are announcing their effect by reviving the complaints against which they are taken, which is as it should be. A great triumph has reached me here: the first French translation, that of the five lectures,[1] published by the *Revue de Genève*. More direct contacts with Paris promise that we shall soon be acquiring an audience in reluctant France.

On July 31 a telegram from Ernst announced the birth of a son[2] of whom (apart from his weight, 7½ pounds) there is no further news. Mother and child are reported to be well. Those in Aussee[3] are not very generous with news; on the fourteenth we are planning to meet on the way to Innsbruck, and on the following day travel on to Seefeld.[4]

I hope to capture you there for several days' conversation. I think it is quite wrong for you to starve yourself. The times demand all our energies, the mobilization of our reserves, and our poor Conrad[5] (our physique) is prepared to cooperate only if he is well treated. On rainy days we will get together for a game of taroc.

Your friendly words about me have done me good although they didn't tell me anything new, because I have been looking upon your friendship for more than a lifetime as an assured possession. I have been able to give something to many people in my life; from you fate has allowed me only to receive. I have confidence in your strength and that you will also prove yourself superior to the present situation.

My latest little book, on group psychology,[6] probably reached you some time ago? At least I gave orders for it to be sent before I left.

---

[1] See note to letter of Nov. 23, 1913.

[2] Ernst Freud's eldest son Stephen Gabriel.

[3] The family.

[4] Resort in the Tyrol.

[5] Allusion to an old German student song, the first line of which runs: "Oh dear! me poor Chorydon, oh dear!" The "poor Conrad" is a bowdlerization of "poor Chorydon."

[6] *Massenpsychologie und Ich-Analyse*, Vienna, 1921. (*Group Psychology and the Analysis of the Ego*, Standard Ed., 18.)

Minna, who is again benefiting very much from Gastein, sends you her greetings. I conclude in the hope of meeting you this month.

Your

Freud

194 *To* ERNST AND LUCIE FREUD

Vienna IX, Berggasse 19

December 20, 1921

Dear Ernst and dear Lux

Frau Lou[1] left this morning, so this evening is the first moment I have had to answer yours of November 30-December 7, with its good news. She was a charming guest and is altogether an outstanding woman. Anna discussed analytic topics and visited a number of interesting people with her, and enjoyed being with her very much. Mama took loving care of her; I myself, working nine hours a day, hadn't much time for her, but she was a discreet and undemanding guest. Apart from all that, there has been a great deal going on just now; I never realized that the older one grows the more there is to do. The idea of peaceful old age seems as much of a legend as that of happy youth. Much of my time is taken up with refusals and information to all corners of the world; everyone wants to be analyzed by me, and at the same time none of my patients is leaving before the end of February.

More important successes to tell you about are the appearance of the French and Italian translations of the lectures (Part I) and my nomination as an honorary member of the Dutch Society of Psychiatry (at the suggestion of an opponent). The present popularity of psychoanalysis may be judged by the fact that within one week two applications for the formation of new local branches of the International Association have been received—one from Calcutta, the other from Moscow! The second letter had on it stamps to the value of 10,000 rubles—a glimpse into our future, perhaps into yours, too.

Early in January the room inhabited by Frau Lou will be occupied successively by Abraham and Ferenczi, who are going to

---

[1] Lou Andreas-Salomé.

lecture to our Americans. On his return Abraham will bring you Mama's Christmas present for Gabriel, a gift that you will then have to refuse from others, a silver feeding spoon, which should be useful in sparing his learned mother. . . .

I don't need any better news from you than that contained in your last letter. I hope the next one will be in a similar vein.

Fond greetings to you all for Christmas and the New Year

Papa

195 *To* MAX EITINGON

Vienna IX, Berggasse 19
January 24, 1922

Dear Max

Abraham is spending this evening with Rank, so I take this opportunity of answering your friendly reference to the fifteenth anniversary of our relationship. You are aware of the role you have acquired in my life and that of my family. I know I was not in a hurry to assign it to you. For many years I was aware of your efforts to come closer to me, and I kept you at bay. Only after you had expressed in such affectionate terms the desire to belong to my family—in the closer sense—did I surrender to the easy trusting ways of my earlier years, accepted you and ever since have allowed you to render me every kind of service, imposed on you every kind of task.

Today I confess that in the beginning I did not appreciate your sacrifices as highly as I did later after recognizing that—burdened with a loving and loved wife who isn't too fond of sharing you with others, and tied to a family who fundamentally has little sympathy for your endeavors—you actually overtaxed your strength by making that offer. But do not conclude from this remark that I am ready to release you. Your sacrifices have become all the more valuable to me; if they have become too much for you, then it is up to you to tell me.

So I suggest that we continue our relationship which has developed from friendship to sonship until the end of my days. Since you were the first to come to the lonely man, you may as well stay with him to the last. Things will no doubt have to remain much as they were before—I shall need something and you will try to produce it. It is your self-chosen fate, for which I pitied you in Berlin, too. But I am familiar from the peripatetic analysis[1] with the pattern of your love life, from which it wasn't possible to release you.

My situation has fundamentally changed in the past fifteen years. I am relieved of material worries, surrounded by popularity which is distasteful to me, and involved in enterprises which rob me of the time and leisure for calm scientific work. What I need now is support for the cultivation of the psychoanalytical movement and above all for the *Verlag*. The most immediate service you can render it is to write an eloquent and detailed annual report of the Polyclinic—the very one you founded—so that with this document I can canvass elsewhere for the support of the institute and encourage the foundation of similar institutes. The prospects in America seem to be not unfavorable.

And now over the distance Vienna-Passau-Berlin please shake your faithful

Freud

warmly by the hand.

P.S. Please tell Mirra[2] how glad we all are to hear the news of her great improvement.

[1] Allusion to Freud and Eitingon walking and talking together.
[2] Eitingon's wife.

196 *To* ERNST FREUD

Vienna IX, Berggasse 19
April 3, 1922

My dear Ernst

It really is not necessary to wish you luck on your thirtieth birthday. You are the only one among my children who already possesses everything a man can want at your age: a loving wife, a splendid child, work, income, and friends. You deserve it all too, and since not all things in life go according to merit, let me express the wish that luck will continue to remain faithful to you.

Love

Papa

197 *To* ARTHUR SCHNITZLER

Vienna IX, Berggasse 19
May 14, 1922

Dear Dr. Schnitzler

Now you too have reached your sixtieth birthday, while I, six years older, am approaching the limit of life and may soon expect to see the end of the fifth act of this rather incomprehensible and not always amusing comedy.

Had I retained a remnant of belief in the "omnipotence of thoughts," I would not hesitate today to send you the warmest and heartiest good wishes for the years that await you. I shall leave this foolish gesture to the vast number of your contemporaries who will remember you on May 15.

But I will make a confession which for my sake I must ask you to keep to yourself and share with neither friends nor strangers. I have tormented myself with the question why in all these years I have never attempted to make your acquaintance and to have a talk with you (ignoring the possibility, of course, that you might not have welcomed my overture).

The answer contains the confession which strikes me as too intimate. I think I have avoided you from a kind of reluctance to meet my double. Not that I am easily inclined to identify myself with another, or that I mean to overlook the difference in talent that separates me from you, but whenever I get deeply absorbed in your beautiful creations I invariably seem to find beneath their poetic surface the very presuppositions, interests, and conclusions which I know to be my own. Your determinism as well as your skepticism—what people call pessimism—your preoccupation with the truths of the unconscious and of the instinctual drives in man, your dissection of the cultural conventions of our society, the dwelling of your thoughts on the polarity of love and death; all this moves me with an uncanny feeling of familiarity. (In a small book entitled *Beyond the Pleasure Principle*, published in 1920, I tried to reveal Eros and the death instinct as the motivating powers whose interplay dominates all the riddles of life.) So I have formed the impression that you know through intuition—or rather from detailed self-observation—everything that I have discovered by laborious work on other people. Indeed, I believe that fundamentally your nature is that of an explorer of psychological depths, as honestly impartial and undaunted as anyone has ever been, and that if you had not been so constituted your artistic abilities, your gift for language, and your creative power would have had free rein and made you into a writer of greater appeal to the taste of the masses. I am inclined to give preference to the explorer. But forgive me for drifting into psychoanalysis; I just can't help it. And I know that psychoanalysis is not the means of gaining popularity.

> With warmest greetings
> Your
> Freud

198 *To* KARL ABRAHAM

> Vienna IX, Berggasse 19
> December 26, 1922

Dear Friend

I have received the drawing which is supposed to represent your head. It is horrible.

I know what [an] excellent person you are, and I am all the more shocked that such a trifling flaw in your character as your tolerance or sympathy for modern "art" has to be so cruelly punished. I learn from Lampl that the artist declares that this is how he sees you! People like him should be the last to have access to analytical circles, for they are the all-too-undesirable illustration of Adler's theory that it is just the people with serious congenital defects of vision who become painters and draughtsmen.

Allow me to forget this portrait while wishing you and your dear family everything good and pleasant for 1923.

> Cordially yours
> Freud

199 *To* ÉDOUARD MONOD-HERZEN

Vienna IX, Berggasse 19
February 9, 1923

Dear Sir

Your recent letter gave me great pleasure. Our times have unfortunately made us so shy and suspicious that we no longer dare take human sympathy in others for granted. So one appreciates it all the more.

Since you are a friend of Romain Rolland, may I ask you to pass on to him a word of respect from an unknown admirer.

To you yourself a cordial handshake from afar!

Your
Freud

200 *To* ROMAIN ROLLAND

Vienna IX, Berggasse 19
March 4, 1923

Dear Sir

That I have been allowed to exchange a greeting with you will remain a happy memory to the end of my days. Because for us your name has been associated with the most precious of beautiful illusions, that of love extended to all mankind.

I, of course, belong to a race which in the Middle Ages was held responsible for all epidemics and which today is blamed for the disintegration of the Austrian Empire and the German defeat. Such experiences have a sobering effect and are not conducive to make one believe in illusions. A great part of my life's work (I am ten years older than you) has been spent [trying to] destroy illusions of my own and those of mankind. But if this one hope cannot be at least partly realized, if in the course of evolution we don't learn to divert our instincts from destroying our own kind, if we continue to hate one another for minor differences and kill each other for petty gain, if we go on exploiting the great progress made in the control of natural resources for our mutual destruction, what kind of future lies in store for us? It is surely hard enough to ensure the perpetuation of our species in the conflict between our instinctual nature and the demands made upon us by civilization.

My writings cannot be what yours are: comfort and refreshment for the reader. But if I may believe that they have aroused your interest, I shall permit myself to send you a small book which is sure to be unknown to you: *Group Psychology and the Analysis of the Ego,* published in 1921. Not that I consider this work to be particularly successful, but it shows a way from the analysis of the individual to an understanding of society.

Sincerely yours
Freud

201 *To* GEORG GRODDECK

Vienna IX, Berggasse 19
March 25, 1923

Dear Dr. Groddeck

First, my congratulations on *The Id*[1] being published at last. I like the little book very much. I believe it is very important to keep bringing home to people the fundamentals of psychoanalysis to which they are so inclined to close their eyes. The work moreover expounds the theoretically important point of view which I have covered in my forthcoming *The Ego and the Id.*[2]

Among the public, of course, it will arouse even more indignation and distaste than the excellent *Seeker of the Soul,*[3] which compensated for its undesirable content by its artistic treatment. But I trust this response will not affect your self-confidence.

Cordial greetings to you and my lady translator.

Your
Freud

---

[1] *Das Buch vom Es,* Psychoanalytischer Verlag, Vienna, 1923. (*The Book of the Id*), Psychoanalytical Press, Vienna, 1923.

[2] *Das Ich und das Es,* Vienna, 1923. (*The Ego and the Id,* Standard Ed., 19.)

[3] See note to letter of Feb. 8, 1920.

202 *To* LOU ANDREAS-SALOMÉ

Vienna IX, Berggasse 19
May 10, 1923

Dearest Lou

Nothing but visits and celebrating—Ernst and Eitingon have been here—are responsible for postponing my answer to your kind and understanding letter. As a result I am able to tell you that I can again speak, chew and work, indeed even smoking is permitted —to a certain moderate, cautious, so to speak *petit bourgeois* degree. The family doctor himself supplied the cigar holder for my birthday; the latter incidentally was celebrated as though I were a music hall star, or as though it were to be my last.

Even after the operation[1] the prognosis is good. You realize that this means no more than a slight lessening of the suspense that is bound to hover over the years to come. Wife and daughter have nursed me tenderly.

I agree completely with your views about our helplessness in the face of physical, above all painful, suffering; like you I also find it—deplorable and, if one could reproach anyone personally, a dirty trick.

I have not done anything to deserve your husband's kind letter and would like you to convey to him my warm and sincere thanks.

With best greetings and wishes
Your
Freud

[1] Freud had undergone an operation for cancer of the jaw.

203 *To* KATÁ AND LAJOS LEVY

Vienna IX, Berggasse 19
June 11, 1923

Dear Katá and dear Lajos

Anna has gone away for the night on a nursing job and has charged me to thank you for suggesting your visit during this month. I also owe Lajos an acknowledgment for his letter about Ferenczi's operation, which was written with his usual clarity and assurance.

About my operation and affliction there is nothing to say that you yourselves couldn't know or expect. The uncertainty that hovers over a man of sixty-seven has now found its material expression. I don't take it very hard; one will defend oneself for a while with the help of modern medicine and then remember Bernard Shaw's warning:

> Don't try to live for ever, you will not succeed.
> (*The Doctor's Dilemma*)

But now there is something else. We brought here from Hamburg Sophie's younger son, Heinele, now aged 4½. My eldest daughter Mathilde and her husband have almost adopted him and have fallen in love with him to an extent that couldn't be foreseen. He was indeed an enchanting little fellow, and I myself was aware of never having loved a human being, certainly never a child, so much. Unfortunately he was very weak, never entirely free of a temperature, one of those children whose mental development grows at the expense of its physical strength. We thought that in Hamburg he lacked the proper kind of care or medical supervision.

This child became ill again two weeks ago, temperature between 102 and 104, headaches, no clear local symptoms, for a long time no diagnosis, and finally the slow but sure realization that he has a miliary tuberculosis, in fact that the child is lost. He is now lying in a coma with paresis, occasionally wakes up, and then he is so completely his own self that it is hard to believe. . . . After each waking and going to sleep one loses him all over again; the doctors say it can last a week, perhaps longer, and recovery is not desirable, fortunately not likely. His father arrived yesterday.

I find this loss very hard to bear. I don't think I have ever experienced such grief; perhaps my own sickness contributes to the shock. I work out of sheer necessity; fundamentally everything has lost its meaning for me. I am due to leave here on the twenty-ninth; if you want to come between the twenty-third and this date (Sunday the twenty-fourth is the only possible day, as I have to work hard to the last moment), I don't know what you will find. It's at your own risk. Under other circumstances I would have been so pleased.

With particularly warm greetings
Your
Freud

204 *To* AMALIE FREUD

Vienna, Auersperg Sanatorium
October 17, 1923

Dear Mother
    Everyone you ask will confirm the news that on the fourth and eleventh of this month I underwent an operation on the upper jaw which, owing to the skill of the surgeon and the excellence of the nursing, has been very successful so far. It will take me some time to grow accustomed to a partial denture that I have to wear. So don't be surprised if you do not see me for a while and I hope you will be in good spirits when we meet again.

Affectionately
Your Sigm.

205 *To* FRITZ WITTELS

Vienna IX, Berggasse 19
December 18, 1923

Dear Mr. Wittels
    Not to acknowledge and express thanks for a Christmas present that deals so copiously with the recipient would be an act of rudeness for which special motives would have to be provided. I declare with satisfaction that in our case no such motives exist. Your book[1] is not unkind, not too indiscreet; it testifies to serious interest and, as was to be expected, to your art of writing and describing.
    Needless to say, I would never have desired or promoted such

---

[1] *Sigmund Freud: der Mann, die Lehre, die Schule*, Leipzig, 1924. (*Sigmund Freud: His Personality, His Teaching, and His School*, London and New York, 1924.)

a book. It seems to me that the world has no claim on my person and that it will learn nothing from me so long as my case (for manifold reasons) cannot be made fully transparent. You think differently about it and as a result were capable of writing this book. Your personal distance from me, which you consider exclusively an advantage, also has great disadvantages. You know too little about your subject and thus cannot avoid the danger of distorting it with your analytical efforts. It is also very doubtful whether you have made the task of finding the right approach to your subject easier by taking Stekel's point of view and assessing me from his angle.

I also make a preconceived idea of yours responsible for the distortions which I seem to recognize. I can guess what this idea is. In your opinion a great man ought to show such and such advantages, faults and extremes; you consider me a great man and as a result you are entitled to ascribe to me all these, often contradictory, qualities. Many interesting and generally significant things could be said about this, but unfortunately your relationship to Stekel rules out any further effort on my part toward a mutual understanding.

On the other hand I willingly concede that your penetrating observation has divined quite correctly in me features well known to myself; for instance, that I am compelled to go my own way, often a roundabout way, and that I cannot make any use of ideas that are suggested to me when I am not ready for them. Also, in my relation to Adler you have, to my great satisfaction, done me justice. What you don't realize is that I behaved with equal tolerance and patience toward Stekel. In spite of his unbearable manners and his hopelessly unscientific approach, I stood by him for a long time in the face of attacks from all sides, forced myself to ignore his far-reaching lack of self-criticism and truthfulness— an outer as well as an inner truthfulness—until finally on a certain occasion which revealed his treachery and ugly dishonesty "all the buttons of [even my] trousers of patience snapped off."[2] (Moreover, you have failed to defend me against the misconception that I deny things merely because I can't as yet judge or digest them.)

Perhaps you know that I have been seriously ill and that even if I recover I have reason to look upon the experience as a warn-

[2] Quotation from Heine's "Romanzero."

ing of a not-too-distant end. From this partial eclipse I may perhaps ask you to acquit me of any desire to interfere in your relationship with Stekel. I only regret that it should have exerted such a decisive influence on your book about me. It strikes me as not out of the question that you may have to revise this book for a second edition. In view of this possibility I am putting at your disposal the enclosed list of corrections. It contains thoroughly reliable data, quite independent of my subject opinions, a few of them unimportant, others perhaps capable of shaking or modifying some of your assumptions. Please look upon this communication as a sign that although I cannot approve of your effort I by no means wish to underestimate it.

Yours faithfully
Freud

206 *To* WILHELM STEKEL

Vienna IX, Berggasse 19
January 13, 1924

Dear Dr. Stekel

I acknowledge receipt of your letter of December 31, 1923, and thank you for your good wishes regarding the improvement of my health. But I cannot refrain from contradicting you on a few important points.

You are mistaken if you think that I hate or have hated you. The facts are that after an initial sympathy—perhaps you still remember how our relationship began—I had reason for many years to be annoyed with you while at the same time having to defend you against the aversion of everyone around me, and that I broke with you after you had deceived me on a certain occasion[1] in the most heinous manner. (You never mentioned this occasion—*Zentralblatt*—in your letters.) I lost confidence in you at that time and since then you have not provided me with any experience that could help me to regain it.

I also contradict your often repeated assertion that you were rejected by me on account of scientific differences. This sounds

---

[1] After leaving the Psychoanalytical Association, Stekel had refused to resign his editorship of the *Psychoanalytische Zentralblatt*.

quite good in public, but it doesn't correspond to the truth. It was exclusively your personal qualities—usually described as character and behavior—which made collaboration with you impossible for my friends and myself. Since you most certainly will not change —you don't need to, for nature has endowed you with an unusual degree of self-complacency—our relationship stands no chance of becoming any different from what it has been during the past twelve years. It will not annoy me to learn that your medical and literary activities have earned you success; I admit that you have remained loyal to psychoanalysis and have been of use to it; you have also done it great harm.

My friends and pupils will find it easier to value your publications objectively when you begin to voice your criticism and polemics in a more polite tone.

> With good wishes
> Yours faithfully
> Freud

207 *To* STEFAN ZWEIG

Vienna IX, Berggasse 19
May 11, 1924

Dear Stefan Zweig

On reading in the paper that Romain Rolland is in Vienna I immediately felt the desire to make the personal acquaintance of the man I have revered from afar. But I did not know how to approach him. I was all the more pleased to hear from you that he wants to visit me, and I hasten to submit to you my suggestions. My day allows me to be free from two to four, so I could receive you both during this time from Tuesday on, provided I know in advance. But it would suit me even better if you both would give me the pleasure of taking a cup of tea at nine (after supper) with my family. No one will be here except the women of my household. A visit at this hour would be possible on Monday.

I am very sorry to hear that Rolland himself has to be careful of his health. I am counting on your presence all the more as during the past six months my speech has been seriously affected;

above all my French would hardly be adequate for conversation.
I also intend to ask you a personal favor on this occasion.

> With cordial greetings to you and your great friend
> Your
> Freud

208 *To* LOU ANDREAS-SALOMÉ

Vienna IX, Berggasse 19
May 13, 1924

Dearest Lou

I have admired your art more than ever this time. Here
is a person who, instead of working hard into old age (see the
example[1] beside you) and then dying without preliminaries, con-
tracts a horrible disease in middle age, has to be treated and
operated on, squanders his hard-earned bit of money, generates
and enjoys discontent, and then crawls about for an indefinite
time as an invalid. In *Erewhon* (I trust you know Samuel Butler's
brilliant phantasy) such an individual would certainly be punished
and locked up. And you can still praise me for bearing my suffering
so well. Actually, it isn't quite like this; although I have weathered
the awful realities fairly well, it is the possibilities that I find hard
to bear; I cannot get accustomed to life under sentence.

It is now six months since my operation and the attitude of my
surgeon, who allows me to travel far afield in summer, ought to lull
me into something like security—so far as such a feeling is admiss-
ible, considering the *di doman non c'è certezza*[2] which affects us
all. But it has no effect on me; perhaps partly because the extent
to which the prosthesis has restored both functions of the mouth
is a very modest one. In the beginning it promised to be more successful, but the promise was not fulfilled.

Six hours of psychoanalysis, this is all I have retained of my
capacity for work; everything else, especially social contact, I keep
at bay. (Of course Romain Rolland, who has announced himself for
tomorrow, I cannot refuse. . . .)

[1] Friedrich Carl Andreas (1846-1930), Professor of Oriental Languages at the University of Göttingen, Lou Andreas-Salomé's husband since 1887.
[2] Last line of the poem "Il Trionfo di Bacco e di Arianna," by Lorenzo de Medici
(1449-1492): "Of tomorrow there is no certainty."

Well, I have written this off my chest because we both seem to be prevented from meeting. The number of things one has to renounce! And instead one is overwhelmed with honors (such as the Freedom of the City of Vienna), for which one would never have lifted a finger.

This brings you my warmest greetings. I am adding a few words of thanks on a card for your husband.

Your
Freud

209 *To* FRITZ WITTELS

Semmering, Villa Schüler
August 15, 1924

Dear Mr. Wittels

Today I received the English translation of your book about me and looked into it here and there. This is the reason for my writing.

You know my attitude to this book; it has not become friendlier.[1] I still maintain that someone who knows as little about a person as you do about me is not entitled to write a biography about that person. One waits till the person is dead, when he cannot do anything about it and fortunately no longer cares.

I cannot compare the English edition with the German, for I didn't bring the latter with me (and the same applies to the Nietzsche). You have apparently made use of my corrections. Certain passages strike me as additions, but this may be the result of a faulty memory. Other passages again gave me the opportunity to admire, if not exactly to envy, your facility.

A biographer should at least attempt to be as conscientious as a translator. *Traduttore = traditore,*[2] says the proverb! I realize that the circumstances have made this especially difficult for you, with the result that omissions occur which distort certain facts, lead to outright mistakes, and so on.

For instance, in the cocaine story by which, for reasons unknown to me, you set such great store. The whole analogy with Brücke's

---

[1] See letter of Dec. 18, 1923.
[2] "Translator = traitor."

eye discovery collapses if one takes into account what you didn't
know (but perhaps should have known?)—i.e., that I guessed its
usefulness for the eye, but for private reasons (in order to travel)
had to drop the experiment and personally charged my friend
Königstein[3] to test the drug on the eye. On my return I realized
that he had done a bad job, had dropped the project, and that another man, Koller, had become the discoverer.

The reader would also have gained a different impression of my
attitude to Koller's discovery had he been told, which indeed you
couldn't have known, that Königstein (it was *he,* not I, who so
deeply regretted having missed winning these laurels) then claimed
to be considered the codiscoverer, and that both Königstein and
Koller chose Julius Wagner[4] *and myself* as arbitrators. I think it
did us both honor that each of us took the side of the opposing
client. Wagner, as Koller's delegate, voted in favor of recognizing
Königstein's claim, whereas I was wholeheartedly in favor of award-
ing the credit to Koller alone. I can no longer remember what compromise we decided on.

In addition to knowing too little, there is also the question of
knowing too much. Anyone presuming to pass sentence in public
on the intimate feelings of a living person is obliged to be very
conscientious and reliable. In discussing the Fliess episode,[5] which
indeed affected me deeply, the English text says: "To him (a
friend of Weininger's) Freud had blabbed." This is somewhat
grandiloquently and impudently expressed and gives a completely
distorted picture of the facts. The notion of general homosexuality
and bisexuality had been accepted for quite some time by then,
and *had* to be emphasized in the treatment of any patient, just as
it is today. W.'s friend Swoboda came to learn of it as a patient. I
couldn't guess that he had a friend (completely unknown to me at
that time) to whom he would pass on the information and who
would thus be in a position to publish the idea before Fliess himself. In my answer to Fliess I reproached myself and expressed my
regret at the chain of events in a rather exaggerated way, including a

[3] Dr. Leopold Königstein (1850-1924), Professor of Ophthalmology at the University of Vienna, Freud's lifelong friend.

[4] Dr. Julius von Wagner-Jauregg (1857-1940), Professor of Psychiatry at the University of Vienna; Nobel Prize winner in 1927 for his discovery of the treatment of
syphilis with the malaria bacillus.

[5] See letter of Jan. 12, 1906.

self-tormenting reference to the role played by the unconscious. But I also read in your book that at first I forgot the conversation with W.'s friend during which the secret was passed on, and only later admitted it. This is so impossible—for it was not an isolated conversation but a substantial part of a treatment—that I should like to ask you where you acquired this information. If you cannot establish this and have misunderstood or confused some facts, then you cannot be spared the reproach of having committed a serious offense against an ethical obligation. It just will not do to say: "In my opinion you are a great man and a genius; as a result you have to allow yourself to be exposed in every way. I have flattered you so outrageously that I am entitled to your complete tolerance."

Neither would a little more truthfulness have done your biography any harm. After I had told you my reasons for my break with Stekel I of course didn't expect that you would inform the world about it, but I did expect that you would choose another interpretatioin for my attitude toward him. This you did not do in spite of your better knowledge. Well, it's done now! But your relationship with Stekel remains the blot that depreciates the value of your book in personal as well as in factual respects. You even follow him on scientific issues. One day when I am no more—my discretion will also go with me to the grave—it will become manifest that Stekel's assertion about [my alleged claim of] the harmlessness of unrestrained masturbation is based on a lie. It is a pity that—but this is enough. You will have realized long ago that I am not pleased with the success of your book about me. But there it is, one is a "great man," therefore helpless.

I greet you with the respect due to your powerful position as a biographer and with something of the old sympathy.

Your
Freud

210 *To* OTTO RANK

Semmering
August 25, 1924

Dear Mr. Rank

Today's mail brings a letter from you; but it contains nothing but an introduction for a superfluous Dr. W.

I cannot help noticing that during these months of your absence in situations critical for us, for you and me, you haven't shown more desire to let me know what is going on in and with you, and it worries me.

Although nowadays I view most events *sub specie aeternitatis* and can no longer devote to them the same passionate intensity as in former years, I am not indifferent to a change in our relationship.

My physical condition seems to suggest that I still have a little while to live, and it is my ardent wish that during this time I won't have to suffer your loss. I am told that you left Europe in an agitated and distrustful mood. The realization that I partly altered my opinion of your last work[1] may have increased your resentment. You probably overestimate the affective significance of this theoretical difference and believe that during your absence I have been exposed to the influences of people hostile to you. The object of this letter is to assure you that this is not the case. I am not so easily influenced, and others—Eitingon and Abraham have just been here for several days—are just as sincere in their recognition of your extraordinary merits as they are full of regret about the abrupt way with which you shut yourself off. There is no hostility toward you, neither with us nor with my New York family. There is just enough time before your return for an exchange of letters. I would like you to enlighten and reassure me about your condition at the moment.

I don't take the difference of opinion concerning the birth trauma seriously. Either you will convince and correct me in the course of time, provided there is still time enough, or you will correct yourself and sift what is a permanent new discovery from what has been added by the bias of the discoverer. I know that your discovery does not lack applause, but you must remember how few people are capable of passing judgment and how strong in most of them is the desire to get away from the Oedipus whenever there is a chance. In any case, even if it contains many misconceptions, you don't have to be ashamed of your brilliant and substantial production, which brings new and valuable ideas even to the critics. And you certainly should not assume that this work of yours interferes with our long and intimate relationship.

[1] *Das Trauma der Geburt und seine Bedeutung für die Psychoanalyse,* Intern. Psychoanal. Verlag, Leipzig, Vienna, Zürich, 1924. (*The Trauma of Birth,* New York, Harcourt, Brace, 1929.)

I add to my customary cordial greetings the hope of seeing you soon.

Your
Freud

211 *To* FRANKLIN HOOPER

Semmering
September 4, 1924

Dear Mr. Hooper

This is a case of an interesting contrast between American and European practices in publishing. Over here it would not occur to anyone that a book such as *These Eventful Years*,[1] published by you, would require any publicity, least of all by its contributors. I will, however, comply with your wish and declare that in my opinion too this book is among the strangest, most deserving, and most informative works that has come my way during this century. I am studying it assiduously so as to orient myself in this world in which, considering my age, I shall not be spending much more time.

My complete admiration goes to the introductory essay by Garvin,[2] and my warmest sympathy to the efforts to correct the opinions distorted by partisan passions about the origins of the war and the consequences of the peace.

I am very proud that you have granted psychoanalysis a chapter to itself. I hope that the future will justify your assessment. If my essay[3] has turned out longer than you wished it to be, my excuse is that a shorter description of the difficult topic would have offered nothing comprehensible to the reader.

Yours sincerely
Freud

[1] Published by the Encyclopaedia Britannica Co., London and New York, 1924.

[2] James Louis Garvin (1868-1947), English journalist, editor of *The Observer*.

[3] "Psychoanalysis: Exploring the Hidden Recesses of the Mind," Chap. 73, Vol. II of *These Eventful Years*, London and New York, 1924. Standard Ed., 19.

212 *To* GEORG GRODDECK

Vienna IX, Berggasse 19
December 21, 1924

Dear Dr. Groddeck

There is one trait in your character which annoys me and which I would love to influence, although I realize that I wouldn't get very far. I am sorry that you try to erect a wall between yourself and the other lions in the Congress menagerie. It is difficult to practice psychoanalysis in isolation; it is an exquisitely sociable enterprise. It would be so much nicer if we all roared or howled in chorus and in the same rhythm, instead of each one growling to himself in his corner. You know how much store I set by your personal sympathy, but it is time that you transferred some of it to the others. It could only benefit psychoanalysis.

The mention of Ferenczi allows me to guess the reproach of not yet having visited you in your beautiful native place.[1] I would love to do so, but please realize my present situation and how difficult traveling is for me at the moment, and perhaps forever.

With cordial greetings to you both
Your
Freud

[1] Baden-Baden.

1925

1931

213 *To* JULIUS TANDLER

Vienna IX, Berggasse 19
March 8, 1925

Dear Professor

I have been informed by Mr. Theodor Reik,[1] one of my best-trained nonmedical pupils, that he has been prohibited from practicing psychoanalysis by the Municipal Council of Vienna as of February 24, 1925.

I remember a conversation I had with you on this topic which resulted in a conformity of opinion much welcomed by myself. You seemed to approve of my statement that "in psychoanalysis everyone has to be considered a layman who cannot prove a satisfactory training in the theory and technique of it," no matter whether he has a medical degree or not.

The reasoning in the document of the Vienna Council leaves room, in my opinion, for serious objections. They ignore above all two undeniable facts—first, that psychoanalysis is not a purely medical matter, either as a science or as a technique; second, that it is not taught to students of medicine at the university.

I look upon the injunction of the Council as an unjustified encroachment in favor of the medical profession and to the detriment of patients and science.

The therapeutic interest is protected as long as the decision whether a certain case should undergo psychoanalytical treatment or not is made by a physician. In all Mr. Reik's cases I myself have made decisions of this kind. After all, I also look upon it as my right to send a patient complaining of pains in the feet and difficulties in walking to an orthopedic shoemaker (if I can diagnose the trouble as flatfoot), instead of prescribing antineuralgic and electrical treatment.

[1] See note 2 to letter of Jan. 2, 1912.

359

If the official authorities, to whom hitherto psychoanalysis has had so little to be grateful for, are now willing to recognize it as an effective, under certain circumstances, dangerous treatment, then they should create guarantees that such treatment cannot be undertaken recklessly by the untrained, whether physicians or not. The Vienna Psychoanalytical Society could without difficulty function as a supervisory committee of this kind.

I ask you herewith to grant Mr. Reik an interview on his behalf. He will also be the bearer of a recently published "Ergography"[2] by me.

I beg to remain
Yours sincerely
Freud

[2] *An Autobiographical Study,* which first appeared in *Die Medizin der Gegenwart in Selbstdarstellungen,* L. R. Grote, editor. Ergography was the autobiographical sketch containing not only a biographical description of the author's personal life but also a description of his work.

214 *To* LOU ANDREAS-SALOMÉ

Vienna
May 10, 1925

Dearest Lou

Sunday afternoon and quiet! This morning with the help of Anna and the typewriter I dealt with my correspondence, which had accumulated as a result of my superannuated imprudence,[1] and [now] I can thank you and have a talk with you.

First of all let me thank your dear old gentleman for the charming lines he wrote to someone unknown to him. May he keep well as long as he himself wants to.

As for me, I no longer want to ardently enough. A crust of indifference is slowly creeping up around me; a fact I state without complaining. It is a natural development, a way of beginning to be inorganic. The "detachment of old age," I think it is called. It must be connected with the decisive crisis in the relationship of the two instincts[2] stipulated by me. The change taking place is perhaps not very conspicuous; everything is as interesting as it was before;

[1] Allusion to Freud's sixty-ninth birthday.
[2] Libido and death instinct.

neither are the qualities very different; but some kind of resonance is lacking; unmusical as I am, I imagine the difference to be something like using the pedal or not. The never-ceasing tangible pressure of a vast number of unpleasant sensations must have accelerated this otherwise perhaps premature condition, this tendency to experience everything *sub specie aeternitatis*.

Otherwise existence is still bearable. I even think I have discovered something of fundamental importance for our work, which I am going to keep to myself for a while. It is a discovery of which one ought almost to be ashamed, for one should have divined these connections from the beginning and not after thirty years. One more proof of the fallibility of human nature. . . .°

It isn't right that our efforts to meet remain so unsuccessful. I very probably won't go to Homburg,[3] but I shall insist that Anna goes, and then she is bound to meet you there. In any case you are continually informed about the details of our existence.

With fond greetings
Your Freud

° *überall mit Wasser gekocht wird.* . . .
[3] The Seventh Psychoanalytical Congress in Bad Homburg.

215 *To* KARL ABRAHAM

Semmering, Villa Schüler
July 21, 1925

Dear Friend

No, writing is really no hardship for me. I was glad to get news from you so soon, for my thoughts often stray from the self (which has become tiresome with all its demands) to others, to friends. When you describe your unusual preoccupation with your "poor Conrad,"[1] then I, the experienced one, am comforted by the certainty that you will have to submit to this adaptation only for a limited time. A permanent new adjustment is far more difficult.

A strange thing occurred when, on looking through the list of the Congress lectures, I actually felt relief at *not* finding your name among the lecturers. To curb your mental activity with such a

[1] See note 5 to letter of Aug. 4, 1921.

task ahead of you would be impossible. But we all hope that by the first week of September our president[2] will have regained his ability to breathe and act freely.

We are living this year very comfortably in the Semmering milieu with which you are familiar. The summer has assumed another, friendlier character and even the modest attractions of the neighborhood do, after a prolonged intimacy, cast their spell. The housekeeping is very easy for the women and life in this beautifully furnished house could almost be called ideal.

The day has passed before one has quite realized it. When thinking it over in the evening it turns out to have had little content. Some ruminating at the writing table, an hour with the American patient . . . , adventures with Wolf[3] (as yet unknown to you), who with his passionate affection and jealousy, his suspicion of strangers, his combination of wildness and docility, has become the center of public interest; a few letters, corrections, family visits from America, etc.

The drastic operation with which my doctor dismissed me three weeks ago has changed the character of my complaint very much to its advantage. All the parentheses which had so tyrannically forced themselves upon one's attention have disappeared and once more set free the human being, albeit one who can still complain of clumsy speech and an endless nasal infection if he feels like it. Thus life is fairly bearable, but what is regular work going to taste like in October after this period of being idle and spoiled?

I have written several brief essays, but they are not very seriously meant. Perhaps I will tell you about them later, when I have decided to recognize them. I will let you know the titles: "Negation"—"Inhibition and Symptom"—"Some Psychological Consequences of the Anatomical Distinction between the Sexes."[4]

Now try to keep well and get out of Wengen[5] everything needed to finish with this episode of sickness. Our warmest thanks go to

[2] Dr. Abraham.

[3] Anna's wolfhound.

[4] "Die Verneinung," Vienna, 1925 ("Negation," Standard Ed., 19); "Hemmung, Symptom und Angst," Vienna, 1926 ("Inhibitions, Symptoms and Anxiety," Standard Ed., 20); "Einige psychische Folgen des anatomischen Geschlechtsunterschieds," Vienna, 1925 ("Some Psychological Consequences of the Anatomical Distinction between the Sexes," Standard Ed., 19).

[5] Mountain resort in Switzerland.

your wife for the share she is taking in your recovery. I have similar things to remember.

> Fond greetings to you all!
> Your
> Freud

216 *To* ERNEST JONES

Vienna IX, Berggasse 19
December 30, 1925

Dear Jones

I can only repeat what you have said: Abraham's death is perhaps the greatest loss that could have struck us—and it has struck us! In my letters I used to call him jestingly my *"rocher de bronze";* I felt safe in the absolute confidence he inspired in me, as in everyone else. In the brief obituary[1] which I wrote for the *Zeitschrift* (Issue I, 1926)—a detailed appreciation of him will probably come from another side; Issue 2 is being reserved for it— I applied to him Horace's line: *"Integer vitae scelerisque purus."*[2]

I have always found exaggerations on the occasion of death particularly distasteful and I have been careful to avoid them, but in citing this quotation I feel I have been sincere.

Who would have thought when we met that time in the Harz mountains[3] that he would be the first to leave this unreasonable life!—We must continue to work and stand together. As a human being no one can replace this loss, but for psychoanalysis no one is allowed to be irreplaceable. I shall soon drop out, the others not till much later, I hope, but our work, compared to whose importance we are all insignificant, must continue.

I am certainly in favor of Eitingon as president; we have owed him this honor for a long time, but he doesn't intend to return till March; his ailing wife makes too many claims on him. What is the

---

[1] "Karl Abraham," *Internationale Zeitschrift für Psychoanalyse.* (International Journal of Psycho-Analysis, 1926, and Standard Ed., 20.)

[2] Horace, *Odes,* I, 22, 1: "He whose life is blameless and free of guilt."

[3] Where the "Committee" met in September, 1921.

good to us of an absent president? I hope that in Berlin you are arranging everything for the best with him, Ferenczi, and Emden. I clasp your hand in a warm feeling of fellowship.

Your
Freud

217 *To* ROMAIN ROLLAND[1]

Vienna IX, Berggasse 19
January 29, 1926

Unforgettable man, to have soared to such heights of humanity through so much hardship and suffering!

I revered you as an artist and apostle of love for mankind many years before I saw you. I myself have always advocated the love for mankind not out of sentimentality or idealism but for sober, economic reasons: because in the face of our instinctual drives and the world as it is I was compelled to consider this love as indispensable for the preservation of the human species as, say, technology.

When I finally came to know you personally I was surprised to find that you hold strength and energy in such high esteem, and that you yourself embody so much will power.

May the next decade bring you nothing but fulfillment.

Most cordially
Your

Sigm. Freud, *aetat.* seventy

[1] Freud's contribution to the *Liber Amicorum Romain Rolland,* edited by Maxim Gorki, Georges Duhamel, and Stefan Zweig. Rothapfel-Verlag, Zürich, 1926. ("To Romain Rolland," Standard Ed., 20.)

218 *To* HANS AND JEANNE LAMPL DE GROOT

Vienna IX, Berggasse 19
February 11, 1926

My dear Friends

It was to be expected that Jeanne would pass the great test with flying colors and fulfill the consequent duties admirably, but we are scientists after all and the confirmation by experience must be more important for us than the certain expectation. I was of course very pleased to hear that everything went so well and am even inclined to think that with today's attitude toward the sexes it doesn't make a great difference whether the baby is manifestly male or female. Especially as a clear predominance in one direction can be compensated to your liking by the result of future experiments.

My warmest greetings to all three members of the happy union!
Your
Freud

219 *To* ENRICO MORSELLI

Vienna IX, Berggasse 19
February 18, 1926

Dear Dr. Morselli

While reading your important work[1] on psychoanalysis I noticed with regret that you cannot accept our youthful science without great reservations, and I have to comfort myself with the divergence of opinion inevitable in such difficult topics as well as with the certainty that your book will contribute a great deal toward arousing the interest of your compatriots for psychoanalysis. But your brief pamphlet on the Zionist question I was able to read without any mixed feelings, with unreserved approval, and I was pleased to see with what sympathy, humaneness, and understanding you were able to choose your point of view concerning this matter which has been so distorted by human passions. I feel as though obliged to send you my personal thanks for it. I am not sure that your opinion, which looks upon psychoanalysis as a direct product of the Jewish mind, is correct, but if it is I wouldn't be ashamed. Although I have been alienated from the religion of my forbears for a long time, I have never lost the feeling of solidarity with my people and realize with satisfaction that you call yourself a pupil of a man of my race—the great Lombroso.

In the past I wouldn't have hesitated to ask your permission to

---

[1] *La Psicanalisi,* Fratelli Bocca, Turin, 1926.

visit you on my next trip to Italy. Unfortunately I cannot consider traveling at the moment.

Yours sincerely

Freud

220 *To* MEMBERS OF THE B'NAI B'RITH LODGE

May 6, 1926

Right reverend Grand Master, reverend Master, dear Brothers!

I thank you for the honor shown to me today. You know why I cannot reply with the sound of my own voice. You have just heard one of my friends[1] and pupils talk about my scientific work, but to pass judgment on these things is difficult and perhaps for some time to come will not be possible with any degree of certainty. Permit me to add something to the speech of my predecessor, who happens to be my friend and painstaking physician. I would like to tell you briefly how I became a B.B. and what I sought with you.

It happened that in the years after 1895 two strong impressions coincided to produce the same effect on me. On the one hand I had gained the first insight into the depths of human instinct, had seen many things which were sobering, at first even frightening; on the other hand the disclosure of my unpopular discoveries led to my losing most of my personal relationships at that time; I felt as though outlawed, shunned by all. This isolation aroused in me the longing for a circle of excellent men with high ideals who would accept me in friendship despite my temerity. Your Lodge was described to me as the place where I could find such men.

That you are Jews could only be welcome to me, for I was myself a Jew, and it has always appeared to me not only undignified, but outright foolish to deny it. What tied me to Jewry was—I have to admit it—not the faith, not even the national pride, for I was always an unbeliever, have been brought up without religion, but not without respect for the so-called "ethical" demands of human civilization. Whenever I have experienced feelings of national exaltation, I have tried to suppress them as disastrous and unfair, frightened by the warning example of those nations among which

---

[1] Dr. Ludwig Braun, Professor of Medicine at the University of Vienna.

we Jews live. But there remained enough to make the attraction of Judaism and the Jews irresistible, many dark emotional powers all the stronger the less they could be expressed in words, as well as the clear consciousness of an inner identity, the familiarity of the same psychological structure. And before long there followed the realization that it was only to my Jewish nature that I owed the two qualities that have become indispensable to me throughout my difficult life. Because I was a Jew I found myself free of many prejudices which restrict others in the use of the intellect: as a Jew I was prepared to be in the opposition and to renounce agreement with the "compact majority."[2]

So I became one of you, took part in your humanitarian and national interests, made friends among you and persuaded the few friends who had remained with me to join our Lodge. There was, after all, no question of my trying to convince you of my new doctrines, but at a time when no one in Europe would listen to me and I had no pupils in Vienna, you offered me your sympathetic attention. You were my first audience.

Some two-thirds of the long time since my entrance, I conscientiously kept up with you, finding recreation and stimulation in my contact with you. You were kind enough not to reproach me today for having stayed away from you during the last third of this period. By that time I was up to my eyes in work, demands connected with it took precedence, the work no longer allowed me to prolong the day by attending your meetings, and soon the body would not permit postponement of the evening meal. Finally there came the years of that illness which also prevents me from appearing among you today.

Whether I have been a proper B.B. in your sense I do not know. I am inclined to doubt it; there were too many special conditions affecting my case. But I can assure you that during the years I have belonged to you, you have meant much to me and done much for me. And so please accept for those years as well as for today my warmest thanks.

In benevolence, brotherly love and harmony

Your

Sigm. Freud

[2] Quotation from *An Enemy of the People,* by Henrik Ibsen (1828-1906).

221 *To* MARIE BONAPARTE

Vienna IX, Berggasse 19
May 10, 1926

My dear Princess

. . . The days of celebration[1] are already over, only letters still keep trickling in. It was fairly exhausting, but I suffered from it only on the first day (May 5). All that happened later I stood very well. There was no false note; I got all kinds of flattering things to hear and to read. It was concluded by a celebration of the Jewish Lodge to which I have belonged for twenty-nine years. My doctor, Prof. Ludwig Braun, made a speech there of a kind which cast a spell over the whole audience, including my family. I had asked to be excused from attending. It would have been embarrassing and in poor taste. As a rule when I am attacked I can defend myself; but when I am praised, I am helpless. All the Viennese papers, and many German ones, printed articles on *feuilletons,* most of them very appreciative, some rather artificial and forced. The Jewish societies in Vienna and the University in Jerusalem (of which I am a trustee), in short the Jews altogether, have celebrated me like a national hero, although my service to the Jewish cause is confined to the single point that I have never denied my Jewishness. The official world—the University, Academy, Medical Association—completely ignored the occasion. Rightly, I think; it was only honest. I could not have looked upon their congratulations and honors as sincere. Only the (social-democratically ruled) City of Vienna sent me the diploma of the Freedom of the City which was bestowed on me two years ago; the Lord Mayor[2] brought it to me himself. My closest friends in London, Berlin, and Budapest were here all the time; on the seventh we had 7½ hours of discussions and conferences. Four-fifths of the 30,000 marks given to me I have turned over to the *Verlag,* one-fifth to the Vienna Clinic. I now have to thank the donors. *One of them I am thanking in this letter.*

Among the written congratulations those that pleased me most came from Einstein, Brandes, Romain Rolland and Yvette Guil-

[1] For Freud's seventieth birthday.
[2] Dr. Karl Seitz.

bert; the best newspaper articles were by Bleuler (Zürich) and Stefan Zweig in the Viennese *Neue Freie Presse.*

General impression: the world has acquired a certain respect for my work. *But so far analysis has been accepted only by analysts.*

My wife, who is fundamentally quite ambitious, has been very satisfied by it all. Anna, on the other hand, shares my feeling that it is embarrassing to be publicly exposed to praise. Mathilde has taken endless trouble with everything, my two sons[3] from Berlin have used the occasion for a visit with their wives; my son Martin, who possesses more wit and humor than he can dispose of as a bank official, has fabricated a wonderful group: Oedipus and the Sphinx, out of the most amusing material.

The document about your beautiful bronze has not completely explained to me its connection with Giovanni da Bologna,[4] especially as I don't possess the work by Bode.[5] Further information would be welcome. An etching which Schmutzer[6] completed for the birthday strikes me as excellent. Others find its expression too severe, almost angry. Inwardly this is probably what I am. I now wonder whether I can send a copy of it to St. Cloud or whether you would rather accept it on your arrival on the Semmering.

> With warmest greetings
> Your barely 70-year-old
> Freud

[3] Oliver and Ernst.

[4] Giovanni da Bologna (1524-1608), Italian sculptor.

[5] Wilhelm von Bode (1845-1929), German art historian and Director of the National Museums in Berlin.

[6] Ferdinand Schmutzer (1870-1928), Austrian engraver.

222 *To* MATHILDE BREUER

Vienna IX, Berggasse 19
May 13, 1926

My dear Frau Breuer

The lines with which you congratulate me on my seventieth birthday have moved me most deeply. The black-edged notepaper reminded me with lightning speed of everything from the moment when, glancing through the door of the consulting room, I first saw you sitting at the table with a barely two-year-old daughter, through all the years when I could almost count myself a member of your family, and the various vicissitudes my life has known since then. Please accept from me in remembrance of this past my most respectful thanks.

Most cordially

Your

Freud

223 *To* ROMAIN ROLLAND

Vienna IX, Berggasse 19

May 13, 1926

Dear Friend

Your lines are among the most precious things which these days have brought me. Let me thank you for their content and your manner of address.

Unlike you I cannot count on the love of many people. I have not pleased, comforted, edified them. Nor was this my intention; I only wanted to explore, solve riddles, uncover a little of the truth. This may have given pain to many, benefited a few, neither of which I consider my fault or my merit. It seems to me a surprising accident that apart from my doctrines my person should attract any attention at all. But when men like you whom I have loved from afar express their friendship for me, then a particular ambition of mine is gratified. I enjoy it without questioning whether or not I deserve it, I relish it as a gift. You belong to those who know how to give presents.

With my warmest wishes for your well-being

Your devoted

Freud

224 *TO* HAVELOCK ELLIS

Semmering

September 12, 1926

Dear Friend

It is not a bad thing once in a while to go against one's principles and habits. Had I suppressed that bit of annoyance over your essay in the *Forum*,[1] as I usually do, I would not feel entitled to address you so intimately today and probably would not have been presented with your biography, which fulfills an old wish of mine. For I have often been anxious to know more personal things about you. *Well aware of the narrowness and the limitations of my own being,** I have always liked to look out for richer and happier natures, and for a long time I have suspected that you have not been a *beast of burden** as I was, but instead could afford to lead a many-sided and harmonious life. Although I cannot imagine being you—I just don't possess your goodness and I am far more urged than you to make decisions—I nevertheless could not help looking for similarities and was glad to find one in the first chapter. The etching of St. Hieronymus[2] in his study is also a favorite of mine and has been hanging in front of me in my room for years. And perhaps some of the ideals you have realized in life have also been mine.

I also see a sign of your magnanimity in your readiness to place so much personal material at the disposal of a biographer. For this I would lack all incentive. But there is no point in continuing the comparison.

I learned with great regret that an internal complaint has compelled you to husband your energy. But I am convinced that it will not interfere with the harmony of your life. As for myself I am of course reminded every hour of the mutilation caused by the operation on my jaw, but after almost three years a relapse is not expected to occur. It looks as though I shall have to wait for another way out of this life; in any case, the heart has already declared it has had enough. I am altogether not very fond of Bernard Shaw, but his *Back to Methuselah* struck me as all but *insipid.**

Don't be surprised if I avail myself of the opportunity to write to you again, and accept the warm greetings of

Your

Freud

[1] Journal edited by Wilhelm Herzog of Munich and Berlin.
* Written in English.
[2] By Albrecht Dürer (1471-1528).

225 *To* ERNEST JONES

Vienna IX, Berggasse 19
November 20, 1926

Dear Jones

Can it possibly be twenty years since you joined the psycho-analytical movement? It has really become entirely yours, for you have achieved everything that could be gained from it: a society, a journal, and an institute.[1] What you have meant to it we may leave to the historians to establish. That you can mean still more I feel quite sure, once the many business activities you complain of have turned into a smooth routine. Then you will find the leisure to offer more of your accumulated experience to your colleagues and to posterity.

We have reason to be well satisfied with each other. I myself have the impression that you sometimes overestimate the significance of the dissensions that have cropped up even between us. It is hard, after all, to succeed in satisfying each other completely; one misses, and is critical of, something in everyone. You yourself have pointed out that there were also differences of opinion between Abraham and myself—it is much the same thing as with one's wife and children; only funeral orators deny these traces of reality; the living have the right to assert that such criticism of an ideal picture doesn't interfere with their enjoyment of reality.

You will be surprised when I reveal the reasons that hamper my correspondence with you. It is a classic example of the petty restrictions to which our nature is subject. I find it really very difficult to write German in Latin characters as I am doing today. All fluency—on a higher level it is called inspiration—leaves me at once. But you have often told me that you cannot read Gothic script, which leaves me with but two means of communication, both of which impair the sense of intimacy: either to dictate the letter to Anna on the typewriter, or to use my clumsy English. . . .

My general condition is turning me away from work, I am inclined to think for good. It is just as well not to deceive oneself. I

[1] The British Psycho-Analytical Society, *The International Journal of Psycho-Analysis*, The Institute of Psycho-Analysis.

feel as though I shall be permitted to live a little longer on bounty.*
Your wife, to whom I send warm greetings, can translate this word
for you.

Cordially
Your
Freud

* *Gnadenbrot.*

226 *To* OSKAR PFISTER

Vienna IX, Berggasse 19
November 21, 1926

Dear Pfister

    I seize the opportunity to congratulate you; I would be very
happy to set aside a special shelf in my bookcases for the translations
of your books if you would like to present me with them. Of all the
applications of psychoanalysis the only one actually thriving is
that which you inaugurated, the branch applied to pedagogy. I
am very pleased that my daughter is beginning to achieve something in this field.

    The day after receiving your book I dreamed I was in Zürich. I
probably wanted to visit you. The dream of a travel cripple! At
Christmas I would like to risk the attempt of a trip to Berlin—to
make the acquaintance of four little grandchildren.[1]

Cordially
Your
Freud

[1] Ernst's three sons and Oliver's daughter.

227 *To* ROZSI VON FREUND

Vienna IX, Berggasse 19
December 17, 1926

Dear Roszi

    I have purposely allowed your first period of mourning[1] to
pass before sending you a few words of sympathy. I feel that some-

[1] Frau von Freund's mother had died.

one who has suffered a great loss is entitled to be left in peace. In fact, I think this period of silence ought to be extended for a long time if the fear of appearing unsympathetic did not compel one to communicate. As it is, anything one can say must sound to the bereaved like so many empty words. The "work of mourning" is an intimate process which cannot stand any interference.

My thoughts go back to another occasion[2] of painful mourning which fate brought to you—but to me, too; and I cannot get over it that so much sorrow has been inflicted on you, undeserved as usual in such cases and out of proportion to the strength with which nature has endowed you. Where are we to look for justice? No one inquires after our wishes, our merits, or our claims. But if the wishes of your friends had any power, your life would have taken a happier course.

Old and ailing myself, I can only ask you to believe that my sympathy for you and your family has not faded and will not fade.

Most cordially
Your
Freud

[2] Anton von Freund's death (see letter of May 14, 1920).

228   *To* WERNER ACHELIS

Vienna IX, Berggasse 19
January 30, 1927

Dear Mr. Achelis

With politeness rare among German scholars you have sent me an essay[1] discussing my work, which you rightly assume would arouse my interest. I thank you for it as well as for the accompanying letter, and will return the manuscript in the immediate future.

What I have to say about your argument will not surprise you, as you seem to be familiar with my attitude to philosophy (metaphysics). Other defects in my nature have certainly distressed me

[1] "Das Problem des Traumes, eine philosophische Abhandlung" ("The Problem of Dreams, a Philosophical Essay"), Stuttgart, 1928.

and made me feel humble; with metaphysics it is different—I not only have no talent for it but no respect for it, either. In secret—one cannot say such things aloud—I believe that one day metaphysics will be condemned as a nuisance, as an abuse of thinking, as a survival from the period of the religious *Weltanschauung*. I know well to what extent this way of thinking estranges me from German cultural life. Thus you will easily understand that most things I read in your essay have remained unappreciated by me, although I several times felt that the essay contained quite "brilliant" thoughts. At other times, for instance when you invite the reader to admire Blüher's[2] genius, I had the impression of being faced with two worlds separated by an unbridgeable gulf.

However this may be, it is certainly simpler to find one's way in "this world" of facts than in the "other world" of philosophy. So please allow me a correction which can easily be confirmed in my various writings. I have never maintained that all dreams have a sexual content, nor that sexual impulses are the mainspring of all dreams. On the contrary, I have always forcefully contradicted this opinion whenever it has been ascribed to me. So I have reason to be dissatisfied when you too repeat this error.

Finally, a word about the translation of the motto above *The Interpretation of Dreams,* also about the interpretation of the motto. You translate *"acheronta movebo"* as "moving the citadels of the earth." But it means to "stir up the underworld." I had borrowed the quotation from Lassalle,[3] in whose case it was probably meant personally and relating to social—not psychological—classifications. In my case it was meant merely to emphasize the most important part in the dynamics of the dream. The wish rejected by the higher mental agencies (the repressed dream wish) stirs up the mental underworld (the unconscious) in order to get a hearing. What can you find "Promethean" about this?

Yours sincerely
Freud

[2] Hans Blüher (1888-1955), German writer, author of *The Role of Erotism in a Male Society.*

[3] Ferdinand Lassalle (1825-1864), one of the founders of the socialist movement.

229   *To* MARGIT FREUD

Vienna IX, Berggasse 19
March 4, 1927

My dear Gretel

I too have been very saddened by Brandes' death. Not that I begrudge him his eternal rest at the age of eighty-five, but I made his acquaintance so late and would like to have met him again. Our meeting that time as he lay in the hotel room in Vienna was a very special experience. I didn't know anything about his hostility to psychoanalysis and approached him quite ingenuously, with unclouded respect. When the conversation finally turned to psychoanalysis, his conversion was seemingly a matter of two minutes. I cleared up the most obvious misconception, his failure to recognize the distinction between conscious and unconscious, and he appeared pacified at once. Probably it wasn't even a question of changing his mind, rather that psychology [psychoanalysis] had always been very alien to him and that, prepared by your influence and disarmed by my guilelessness, he was ready to relinquish a prejudice where he was unable to form an opinion. He could not fail to realize how highly I respected him. When he modestly tried to take second place behind the "scientist," I pointed out to him his position among the descendants of our prophets.

I am fond of looking for resemblances. At that time I was struck by his likeness to Wallenstein, which however didn't signify anything in particular. After I had talked to Einstein in Berlin and seen the bust of Popper-Lynkeus in our Rathauspark, I discovered a more significant resemblance between these three great Jewish personalities.

I was very pleased to hear from you and from other sides that you are beginning to prosper and that your industry and ability are at last bearing fruit. Earning brings so much more pleasure than owning. Anna will be in Berlin on the nineteenth and twentieth of March. If you are thinking of coming to Vienna, I hope you will not choose just these days.

With fond greetings to you and your two men
Your uncle
Sigm.

230 *To* ARNOLD ZWEIG

Vienna IX, Berggasse 19
March 20, 1927

Dear Sir

I accept the offer of the author of *Claudia*[1] to dedicate to me his new book with thanks and full appreciation of the honor shown to me.

This I would have done anyway, but I do it all the more now that I have learned from your letter that you appreciate psychoanalysis and that you have established a personal relationship with it.

Please fulfill the promise to visit me one day. (Don't wait too long, I am soon going to be 71.)

Yours very sincerely
Freud

[1] *Novellen um Claudia*, Kurt Wolff Verlag, Leipzig, 1914.

231 *To* OSKAR PFISTER

Vienna IX, Berggasse 19
June 1, 1927

Dear Pfister

I have just accomplished the hangman's job you asked me to do via Frau H. The letters of the year 1912 have been destroyed; a few of impersonal content still lie in front of me.

I have done what you asked me to do, but I didn't like doing it. I was sorry for the letters, which I re-read for the first time for many years. Your picture rose before me as you had been then, with all your winning features, your enthusiasm, your exuberant gratitude, your courage to face the truth, your blossoming out after the first contact with psychoanalysis, also the blind confidence you placed in human beings who were to disappoint you so soon. And although I realized that you had been in danger of committing some follies at that time and that since then your fate has taken a gentler course, I nevertheless couldn't help regretting that that battle passed you by. And perhaps it wasn't just my fanaticism or the spectator's desire for sensation that made me feel like that. Perhaps friendship had something to do with it. . . .

I wonder if you would write today what I have just read: "Don't speak of your age, you are the youngest of us all"?

I greet you cordially
Your
Freud

232 *To* JULIE BRAUN-VOGELSTEIN

Vienna IX, Berggasse 19
October 30, 1927

Dear Madam

It does not require a mediator to induce me to fulfill your wish.[1] I am only afraid that I cannot offer as much as you expect. Our relationship, so close during school years, had already come to an end while we were at the university.

In the last fifty (!) years I have heard hardly anything about him. In all that time I have run into him on several occasions by chance in the street in Vienna; he may once have come to visit me in my home. On another occasion I received a letter from him asking me to analyze one of his sons who was causing him difficulties. (It must have been during the first decade of psychoanalysis.) This I declined because I was not ready for the treatment of young people and thought—rightly—that personal relations were not favorable for the psychoanalytical situation.

I am afraid I don't have any letters from him. The last memorable encounter we had may have taken place in 1883 (?) or 1884 (?). At that time he came to Vienna and invited me to lunch at the house of his brother-in-law, Victor Adler.[2] I still remember that he was a vegetarian at the time, and that I got a glimpse of

[1] Julie Braun-Vogelstein had asked Freud for memories of her late husband, the German Socialist politician Heinrich Braun (1854-1927), whose biography she was about to write.

[2] Victor Adler (1852-1918), leader of the Austrian Socialists, had married Braun's sister.

little Fritz,[3] who was between one and two years old. (It strikes me as strange that this took place in the same rooms as those in which I have been living for the past thirty-six years.)

I know that I made Heinrich Braun's acquaintance during the first school year on the day of the first annual school report, and that we soon became inseparable friends. I spent every hour not taken up by school with him, mostly at his place—especially as long as his family were not yet in Vienna and he was living with his elder brother Adolf[4] and a tutor. This brother tried to upset our friendship, but we ourselves got along very well. I can hardly remember any quarrels between us or periods when we were "cross" with each other, as usually happens in these early friendships. What we did all these days and what we talked about is hard to imagine after so many years. I think he encouraged me in my aversion to school and what was taught there, aroused a number of revolutionary feelings within me, and we encouraged each other in overestimating our critical powers and superior judgment. He directed my interest toward books like Buckle's[5] *History of Civilization* and a similar work by Lecky,[6] which he very much admired. I admired him, his energetic behavior, his independent judgment, compared him secretly with a young lion and was deeply convinced that one day he would fill a leading position in the world. A learner he was not, but although I soon became and remained head boy I did not hold this against him; with the vague perception of youth I guessed that he possessed something which was more valuable than any success at school and which I have learned since to call "personality."

Neither the aims nor the means of our strivings were very clear to us. I have since come to suspect that his aims were mainly negative; but it was understood that I would work with him and never let down his side. Under his influence I also decided at that time to study law at the university.

The first interruption of our relationship occurred when he—I

[3] Friedrich Adler assassinated the Austrian Prime Minister Graf Stürgkh in 1916. Later he was made Secretary of the Second International (Association of European Workers).

[4] Actually his name was Anton.

[5] Henry Thomas Buckle (1821-1862), English historian.

[6] W. E. H. Lecky (1838-1903), Irish historian.

think it was in the "Septima," the highest class but one—left school, unfortunately not of his own free will. He turned up again during my first year at the university. But I was studying medicine, and he law. He soon made the acquaintance of Victor Adler, who later on became his brother-in-law. Our ways slowly parted; he always had more social contacts than I had, always found it easy to make new ones. He probably no longer felt any need to associate with me. So it came about that during the later years at the university I lost sight of him completely. I don't even know whether he took his degree in Vienna. I doubt it. One day, in 1881 or 1882, after a long estrangement, we met in the street. It was still natural for both of us to begin talking immediately about personal things. He showed me the picture of the girl he was engaged to, and I confessed that I was in the same situation. Perhaps it was then that he invited me to lunch at Adler's. After this, we didn't meet for a long time and there was never a relationship between us again.

Yours respectfully
Freud

233 *To* HAVELOCK ELLIS

Vienna IX, Berggasse 19
May 12, 1928

Dear Friend

My cordial thanks for sending me your most recent work.[1] I now possess the complete series of your valuable studies, most of them with your inscription, except for Volume IV (*Seven Selections*). Does it make any sense for a man of my age to try and fill his library? At best in the interest of his heirs. Life already lies heavy on my shoulders and I feel in no way privileged to have reached an age attainable only by a few. I continue with my professional work partly because I have to, but the material no longer shapes itself into coherent systems, no longer submits to ideas which used to appear out of the unknown. *Im Austragstüberl,*[2] it is called in our Alpine regions.

---

[1] *Studies in the Psychology of Sex*, Random House, New York.
[2] A room set aside for the old and infirm members of the family.

When one knows little of the other person one is at liberty to imagine him in undiminished strength and health. I know I won't be able to congratulate you on having reached the traditional age limit until 1929. If I survive I won't fail to do so.

Yours very sincerely
Freud

234 *To* GEORGE SYLVESTER VIERECK

Semmering
July 20, 1928

Dear Mr. Viereck

I have owed you an acknowledgement of your friendly communications for a long time, and there is no justification for keeping you waiting any longer. I did not write about the *Days in Doorn* at once, because . . . yes, because something in this book rather annoyed me. I should like to spare your feelings, although I know from personal contact that your feelings are not easily hurt. Proof of great tolerance and an amiable nature.

I liked your "Clemenceau"[1] very much indeed. One feels that this is how that strange man is, magnificent and indifferent to what mankind calls greatness. In a long conversation I had with Georg Brandes the year before his death, our talk turned to "The Tiger." I let myself be carried away into calling him the greatest living offender against humanity. I didn't know at the time that he and Brandes had been intimate friends until the Great War (if I am not mistaken) tore them apart. The wise old man did not take me to task; with a friendly look he clasped my hand and changed the subject. I took this to mean: "Never mind, my friend, this is something you don't understand." While reading your interview I realized to my surprise that I could feel a deep sympathy for this hated enemy, as though it were not at all difficult for me to put myself in his place, something I have never succeeded in doing with other despots like Lenin and Mussolini. I think I was overwhelmed by discovering a common outlook on the world. I bear him no grudge for maintaining that he never heard of psychoanalysis. . . .

---

[1] Georges Clemenceau (1841-1929), commonly referred to as "The Tiger."

Your memoirs of the "Eternal Jew" are assured of my greatest interest. I will study them expressly from the point of view which you suggest. My judgment of them will not be worth much, for first of all I find it difficult to make literary evaluations, and secondly my sympathy for the author is bound to falsify my criticism.

I am sorry you won't be paying me a visit this year. I would like to have presented your wife with a bouquet of the most beautiful roses in our garden. You could have convinced yourself of the progress which age has achieved in the destruction of my precious personality. My womenfolk, even your alleged enemy,[2] would have given you a hearty welcome.

With kind regards

Yours sincerely

Freud

[2] Viereck thought that Anna disliked him.

235 *To* SANDOR FERENCZI

Berlin-Tegel

October 12, 1928

Dear Friend

I don't want to deny cordial greetings to the former traveling companion who is now allowing himself to fulfill my ungratified traveling wishes on his own, greetings which spring from an envious but sympathetic heart. I myself am definitely moving toward an improvement which I am hoping to take back home with me during this month.

Many cordial greetings to you as well as to Frau Gisela, also from my faithful Antigone—Anna.

Your

Freud

236 *To* ERNST SIMMEL

Vienna IX, Berggasse 19

November 11, 1928

Dear Dr. Simmel

You are right. Once upon a time these rings were a privilege and a mark distinguishing a group of men who were united in their

devotion to psychoanalysis, who had promised to watch its development as a "secret committee," and to practice among themselves a
kind of analytical brotherhood. Rank then broke the magic spell; his
secession and Abraham's death dissolved the Committee.

When on leaving Tegel[1] I expressed the desire to acknowledge by
a token the quite extraordinary kindness with which you managed
to transform for me a time of personal difficulty into a time of comfort, my daughter suggested that I renew the old custom with you.
And indeed, apart from my personal indebtedness to you, I don't
know anyone in Berlin who, by the purity and intensity of his al-
legiance, would be more worthy of inclusion in that circle—if it
still existed.

Forms may pass away, but their meaning can survive them and
seek to express themselves in other forms. So please don't be disturbed by the fact that this ring signifies a regression to something
that no longer exists, and wear it for many years as a memory of
your cordially devoted

Freud

[1] The doctor's house at the Tegel Psychoanalytical Sanatorium had been put at
Freud's disposal by Dr. Simmel, the Director, while Professor Schröder, Berlin
dentist, treated Freud.

237 *To* RICHARD S. DYER-BENNET

Vienna IX, Berggasse 19
December 9, 1928

Dear Major*

Let me confine my English to this address. The idea that
Havelock Ellis will help you with the translation of this letter makes
the writing of it easier for me. There are few people with whom
I prefer—or would prefer—to correspond.

I conclude from your pamphlet as well as from your letters that
you have an enthusiastic nature, which makes me realize that I
shall have to disappoint you.

I have little fault to find with your *Gospel of Living;* so allow me
to express that little. To turn human beings into gods and the earth
into heaven would not be an aim of mine. This is too reminiscent of

* Salutation written in English.

vieux jeu, nor is it quite feasible. We human beings are rooted in our animal nature and could never become godlike. The earth is a small planet, not suited to be a "heaven." We cannot promise those willing to follow us full compensation for what they give up. A painful piece of renunciation is inevitable. I also object that you, unlike us, look upon "ignorance" as the obstacle preventing the transformation of life that we aspire to. I don't believe that intellect can be assigned such an important role. It is rather in the instinctual drives and the interests of mankind that the obstacles should be sought.

What also seems overoptimistic to me is your opinion that humanity has progressed far enough to react to an appeal such as yours. A very thin layer may come up to your expectations, otherwise all the old cultural levels—those of the Middle Ages, of the Stone Age, even of animistic prehistory—are still alive in the great masses. You are also optimistic in the only positive suggestion in your gospel. The means of controlling human fertility are as yet far from perfected—i.e., safe or psychologically harmless.

So what can be done to improve life? I believe that one must have patience and accept the fact that there is still a long way to go. Meanwhile one should apply one's energies in that place to which one is most suited; thus one should either fight ignorance and prejudice or attempt to increase man's control over nature, etc. Anyone not blind and tough enough to take part in political experiments on the human masses shouldn't force himself to do so. On the other hand it is probably just as well that there are such men of action undeterred by scruples and compassion. But it is equally probable that for the time being these men will offer us nothing but disappointments.

I understand very well that you don't expect anything from the publication of your philanthropic pamphlet. My *Future of an Illusion*[1] has brought me almost only negative response, frequently indignant rejections.

Yours very sincerely
Freud

[1] *Die Zukunft einer Illusion,* Vienna, 1927. (*The Future of an Illusion,* Standard Ed., 21.)

238 *To* ERNEST JONES

Vienna IX, Berggasse 19
January 1, 1929

Dear Jones

As an enemy of all celebrations and formalities, I am writing to you *on* your fiftieth birthday rather than *for* it. Since I cannot postpone the letter I have to apologize for it because I am suffering from an (unfeverish) influenza, feel rather weak and helpless, and the increased narcissism of this condition will have betrayed itself to you by my having relapsed into the Gothic script at the beginning of this letter.

Thus many things that I could have said to you in response to your letter must remain unwritten. Not, however, the assurance that I have always looked upon you as a member of my intimate family circle and will continue to do so, which points (beyond all disagreements that are rarely absent within a family and also have not been lacking between us) towards a fount of affection from which one can always draw again. I think it began on the occasion when I once accompanied you to the station (in Worcester?). It is not in my nature to give expression to my feelings of affection, with the result that I often appear indifferent but my family knows better.

No, I am not going to write about youth and age. Perhaps I won't write anything at all, for the ease of conceiving ideas, which once I could count on, has deserted me in old age, and I am sensible enough not to try to force anything out of myself. I willingly admit the great improvement[1] in my existence since Berlin, but something in me longs for further improvement and I hope this will soon take place. . . .

With my best wishes for your next decade

Your faithful
Freud

[1] Thanks to the dentist's work.

239 *To* LUDWIG BINSWANGER

Vienna IX, Berggasse 19
April 11, 1929

Dear Dr. Binswanger

I don't know whether it was in 1912 or 1913 that I paid you a visit and was so impressed[1] by your courage that it has gained for you ever since a high place in my estimation. The intervening years, as you know, have made a rather frail old man of me. I can no longer travel to shake your hand.

April 12, 1929

My daughter[2] who died would have been thirty-six years old today.

Yesterday I nearly committed a serious blunder. I began reading your letter, deciphered some isolated friendly words which I would have been sorry to miss, but I was unable to fit them into a sentence, and the further I got the more enigmatic your handwriting became. I considered returning the letter to you with a jocular expression of my indignation and the suggestion that you send it back to me re-written. Then my sister-in-law offered her assistance and gave me the deeply moving news contained in the latter part of the letter, whereupon I understood why you had not dictated it into the machine.

Although we know that after such a loss the acute state of mourning will subside, we also know we shall remain inconsolable and will never find a substitute. No matter what may fill the gap, even if it be filled completely, it nevertheless remains something else. And actually this is how it should be. It is the only way of perpetuating that love which we do not want to relinquish.

Please give my warm regards to your wife.

In old friendship
Your old
Freud

---

[1] See letter of Apr. 14, 1912.
[2] Sophie Halberstadt.

240 *To* ISRAEL SPANIER WECHSLER

Vienna IX, Berggasse 19
May 8, 1929

Dear Dr. Wechsler

I am looking forward to the arrival of your new book in the expectation that it will fulfill the promise given in that chapter of your earlier book.

As regards your wish that I should leave the manuscripts of my publications to our university in Jerusalem, it is not so easy for me to answer. Your assumption that this university is dear and important to me is correct, but it may be that we do not agree about the assessment of such manuscripts. To me they do not mean anything; it actually would never have occurred to me to offer them to the university as a gift. I used to cast them into the wastepaper basket after they had been printed until one day someone suggested that they could be put to another use. This man told me that among the rich there are fools who, in the event of my becoming famous, would be willing to pay cash for such pages covered by my hand. Since then I have been preserving them and am awaiting such pleasant consequences of my fame, bearing in mind that there is no other way of my leaving or presenting anything to our own institutions such as the *Verlag,* the Viennese Institute, or the Berlin Sanatorium; indeed, such a legacy might be quite welcome to my seven grandchildren. Before the war I was told that a well-known collector of waste paper was actually considering the acquisition of my rubbish. But then came the war and no other suggestion of this kind has cropped up again. As it doesn't cost anything I am still waiting and wondering if perhaps one day after my death these treasures may rise in value to the extent that their preservation will have been worth while.

Thus I cannot really bestow upon these manuscripts a *pretium affectionis,* nor do I feel that I am depriving the University of Jerusalem of anything by not presenting it with these papers. Or rather only of some commodity whose value at the moment is still very doubtful. Should the University of Jerusalem have a different opinion, then I can make allowance by stating in my will that the manuscripts should be handed over to it if by a certain date after my death no purchaser has appeared.

Yours sincerely
Freud

241 *To* ROMAIN ROLLAND

Berchtesgaden, Haus Schneewinkl
July 14, 1929

My dear Friend

Your letter of December 5, 1927, containing your remarks about a feeling you describe as "oceanic" has left me no peace. It happens that in a new work[1] which lies before me still uncompleted I am making a starting point of this remark; I mention this "oceanic" feeling and am trying to interpret it from the point of view of our psychology. The essay moves on to other subjects, deals with happiness, civilization and the sense of guilt; I don't mention your name but nevertheless drop a hint that points toward you.

And now I am beset with doubts whether I am justified in using your private remark for publication in this way. I would not be surprised if this were to be contrary to your wishes, and if it is, even in the slightest degree, I should certainly refrain from using it. My essay could be given another introduction without any loss; perhaps it is altogether not indispensable.

So if you have any compunction about my quoting this remark I ask you to prevent me from abusing it by dropping me a friendly line.

Please bear in mind that I always think of you with feelings of most respectful friendship.

Very sincerely yours
Freud

---

[1] *Das Unbehagen in der Kultur,* Vienna, 1930. (*Civilization and its Discontents,* Standard Ed., 21.)

242 *To* ROMAIN ROLLAND

Berchtesgaden, Haus Schneewinkl
July 20, 1929

Dear Friend

My best thanks for your permission! But I cannot accept it
before you have reread your letter of the year 1927 which I enclose
herewith. I possess so few letters from you that I do not like the
idea of renouncing the return of this, your first one. I am not normally a hunter of relics, so please forgive this weakness.

I was glad to hear that your book will appear before my small
effort, which is unlikely to be in print before February or March.
But please don't expect from it any evaluation of the "oceanic" feeling; I am experimenting only with an analytical version of it; I am
clearing it out of the way, so to speak.

How remote from me are the worlds in which you move! To me
mysticism is just as closed a book as music. I cannot imagine reading
all the literature which, according to your letter, you have studied.
And yet it is easier for you than for us to read the human soul!

With warm wishes for your well-being

Very sincerely yours
Freud

243 *To* LOU ANDREAS-SALOMÉ

Schneewinkl
July 28, 1929

Dearest Lou

You will have guessed with your usual perspicacity why I
have taken so long to answer you. Anna has already told you that
I am working on something, and today I wrote the last sentence,
which—as far as is possible without a library—finishes the work.[1] It
deals with civilization, sense of guilt, happiness, and similar exalted
subjects, and strikes me, no doubt rightly, as quite superfluous in

[1] See note 1 to letter of July 14, 1929.

contrast to earlier works which always sprang from some inner necessity. But what else can I do? One cannot smoke and play cards all day; I am no longer much good at walking, and most of what I read doesn't interest me any more. I wrote, and in doing so the time passed quite pleasantly. While engaged in this work I have discovered the most banal truths.

Thomas Mann's essay[2] is no doubt quite an honor. He gives me the impression of having just completed an essay on romanticism when he was asked to write about me, and so he applied a veneer, as the cabinetmaker says, of psychoanalysis to the front and back of this essay: the bulk of it is of a different wood. Nevertheless, whenever Mann says something it is pretty sound.

I am very interested in your analysis of my production, but I find myself without an opinion. All I know is that I worked terribly hard; the rest followed as a matter of course. It could also have been very much better. I was aware only of the objective, not of myself. My worst qualities, among them a certain indifference to the world, probably had the same share in the final result as the good ones—i.e., a defiant courage about truth. In the depths of my heart I can't help being convinced that my dear fellow men, with a few exceptions, are worthless.

I would love to have had a leisurely talk with you about all this in our idyllically quiet and beautiful Schneewinkl, if only it had been possible to invite you here. But in the house itself there is no room, and in Berchtesgaden there isn't so much as an attic to be had. We have had all kinds of visitors—including some uninvited ones, all three sons in turn, two of whom finally found accommodation at a considerable distance. Ernst and Lux took advantage of Anna's absence and are living with us. According to her telegraphic reports, Anna is having a hard time in Oxford; this evening she will have delivered her lecture and from then on I hope she will take things easier. As regards accommodation she writes significantly: "More tradition than comfort." I expect you know that the English, after having created the notion of comfort, refused to have anything more to do with it. Like Wolf, I can hardly wait for her return. I write, and he spends half the day lying apathetically on his couch.

---

[2] "Die Stellung Freuds in der modernen Geistesgeschichte" ("Freud's Position in the History of Modern Culture"), *Almanach der Psychoanalyse*, Vienna, 1929.

Cordial greetings to you and your old gentleman and let us hope we can meet after all. Perhaps in Berlin, if I have to see Schröder.

Your old
Freud

244 *To* EDWARD L. BERNAYS

Berchtesgaden
August 10, 1929

Dear Edward

That is of course quite an impossible suggestion. A biography is justified under two conditions. First, if the subject has had a share in important, generally interesting events; second, as a psychological study. Outwardly my life has passed calmly and uneventfully and can be covered by a few dates. A psychologically complete and honest confession of life, on the other hand, would require so much indiscretion (on my part as well as on that of others) about family, friends, and enemies, most of them still alive, that it is simply out of the question. What makes all autobiographies worthless is, after all, their mendacity.

Incidentally, it is American naïveté on the part of your publisher to expect a hitherto decent person to commit so base a deed for $5,000. For me temptation might begin at a hundred times that sum, but even then it would be turned down after half an hour.

In the hope that you and your wife and daughter are in good health, I greet you cordially.

Your uncle
Sigm.

245 *To* MAX EITINGON

Vienna IX, Berggasse 19
December 1, 1929

Dear Max

We received your sad news yesterday with that deep sorrow which finally turned into the realization that a sudden end can only

391

be looked upon as a blessing, considering the hopelessness of the outlook. I have been charged to express to you and your family the deep sympathy of us all, and I am not going to try to console you. The loss of a mother must be something very strange, unlike anything else, and must arouse emotions that are hard to grasp. I myself still have a mother, and she bars my way to the longed-for rest, to eternal nothingness; I somehow could not forgive myself if I were to die before her. But you are young, you actually have the best and most eventful decade, from fifty to sixty, still ahead of you, and your friends have reason to hope that you will soon be reconciled to a misfortune which does not surpass the bounds of the average human fate. I trust that your Mirra, with whom (as you know) I sometimes remonstrate in my mind, will use this occasion also to show her ability as helper and comforter, as she invariably does in an emergency.

With cordial greetings and in the hope of seeing you soon

Your

Freud

246 *To* ROMAIN ROLLAND

Vienna IX, Berggasse 19
January 19, 1930

Dear Friend

My warm thanks for the gift of your twin-headed, three-volume work![1] Contrary to my calculation, my "discontented" little book preceded yours by several weeks. I shall now try with your guidance to penetrate into the Indian jungle from which until now an uncertain blending of Hellenic love of proportion ($\sigma\omega\phi\rho\sigma\acute{\nu}\nu\eta$), Jewish sobriety, and philistine timidity have kept me away. I really ought to have tackled it earlier, for the plants of this soil shouldn't be alien to me; I have dug to certain depths for their roots. But it isn't easy to pass beyond the limits of one's nature.

Of course I soon discovered the section of the book most interesting to me—the beginning, in which you come to grips with us extreme rationalists. That you call me "grand" here I have taken

[1] *Essai sur la mystique et l'action de l'Inde vivante: La vie de Rama Krishna,* Stock, Paris, 1929; *La vie de Vivekananda et l'Evangile Universel* (2 volumes), Stock, Paris, 1930.

quite well; I cannot object to your irony when it is mixed with so much amiability.

Concerning the criticism of psychoanalysis, you will permit me a few remarks: the distinction between *extrovert* and *introvert* derives from C. G. Jung, who is a bit of a mystic himself and hasn't belonged to us for years. We don't attach any great importance to the distinction and are well aware that people can be both at the same time, and usually are. Furthermore, our terms such as *regression, narcissism, pleasure principle* are of a purely descriptive nature and don't carry within themselves any valuation. The mental processes may change direction or combine forces with each other; for instance, even reflecting is a regressive process without losing any of its dignity or importance in being so. Finally, psychoanalysis also has its scale of values, but its sole aim is the enhanced harmony of the ego, which is expected successfully to mediate between the claims of the instinctual life (the "id") and those of the external world; thus between inner and outer reality.

We seem to diverge rather far in the role we assign to intuition. Your mystics rely on it to teach them how to solve the riddle of the universe; we believe that it cannot reveal to us anything but primitive, instinctual impulses and attitudes—highly valuable for an embryology of the soul when correctly interpreted, but worthless for orientation in the alien, external world.

Should our paths cross once more in life, it would be pleasant to discuss all this. From a distance a cordial salutation is better than polemics. Just one more thing: I am not an out-and-out skeptic. Of one thing I am absolutely positive: there are certain things we cannot know now.

With my warmest wishes for your well-being

Your devoted
Freud

247 ANONYMOUS

Vienna IX, Berggasse 19
February 8, 1930

Dear Madam

You have guessed my attitude toward autograph hunters correctly. If by a stroke of the pen one can really do something for a worthy person in need, then one has no reason to hesitate, and one can even discover something useful in human folly. Please do not omit in the interest of your protégé to drive home to the wealthy lady that a specimen such as that enclosed is otherwise difficult to obtain.

Sincerely yours
Freud

248 *To* A. A. ROBACK

Vienna IX, Berggasse 19
February 20, 1930

Dear Dr. Roback
　　I hasten to acknowledge receipt of your book, *Jewish Influence, etc.*,[1] and to thank you for it. Nor have I delayed reading the enclosed reprint and at least glancing through the book.

I cannot refrain from confessing to a certain disappointment. You do me great honor in the book; you mention my name among the greatest of our people (which goes far beyond my ambition) and so on; [but] in the essay ("Doctrine of Lapses") you express disbelief about just that part of psychoanalysis which has most readily found general recognition. How then are you going to judge our far less attractive discoveries? My feeling is that if your objections to my interpretation of lapses is justified, then I have very little claim to be mentioned with Bergson and Einstein among the intellectual giants. You realize what I am aiming at. I wish neither to be crowned with the Nobel Prize nor to be discussed in every newspaper; I hope to have gained a useful piece of new insight.

I haven't yet read the paragraph about psychoanalysis in your book; I am afraid to find in it incorrect statements which I shall regret. In some of your assertions I recognize myself as little (for instance, no one has so far reproached me with "mystical leanings," and over the question of hypnosis I sided against Charcot, even if not entirely with Bernheim)[2] as I do in the appalling picture of my physical appearance which you have included.

---

[1] *Jewish Influence in Modern Thought,* Science Art Publication, Cambridge, Mass.
[2] Henri Bernheim (1837-1919), Professor at Strassburg University, practiced psychiatry at Nancy.

It may interest you to hear that my father did indeed come from a Chassidic background. He was forty-one when I was born and had been estranged from his native environment for almost twenty years. My education was so un-Jewish that today I cannot even read your dedication, which is evidently written in Hebrew. In later life I have often regretted this lack in my education.

With the expression of that sympathy which your courageous defense of our people demands, I beg to remain

Yours sincerely
Freud

249 *To* RICHARD FLATTER

Vienna IX, Berggasse 19
March 30, 1930

Dear Mr. Flatter

I thank you for kindly sending me your translation of *King Lear,* which gave me the opportunity of rereading this powerful work.

As to your question whether one is justified in considering Lear a case of hysteria, I should like to say that one is hardly entitled to expect from a poet a clinically correct description of a mental illness. It should be enough that our feelings are at no point offended and that our so-called popular psychiatry enables us to follow the person described as abnormal in all his deviations. This is the case with Lear; we are not shocked when, in his sorrow, he abandons contact with reality, nor when, clinging to the trauma, he indulges in phantasies of revenge; nor when, in his excess of passion, he storms and rages, although the picture of a consistent psychosis is disrupted by such behavior. As a matter of fact I am not sure whether such hybrid formations of an affective clinging to a trauma and a psychotic turning away from it do not occur often enough in reality. The fact that he calms down and reacts normally when he realizes he is safely protected by Cordelia doesn't seem to me to justify a diagnosis of hysteria.

Sincerely yours
Freud

250 *To* HEINRICH LÖWY

Vienna IX, Berggasse 19
March 30, 1930

Dear Dr. Löwy

Your biography[1] of Popper-Lynkeus which accompanied your letter has pleased me so much by its dignity and truthfulness that I would like to comply with your wish and make a contribution to your collection of solutions[2] to scientific problems. But in trying to find some suitable examples I have encountered strange and almost insuperable obstacles, as though certain procedures that can be expected from other fields of investigation could not be applied to my subject matter. Perhaps the reason for this is that within the methods of our work there is no place for the kind of experiment made by physicists and physiologists.

When I recollect isolated cases from the history of my work, I find that my working hypothesis invariably came about as a direct result of a great number of impressions based on experience. Later on, whenever I had the opportunity of recognizing an hypothesis of this kind to be erroneous, it was always replaced—and I hope improved—by another idea which occurred to me (based on the former as well as new experiences) and to which I then submitted the material.

I am afraid the above will not be of any great use to your collection.

With kind regards
Yours sincerely
Freud

[1] See List of Addressees and letter of Aug. 4, 1916.
[2] The collection planned by Professor Löwy and Professor R. von Mises did not materialize.

251 *To* MAX SCHUR

Berlin-Tegel
June 28, 1930

Dear Dr. Schur

I will not be in Vienna when you get married, so I am writing to you today, several days before the event, to wish you all the hap-

piness you have the right to expect in your married life. Mindful of
the rare kindness and conscientiousness that you have shown in
caring for the remains of my physical self, I should like to endow
my wishes with the power to enforce their fulfillment. This is hardly
the occasion to bother you with medical reports. I only want to
say that I will not forget how often your diagnoses have turned out
to be correct in my case, and that for this reason I am a docile
patient, even when it is not easy for me.

With kind regards
Yours sincerely
Freud

252 *To* ADOLFINE FREUD

Berlin-Tegel
July 22, 1930

Dear Dolfi
   The enclosed is just some pocket money for your birthday
for which I send you my best wishes as well as my most heartfelt
gratitude for your invaluable services throughout all these years.
As you know, we shall be in Mother's immediate neighborhood this
summer, but unfortunately I don't yet know when.

Affectionately
Your
Sigm.

253 *To* ALFONS PAQUET

Grundlsee,[1] Rebenburg
July 26, 1930

Dear Mr. Paquet
   I have not been spoiled by public honors, and have therefore
accustomed myself to getting along without them. I cannot deny,
however, that the award of the Goethe Prize by the City of Frankfort has given me great pleasure. There is something about it that
particularly warms the imagination, and one of its conditions elimi-

---

[1] Resort in the Salzkammergut.

nates the humiliation normally associated with such distinctions.

I owe you special thanks for your letter; it moved and surprised me. Quite apart from [my appreciation of] the trouble you have taken to study my work, [I must tell you that] I have never before found its secret personal intention recognized with so much clarity, and I would have liked to ask you how you happen to have divined it.

I am sorry to learn from your letter to my daughter that I shall not be seeing you in the near future; delay at my time of life is somewhat risky. Of course I shall be very pleased to receive the gentleman (Dr. Michel) announced by you.

Unfortunately I cannot come to the celebration in Frankfort; my health is not reliable enough for such an undertaking. The audience will lose nothing by my absence; my daughter is certainly more pleasant to the eye and ear than I am. She is going to read a few lines dealing with Goethe in relation to psychoanalysis and defending the analysts against the reproach of having impugned the veneration due to the great man by their analytical investigations. I hope it will be possible to interpret the subject given to me: "The relations of the man and scholar to Goethe"[2] in the above sense, or, if this can't be done, that you will be good enough to advise me against it.

Very sincerely yours
Freud

[2] "Ansprache im Frankfurter Goethehaus," 1930. ("Address delivered in the Goethe House at Frankfort," Standard Ed., 21.)

254 *To* SANDOR FERENCZI

Grundlsee, Rebenburg
August 1, 1930

Dear Friend

We returned from Berlin a week ago, and have been here four days. This spot is marvelously beautiful, the house spacious and comfortable in spite of some abominations, the view over the lake enchanting. It rains too often and too much, as usual here; after all we missed the best part of the summer.

My prosthesis promises to improve; in some respects it is a noticeable change for the better.

The day before yesterday I received the official news of my having been awarded the Goethe Prize by the City of Frankfort. This news incidentally came in a quite charming, very intelligent letter from the secretary, Mr. Paquet, who himself must be someone [of importance]. By its connection with Goethe the prize possesses rather more dignity than many others of its kind. There is nothing to spoil my pleasure. It is worth 10,000 marks, more or less covering the expenses of my stay in Berlin. The solemn bestowal of the Prize is scheduled to take place in Frankfort on August 28, Goethe's birthday, in the presence of the man thus honored, who in return has to make a speech about his relationship with Goethe. Needless to say I cannot go, but Anna will represent me and read what I have to say[1] about Goethe in relation to psychoanalysis and its right to use him as a subject for investigation.

With cordial greetings to you and Frau Gisela in your new house and the question whether you still consider traveling this year?

Your
Freud

[1] See note 2 to letter of July 26, 1930.

255 *To* ARNOLD ZWEIG

Grundlsee, Rebenburg
August 21, 1930

Dear Mr. Zweig

Among the many congratulations which the Goethe Prize has brought me, none has moved me so much as that which you have extorted from your poor eyes (although your handwriting doesn't show it); evidently because in hardly any other case do I feel so certain that my friendship meets with such loyal response. I do not deny that the Goethe Prize has pleased me very much. The idea of a closer connection with Goethe is too tempting; the prize itself is more of a bow made to the recipient than an assessment of his achievement. On the other hand at my time of life such recognition has neither much practical value nor great emotional significance. For a reconciliation with my contemporaries it is pretty late; and that psychoanalysis will win through long after my time I have never doubted. In reading your lines I have made the discovery that I wouldn't have been much less pleased had the prize been awarded to you, and in your case it might have been more appropriate. But no doubt many similar honors are in store for you.

My wife and daughter send warm regards to you as well as to your wife, whose letter deserves my special thanks. My daughter will represent me at the Goethe celebration in Frankfort.

>   With every good wish
>   Your
>   Freud

256 *To* SANDOR FERENCZI

>   Grundlsee, Rebenburg
>   September 16, 1930

Dear Friend

Above all my warm thanks for your beautiful words about the death of my mother. It has affected me in a peculiar way, this great event. No pain, no grief, which probably can be explained by the special circumstances—her great age, my pity for her helplessness toward the end; at the same time a feeling of liberation, of release, which I think I also understand. I was not free to die as long as she was alive, and now I am. The values of life will somehow have changed noticeably in the deeper layers.

I did not go to the funeral; Anna represented me there, too. Today she went on a Swiss-Italian tour with her friend Dorothy,[1] for which I wish her better weather.

The gruesome newspaper reports about my health will probably have reached you, too. I find them very interesting as a proof of the difficulty of forcing upon the general public something it doesn't like. For they are the reaction to the Goethe Prize and must warn us against the illusion that the resistance to psychoanalysis has subsided in a practical, tangible way. The same style of reaction is also

[1] Mrs. Dorothy Burlingham, Co-director of the Hampstead Child Therapy Clinic, London.

revealed in Bumke's[2] speech (which I know only from a report in the *Neue Freie Presse*) as well as in the increased activities of the Adlerian buffoons who are now publishing books about the meaning of life (!) and homosexuality. In short, we will have to pay dearly for the Goethe Prize. . . .

I am glad you are working. What with the congratulations for the prize, the letters of condolence concerning my own fatal illness and now the death of my mother, not to mention the discomforts of the continuous abstinence from smoking, I don't find time for anything.

Cordially

Your

Freud

[2] Oswald Bumke, Professor of Psychiatry at the University of Berlin, "Über Psychoanalyse," *Zentralblatt für Psychotherapie*, 1930.

257 *To* ARNOLD ZWEIG

Vienna IX, Berggasse 19

October 22, 1930

My dear Mr. Zweig

I read your little comedy[1] just before I took to my bed for a week (with a temperature). I am far from ascribing my illness to its influence. But perhaps the former was already in me and has colored my opinion of the comedy. For it was not to my taste; above all I didn't find it at all funny and as a result of the differences between north- and south-German I had difficulties in finding my way through the dialogue. I trust you won't take my criticism amiss; at least you will know that next time I praise passionately something you have sent me, it will be seriously meant.

I hope that my wishes for you and your wife's health will reach you at a more propitious moment than was granted to your wish.

Cordially

Your

Freud

[1] *Laubheu und Keine Bleibe*, Kiepenheuer Verlag, Potsdam, 1926.

258   *To* STEFAN ZWEIG

Vienna IX, Berggasse 19
February 7, 1931

Dear Stefan Zweig

I have received your latest work[1] and read it again, taking a more personal interest this time, of course, than I did in your former captivating books. If I may couch my impressions in a critical form, I would say that it is your essay on Mesmer[2] that strikes me as the most harmonious, just, and distinguished. Like you, I believe that the real nature of his discovery, that of suggestion, has so far not been recognized, and that there still remains room for something new here.

What disturbs me about the essay on Mary Baker Eddy[3] is that you lay such stress on her intensity. People like us who cannot free themselves from the pathological point of view are not very impressed by this quality. We know that in a fit of frenzy the raving lunatic releases energies which normally are not at his disposal. The mad and wicked side of the Mary Baker Eddy phenomenon is not sufficiently brought out, nor is the unspeakable grimness of the American background.

That one doesn't like one's own portrait, or that one doesn't recognize oneself in it, is a general and well-known fact. I therefore hasten to express my satisfaction at your having recognized correctly the most important feature in my case. Namely, that in so far as achievement is concerned it was less the result of intellect than of character. This seems to be the core of your opinion, and one in which I myself believe. On the other hand I feel inclined to object to the emphasis you put on the element of *petit-bourgeois* correctness in my person.

The fellow is actually somewhat more complicated: your description doesn't tally with the fact that I too have had my splitting headaches and attacks of fatigue like anyone else, that I was

[1] *Die Heilung durch den Geist: Franz Anton Mesmer, Mary Baker Eddy, Sigmund Freud,* Inselverlag, Leipzig, 1931. (*Mental Healers,* New York, Viking Press; London, Cassell, 1933.)

[2] Franz Anton Mesmer (1734-1815), founder of the doctrine of animal magnetism.

[3] Mary Baker Eddy (1821-1910), founder of Christian Science.

a passionate smoker (I wish I still were), that I ascribe to the cigar the greatest share of my self-control and tenacity in work, that despite my much vaunted frugality I have sacrificed a great deal for my collection of Greek, Roman and Egyptian antiquities, have actually read more archaeology than psychology, and that before the war and once after its end I felt compelled to spend every year at least several days or weeks in Rome, and so on. I realize from my experience with art in miniature that this medium compels the artist to simplify, but the result is often a distorted picture.

I am probably not mistaken in assuming that up to the writing of this book you were unfamiliar with psychoanalytic theory. That you have absorbed so much of it in that time deserves all the more praise. You can be criticized on two counts. You hardly mention the technique of free association, considered by many people the most important contribution made by psychoanalysis, the methodological key to its results. And you make me find my insight into dreams through the dream of childhood; that is historically incorrect, although it has been represented like this for didactic reasons.

Your last doubt whether psychoanalysis can be practiced by average human beings also leads back to ignorance of technique. At the time when the microscope was a new instrument in the hands of physicians, textbooks on physiology declared that a microscopist had to possess very special and rare qualifications. Later on, the same requirements were made of surgeons; today every student learns to use the microscope, and good surgeons are trained in schools. That not everyone carries out his job equally well is something for which there is no remedy in any field.

With sincere greetings and good wishes for your holiday

Your
Freud

259 *To* YVETTE GUILBERT

Vienna IX, Berggasse 19
March 8, 1931

Dear Friend

I would very much like to be with you when your husband[1] translates this letter to you, for owing to my poor state of health

---

[1] Dr. Max Schiller—see List of Addressees and letter of Mar. 26, 1931.

I profited far too little from your last visit to Vienna. It is good to hear that you are again going to write something about yourself. If I understand correctly, you intend this time to elucidate the secret of your achievement and success, and you imagine that your technique consists in relegating your own person completely to the background and replacing it by the character whom you are representing. And now you wish me to tell you whether this process is likely and whether it applies to you.

I wish I knew more about this process; then I would certainly tell you everything I know. Since I don't understand much about it, I must ask you to be content with the following suggestions. I believe that what you consider the psychological mechanism of your art has been claimed very often, perhaps universally. But this idea of the obliteration of one's own person and its replacement by an imagined one has never quite satisfied me. It tells us so little, doesn't inform us how it is brought about, and above all it fails to explain why one person should succeed so much better than another in achieving what every artist allegedly wants. I rather suspect that an element of the opposite mechanism is indispensable for it: that one's own person is not obliterated but that parts of it —repressed desires and traits that haven't had a chance to develop —are employed to represent the chosen character and in this way find expression and give it the stamp of realistic truth. This is less simple than the "transparency" of one's own ego to which you give precedence. I would of course be interested to hear whether you can detect any traces of this other aspect. In any case it is merely one contribution toward solving the beautiful puzzle of why we shudder at the "Voularde"[2] or say yes with all our senses to the question: "Dîtes-moi si je suis belle."[2] But one knows so little about it all.

In affectionate remembrance of everything your letter brought back to me and with friendly greetings to you and Uncle Max

Your

Freud

[2] Two of the songs made famous by Yvette Guilbert.

260 *To* MAX SCHILLER

Vienna IX, Berggasse 19
March 26, 1931

Dear Dr. Schiller

It is a very interesting experience for me to be called upon to defend my theories to Madame Yvette and Uncle Max. Despite my impeded speech and hardness of hearing I wish I didn't have to do it in writing.

Actually, I don't intend to yield much beyond the confession that we know so little. Just recently, for instance, Charlie Chaplin was in Vienna; I almost caught sight of him, but it was too cold for him, and he left in a hurry. He is undoubtedly a great artist— although he always plays one and the same part, the weak, poor, helpless, clumsy boy for whom life turns out all right in the end. Now do you think he has to forget his own self in order to play this part? On the contrary, he invariably plays only himself as he was in his grim youth. He cannot get away from these impressions and even today he tries to compensate himself for the humiliation and deprivation of that time. He is of course an especially simple, transparent case.

The theory that the achievements of artists are conditioned internally by their childhood impressions, vicissitudes, repressions, and disappointments has already clarified many things for us, and we therefore think highly of it. I once dared to tackle one of the very greatest of all, Leonardo da Vinci,[1] of whom unfortunately too little is known. I was able at least to point out that "The Virgin and Saint Anne," which you can see any day in the Louvre, couldn't be understood without some knowledge of Leonardo's peculiar childhood. The same may also apply to other things.

Now you will point out that Madame Yvette doesn't play just one part, that she plays with equal mastery all kinds of characters: saints, sinners, coquettes, the righteous, criminals, and *ingénues*. This is true and testifies to an unusually rich and adaptable psychic life. But I wouldn't hesitate to trace back this whole repertoire to experiences and conflicts of her early youth. It is tempting to continue

[1] See letter of Oct. 1, 1911.

on this subject, but something holds me back. I know that unwarranted analyses call forth antagonism, and I don't want to do anything that could disturb the warm sympathy that dominates our relationship.

With friendly greetings to you and Madame Yvette

Your

Freud

261 *To* ROMAIN ROLLAND

Vienna
May 1931

Dear Friend

You answered my pleasantry[1] with the most precious information about your own person. My profound thanks for it.

Approaching life's inevitable end, reminded of it by yet another operation and aware that I am unlikely to see you again, I may confess to you that I have rarely experienced that mysterious attraction of one human being for another as vividly as I have with you; it is somehow bound up, perhaps, with the awareness of our being so different.

Farewell!

Your

Freud

[1] Allusion to the dedication in the copy of *Civilization and its Discontents* which Freud had sent to Rolland.

262 *To* ALEXANDER LIPSCHÜTZ

Vienna, Pötzleinsdorf
August 12, 1931

Dear Professor Lipschütz

My cordial thanks for your good wishes. I am afraid I am so old that wishes are rather wasted on me.

I am here in the country and far from my library, so cannot confirm my suspicion that you are the same person whom I have quoted several times in my writings and from whose book on the puberty gland I have gained so much knowledge. But I think my assumption is correct. I missed hearing that you had been called to Chile.

What you say about the attitude of scientists toward new scientific discoveries is unfortunately only too true in every respect. One can easily understand the hostility to psychoanalysis which destroys cherished illusions and prejudices, but one may well be surprised by the emotional attitude toward discoveries in your field of endeavor. Probably the reason is that under any circumstances only a few people are constitutionally capable of scientific investigation.

I am very pleased to see from your letter that you are not among those who place psychoanalysis in opposition to endocrinology, as though psychic processes could be explained directly by glandular functions, or as though the understanding of psychic mechanism could replace the knowledge of the underlying chemical process.

Please accept my sincerest good wishes for successful work in your newly-adopted, promising country.

Your
Freud

263 *To* THE MAYOR OF PŘIBOR-FREIBERG

October 25, 1931

I wish to thank the Mayor of Přibor-Freiberg, the organizers of the celebration and all those taking part in it, for the honor they are doing me by distinguishing the house of my birth with this memorial plaque created by the hand of an artist. And this, moreover, during my lifetime and while the contemporary world is not yet agreed on the value of my achievement.

I left Freiberg at the age of three, revisited it as a sixteen-year-old schoolboy on vacation as a guest of the family Fluss,[1] and have never been here since. Many things have happened to me since that time: I have had many difficulties, experienced considerable sorrow as well as happiness and some success, as things are wont to blend in human life. It is not easy for the now seventy-five-year-

---

[1] See letter of June 16, 1873.

old man to recall those early days of whose rich content only a few fragments reach into his memory, but of one thing I am certain: deep within me, although overlaid, there continues to live the happy child from Freiberg, the first-born son of a youthful mother, the boy who received from this air, from this soil, the first indelible impressions. May I therefore be allowed to conclude this vote of gratitude with hearty good wishes for this town and its inhabitants.

Sigmund Freud

1932

1939

264 *To* ARNOLD ZWEIG

Hochroterd
May 8, 1932

Dear Arnold Zweig

Where am I writing from? From a farm cottage on the side of a hill, forty-five minutes by car from the Berggasse, which my daughter and her American friend[1] (she owns the car) have acquired and furnished for their week ends. We had expected that your return home from Palestine would take you through Vienna, and then we would have insisted on your seeing it.

You have been very generous by giving me your picture, which admittedly doesn't show me much more than your forehead, with the photograph of your author's cage[2] which I shall not get to see, with the news that your eyes have improved, and with those few drops from your cauldron in which at the moment so many new and interesting stories are simmering, all of which I should like to read although my days are beginning to be numbered.

You are right, I have just celebrated my birthday and at this moment I am laboriously defending myself against the obligations ensuing from it. But to return to you: how strange this tragic and fantastic country you have just visited must have appeared to you! To think that this strip of our native earth is associated with no other progress, no discovery or invention—the Phoenicians are said to have invented glass and the alphabet (both doubtful!), the island of Crete is said to have given us Minoan art, Pergamon reminds us of parchment, Magnesia of the magnet, and so on *ad infinitum*—but Palestine has produced nothing but religions, sacred frenzies, presumptuous attempts to conquer the outer world of

[1] Mrs. Dorothy Burlingham.
[2] Allusion to Zweig's newly-built house in Berlin.

411

appearances by the inner world of wishful thinking. And *we* hail from there (although one of us considers himself a German also, the other doesn't), our ancestors lived there perhaps for half, perhaps a whole, millennium (but this also only perhaps), and it is impossible to say how much of the life in that country we carry as heritage in our blood and nerves (as is mistakenly said). Oh, life could be very interesting if only one knew and understood more about such things! But the only things we can be sure of are our feelings of the moment! Among them my warm feelings for you and your work!

With greetings to your wife
Your
Freud

265 *To* STEFAN ZWEIG

Vienna, Hohe Warte[1]
June 2, 1932

Dear Stefan Zweig

Whenever a work of mine is published I feel reluctant for a long time to give it any further thought. I should be sorry if this were true for you too, for I intend to draw your attention back to that book[2] of yours, a third of which you devoted to me and my work.

A friend of mine who has recently been in Venice found in a bookshop there the Italian translation of *Mental Healers,* and made me a present of it. As a result I reread parts of your essay and discovered on page 272 an error of representation which cannot be looked upon as unimportant and which, if you don't mind my saying so, actually belittles my merit. It declares that Breuer's patient under hypnosis made the confession of having experienced and suppressed certain *"sentimenti illeciti"* (i.e., of a sexual nature) while sitting at her father's sickbed. In reality she said nothing of the kind; rather she indicated that she was trying to conceal from her father her agitated condition, above all her tender concern. If things had been as your text maintains, then everything else

---

[1] Suburb of Vienna.
[2] See note 1 to letter of Feb. 7, 1931.

would have taken a different turn. I would not have been surprised by the discovery of sexual etiology, Breuer would have found it more difficult to refute this theory, and if hypnosis could obtain such candid confessions, I probably would never have abandoned it.

What really happened with Breuer's patient I was able to guess later on, long after the break in our relations, when I suddenly remembered something Breuer had once told me in another context before we had begun to collaborate and which he never repeated. On the evening of the day when all her symptoms had been disposed of, he was summoned to the patient again, found her confused and writhing in abdominal cramps. Asked what was wrong with her, she replied: "Now Dr. B.'s child is coming!"

At this moment he held in his hand the key that would have opened the "doors to the Mothers,"[3] but he let it drop. With all his great intellectual gifts there was nothing Faustian in his nature. Seized by conventional horror he took flight and abandoned the patient to a colleague. For months afterwards she struggled to regain her health in a sanatorium.

I was so convinced of this reconstruction of mine that I published it somewhere. Breuer's youngest daughter (born shortly after the above-mentioned treatment, not without significance for the deeper connections!) read my account and asked her father about it (shortly before his death). He confirmed my version, and she informed me about it later.

> Very sincerely yours
> Freud

[3] Allusion to an image in Goethe's *Faust*, II.

266 *To* MAX EITINGON

Vienna XVIII, Khevenhüllerstrasse 6
June 5, 1932

Dear Max

In your last letter I miss any mention of the progress of your recovery. Instead you inquire after my health. Now, this is not quite the same thing. I have been passing through a good time which I have managed to use for drafting one of the new lectures, but there are recurring signs of several organs refusing to function—and when one of them becomes too obstreperous the good time is over. And in addition the permanent, never-to-be-ended misery of the prosthesis.

The good progress of our activities on behalf of the *Verlag* is like a ray of light in these gloomy times. I think one can also say already that Martin is proving himself a good choice.[1] I cannot refute the reproach made against you for having tolerated Storfer too long, but I am aware that it applies equally to me. Fortunately there is another side to the story. Storfer reminds me of one of those German princelings who suppressed and exploited his subjects. But after they had been expelled it turned out that their little country possessed a capital with a castle, a theater, and an art collection, which became its most powerful attraction for visiting foreigners. The trouble is, one has to expel them first.

Life here, where my study opens straight into the Park, is pleasant. Yesterday was the first inclement day for three weeks.

Cordial greetings to you and Mirra.

Your
Freud

[1] Martin Freud had become Director of the Internationale Psychoanalytische Verlag, succeeding Albert Josef Storfer (1888-1944).

267 *To* LEON STEINIG

Vienna XVIII, Khevenhüllerstrasse 6
June 1932

Dear Sir

I hasten to answer your letter because you tell me that you intend to use my comments when you meet Professor Einstein at the end of this month.

While reading your letter I have indulged in as much enthusiasm as I am able to muster at my age (seventy-six) and in my state of disillusionment. The words in which you express your hopes and those of Einstein for a future role of psychoanalysis in the life of individuals and nations ring true and of course give me very great pleasure. It has been no little disappointment to me that at a time when we can continue our work only under the greatest social and material difficulties, I haven't seen the slightest sign of interest for our efforts on the part of the League of Nations. Thus practical and idealistic considerations combine to induce me to put myself with all that remains of my energies at the disposal of the Institute for Intellectual Co-operation.

I cannot quite imagine as yet what form my participation is going to take. It will devolve upon Einstein to make suggestions. I would prefer not to hold forth on my own and hope that the character of a discussion can be maintained—in such a way, perhaps, that instead of answering one question put to me by Einstein, I respond from the point of view of psychoanalysis to statements in which he expresses his opinions. I would also prefer not to pick out a single topic from among those enumerated in your letter. It is rather a question of a number of problems of which the most important for practical purposes is the influence of psychoanalysis on education. But as I say, in all these practical details I am ready to follow Einstein's suggestions.[1] When you see him you won't be able to tell him anything more about my personal relationship to him than he knows already, although I only once[2] had the long-desired opportunity of talking to him.

As for yourself, please accept my cordial thanks for your interest in psychoanalysis.

Yours very sincerely
Freud

[1] The result was Freud's essay "Warum Krieg?" 1933. ("Why War?" Standard Ed., 22.)
[2] In Ernst Freud's house in Berlin, in 1927.

268 *To* COUNT HERMANN KEYSERLING

Vienna XVIII, Khevenhüllerstrasse 6
August 10, 1932

Dear Count

I have indeed received as well as read your *South American Meditations*. The book arrived in an impersonal manner, as a reviewer's copy, I think, or with the suggestion that I comment on it for publicity—I can't remember now—certainly not as a gift from the author, otherwise I wouldn't have omitted to thank you.

Everything you write arouses my lively interest, but as a rule I cannot understand it or follow it very far. On this occasion I was surprised by the impression which a glimpse into a society dominated by elemental, instinctual impulses had made on you. To us this picture is familiar; we believe it can also be detected behind the disguise and camouflage of Europe and have become familiar with it by studying our patients and ourselves. I had assumed that you hail from distant philosophical worlds where one doesn't take cognizance of such common but fundamental things, and I now realize that with your great receptiveness you have become responsive to something new.

Yours sincerely
Freud

269 *To* EDOARDO WEISS

Vienna IX, Berggasse 19
April 12, 1933

Dear Dr. Weiss

The Italian "Moses"[1] has given me particular pleasure. My relationship to this work is something like that to a love child. Every day for three lonely weeks in September of 1913[2] I stood in the church in front of the statue, studying it, measuring and drawing it until there dawned on me that understanding which in the essay I only dared to express anonymously. Not until much later did I legitimize this nonanalytical child.

One of the consequences of my failing health difficult to bear is that I can no longer come to Rome (the last time was in 1923). . . .

With cordial greetings
Your
Freud

[1] Weiss had translated Freud's essay into Italian. See letter of Sept. 25, 1912.
[2] It was actually in 1912.

270 *To* GEORGE SYLVESTER VIERECK

Vienna IX, Berggasse 19
Easter Sunday, April 16, 1933

Dear Mr. Viereck

You have sent me the message of the former German Crown Prince and your letter to the New York *Herald Tribune* without adding any personal note. I assume you expect a comment from me. This I would not have withheld and it would have become quite extensive were I not concerned that you might somehow use it for publicity. I will therefore confine myself to a few remarks which, meant for you alone, will defy publication. I will just say I regret that you have debased yourself by siding with those wretched lies as expressed in your royal cousin's letter.

To judge the case it is sufficient to pick out three pronouncements: the Crown Prince's assurance that no one in Germany has "to suffer injustice on account of his religion"; your own admission that "if those who wish to purge their country of such elements invoke racial or religious prejudice, they are guilty of the same fault as those to whose course they take exception," and N. Chamberlain's plain statement that he is not concerned with the alleged atrocities but is going by the official announcements and measures taken by the [National Socialist] movement.

With deep regret
Freud

271 *To* OSKAR PFISTER

Vienna XIX, Hohe Warte 46
May 28, 1933

Dear Pfister

My warm thanks for your letter of condolence on Ferenczi's death. I deserve it, for the loss is very distressing. It was not unexpected, of course. For the last two years our friend had not been himself; he suffered from pernicious anemia with *locomotor ataxia* and psychologically had changed very much; he died finally of asphyxia. We will continue to remember him as he has been for the past twenty years.—I believe some of his achievements, his Genital Theory,[1] for instance, will keep his memory alive for a long time.

I was very pleased to hear such good news of you. Our horizon has become darkly clouded by the events in Germany. Three members of my family, with their families, two sons and a son-in-law,

---

[1] *Versuch einer Genitaltheorie*, Internationale Psychoanalytischer Verlag, Leipzig, Vienna, Zürich, 1924. (*Theory of Genitality*, 1933.)

are looking for a new country and still haven't found one. Switzerland is not among the hospitable countries. My judgment of human nature, above all the Christian-Aryan variety, has had little reason to change. My correspondence with Einstein[2] has been published simultaneously in German, French, and English, but it can be neither advertised nor sold in Germany.

Martin will soon be coming to Zürich again.

I greet you warmly.

Your

Freud

[2] See letter (to Steinig) of June, 1932.

272 *To* JUDAH MAGNES

Vienna IX, Berggasse 19
December 5, 1933

Dear Sir

I thank you for your letter of the twenty-seventh and hasten to answer the question it contains.

The view that it is premature to create a chair for psychoanalysis so long as none for psychology exists invites a discussion of the relationship between the two sciences. My opinion is as follows: psychoanalysis is also psychology in the sense that it is a science of the *unconscious* psychic processes, whereas what is taught as academic psychology is confined to dealing with *conscious* phenomena. There need be no contradiction between the two; psychoanalysis could be presented as an introduction to psychology; in reality, however, the contradiction is produced by the fact that the academic circles don't want to have anything to do with psychoanalysis.

The necessity to start the teaching of psychology with traditional academic psychology doesn't exist. On the contrary, all the applications of psychology to medicine and the arts derive from deepreaching psychoanalysis, whereas academic psychology has proved itself to be sterile.

I see no reason for assuming that Professor Kurt Lewin[1] is the

[1] Dr. Kurt Lewin (1890-1947), Professor of Psychiatry at the Hebrew University.

man to bring about the synthesis between psychology and psycho-analysis. Under these circumstances the intention of creating a chair for psychology suggests a poorly disguised rejection of psycho-analysis, and the University of Jerusalem would have followed the example of other official educational institutions. It does one good to know that Dr. Eitingon is determined to pursue the practice of psychoanalysis in Palestine no matter what the University decides to do.

Yours faithfully
Freud

273 *To* ERNST FREUD

Vienna IX, Berggasse 19
February 20, 1934

Dear Ernst

Thanks to the guiding principle of all journalistic reportage —of making as much noise as possible—it is probably not easy to learn from the papers what is really happening in a city where shooting is going on. What affected us most was that for almost twenty-four hours we were without electric light. (It was some comfort that at least matches still worked.) But on the whole it was civil war and unpleasant. The details of it all are not clear; rumor has it that a certain powerful man insisted on putting an end at last to the conflict which has been smoldering for so long. At some time this was probably bound to happen. Now of course the victors are the heroes and the saviors of sacred order, the others the im-pudent rebels. Had the latter won, however, it wouldn't have been much better, for it would have meant a military invasion of the country. The government shouldn't be judged too harshly; after all, life under the dictatorship of the proletariat, which was the aim of the so-called leaders, would not have been possible either. Needless to say, the victors will not fail to commit every error that can be committed in such a situation. It will not be Dollfuss's[1] fault; he probably won't be able to curb the dangerous fools in the *Heimwehr*.[2]

---

[1] Dr. Engelbert Dollfuss (1892-1934), Austrian Chancellor.
[2] The Austrian National Guard.

The future is uncertain; either Austrian fascism or the swastika. In the latter event we shall have to leave; native fascism we are willing to take in our stride up to a certain point; it can hardly treat us as badly as its German cousin. It wouldn't be pleasant, of course; but life in a foreign country is not so pleasant either—something I don't have to tell you, although you have actually been rather fortunate. Our attitude to the two political possibilities for Austria's future can only be summed up in Mercutio's line in *Romeo and Juliet*: "A plague on both your houses."

. . . Just now—Wednesday morning, February 21—martial law has been repealed. Our government and our Cardinal[3] expect a great deal from God's assistance.

Warm greetings to you and Lux
Papa

[3] Cardinal Archbishop Dr. Theodor Innitzer.

274 *To* OSKAR PFISTER

Vienna XIX, Strassergasse 47
June 13, 1934

Dear Pfister

I congratulate you on having been made a Doctor of Theology, but I cannot allow you to make me responsible for this honor. As a defender of religion against my *Future of an Illusion* you have an exclusive right to it. But that the faculty of Geneva was not deterred by psychoanalysis from bestowing the title is something that at least deserves recognition.

With cordial greetings
Your
Freud

275 ANONYMOUS

Vienna XIX, Strassergasse 47
June 27, 1934

Dear Miss X.

It is not out of the question that an analytic treatment results in an inability to continue artistic creation. If it does, it isn't

the fault of analysis, for this condition would have come about in
any case, and it is only an advantage to find out about it in time.
If, however, the creative urge is stronger than inner resistances,
productivity will only be enhanced, never diminished, by analysis.

> With best wishes
> Freud

276 *To* ARNOLD ZWEIG

> Vienna XIX, Strassergasse 47
> (not much longer!)
> September 30, 1934

Dear Arnold Zweig

I am answering under the immediate impact of alarm that
your play about Bonaparte[1] may have gone astray on the journey.
But it may still turn up.

You think you know why I haven't written to you for so long?
No, you don't. Perhaps you suspect—and not quite wrongly—that
I have decided not to disturb you any more with my continued
objection to your Nietzsche project; but the main reason was another one. For, in a time of relative freedom and at a loss to know
what to do with my surplus leisure, I have written something[2] myself; and this, contrary to my original intention, has taken such a
hold of me that everything else has been left undone. Now, don't
rejoice, for I bet you won't get a chance to read it. But let me
explain what it is all about.

The starting point of my work is known to you; it was the same
as that of your *Bilanz*. Faced with the renewed persecutions, one
asks oneself again how the Jew came to be what he is and why
he has drawn upon himself this undying hatred. I soon found the
formula: Moses created the Jew. And my essay received the title:
The Man Moses, a Historical Novel (with more right than your
novel about Nietzsche). The material is divided into three parts;
the first reads like an interesting novel; the second is laborious and
lengthy, the third substantial and exacting. The enterprise found-

---

[1] *Bonaparte in Jaffa*, 1935.

[2] "Moses—ein Ägypter," Vienna, 1937. ("Moses: an Egyptian," Part I of *Moses
and Monotheism*, Standard Ed., 23.)

ered on the third section, for it contains a theory of religion which, although nothing new to me after *Totem and Taboo,* is nevertheless bound to be something fundamentally new and shattering to the uninitiated. Concern for these uninitiated compels me to keep the completed essay secret. For we live here in an atmosphere of Catholic orthodoxy. It is said that the politics of our country are made by a Father Schmidt who lives in [the monastery] St. Gabriel near Mödling and is a confidant of the Pope. Unfortunately he himself is an ethnologist and a religious scholar who in his books makes no secret of his horror of psychoanalysis and above all of my totem theory. The good Edoardo Weiss[3] has founded a psychoanalytic group in Rome and has published several numbers of a *Rivista Italiana di Psicoanalisi.* All of a sudden this publication has been stopped, and although Weiss has direct access to Mussolini and received from him a favorable promise, the ban could not be lifted. The ban is said to come straight from the Vatican and Father Schmidt is said to be responsible. Now, it stands to reason that a publication of mine would be bound to create a certain sensation and not escape the notice of the inimical priest. This would mean running the risk of having psychoanalysis banned in Vienna and the end of all our publications here. Were this danger confined to myself it would make little impression on me, but to deprive all our members in Vienna of their livelihood strikes me as too great a responsibility. And in addition to all this there is the feeling that the essay doesn't seem to me too well substantiated, nor do I like it entirely. So all in all it isn't quite the proper occasion for martyrdom. Enough for the moment!

Of my so-called health I prefer not to say much. At least it permits me to continue my normal professional activities. When these lovely autumn days have passed we shall be returning to the Berggasse.

The tidbit from your analysis is rather tasty; I hope it won't be confined to such samples. When your play turns up I will pass on to Martin your question about getting it produced. We don't have many connections in the theater world. Your personal presence will probably be indispensable.

We are certainly living through grim times, but when I think

[3] See letter of Apr. 12, 1933.

back to the era in which I grew up I can't muster any real regret that it has passed. It is "all much of a muchness," as they say.

Until I hear from you again I shall be pleased to assume that you and yours are well.

Cordially

Your

Freud

277 ANONYMOUS*

Vienna IX, Berggasse 19

April 9, 1935

Dear Mrs. . . .

I gather from your letter that your son is a homosexual. I am most impressed by the fact that you do not mention this term yourself in your information about him. May I question you why you avoid it? Homosexuality is assuredly no advantage, but it is nothing to be ashamed of, no vice, no degradation; it cannot be classified as an illness; we consider it to be a variation of the sexual function, produced by a certain arrest of sexual development. Many highly respectable individuals of ancient and modern times have been homosexuals, several of the greatest men among them. (Plato, Michelangelo, Leonardo da Vinci, etc.) It is a great injustice to persecute homosexuality as a crime—and a cruelty, too. If you do not believe me, read the books of Havelock Ellis.

By asking me if I can help, you mean, I suppose, if I can abolish homosexuality and make normal heterosexuality take its place. The answer is, in a general way we cannot promise to achieve it. In a certain number of cases we succeed in developing the blighted germs of heterosexual tendencies, which are present in every homosexual; in the majority of cases it is no more possible. It is a question of the quality and the age of the individual. The result of treatment cannot be predicted.

What analysis can do for your son runs in a different line. If he is unhappy, neurotic, torn by conflicts, inhibited in his social life,

---

* Written in English.

analysis may bring him harmony, peace of mind, full efficiency, whether he remains homosexual or gets changed.

If you make up your mind that he should have analysis with me— I don't expect you will—he has to come over to Vienna. I have no intention of leaving here. However, don't neglect to give me your answer.

Sincerely yours with kind wishes
Freud

P.S. I did not find it difficult to read your handwriting. Hope you will not find my writing and my English a harder task.

278 *To* ARNOLD ZWEIG

Vienna XIX, Strassergasse 47
May 2, 1935

Dear Arnold Zweig

I am sitting in my lovely room in Grinzing, before me the beautiful garden with fresh green and reddish brown young foliage (copper beech) and state that the snowstorm with which May introduced itself has ceased (for the time being!), and that a cold sun is dominating the climate. Needless to say, my idea of enjoying spring with you on Mount Carmel was a mere phantasy. Even supported by my faithful Anna-Antigone I could not embark on a journey; in fact, I have recently had to undergo another cauterization in the oral cavity.

I am worried about your poor eyes. The intelligent oculist whom we consulted refuses to give a definite opinion without a detailed report of the condition. Why the symptoms should appear just now, he can't say. On the other hand, according to him there is no doubt that an improvement could be expected from giving the eyes a rest and a general strengthening. I assume your oculist is trustworthy?

I can't say that much is happening in my life. Since I can no longer smoke freely, I no longer want to write—or perhaps I am just using this pretext to veil the unproductiveness of old age. Moses won't let go of my imagination. I visualize myself reading it out to you when you come to Vienna, despite my defective speech. In a report on Tell-el Amarna, which still hasn't been fully excavated, I read a remark about a certain Prince Thothmes of whom nothing else is known. If I were a millionaire in pounds, I would finance the continued excavations. This Thothmes could be my Moses and I would be able to boast that I had guessed right.

At the suggestion of the Fischer Verlag I have composed a brief address for Thomas Mann's sixtieth birthday[1] (June 6) and into it slipped a warning which I trust will not go unnoticed. The times are gloomy; fortunately it is not my job to brighten them.

With kindest regards
Your
Freud

[1] See letter of June 6, 1935.

279 *To* LOU ANDREAS-SALOMÉ

Vienna XIX, Strassergasse 47
May 16, 1935

Dear Lou

If one lives long enough (say about seventy-nine years), one may live to get a letter and even a photo from you—whatever the latter may look like. I refrain from sending you one of myself. What an amount of good nature and humor it takes to endure the gruesome business of growing old! The garden outside and the flowers in the room are beautiful, but the spring is a *Fopperei,** as we say in Vienna. I am learning at last what it is to be cold. My doctor has ordered me to drink sugar water for my subnormal temperatures which makes one feel miserable.

Don't expect to hear anything intelligent from me. I doubt that I can still produce anything—I don't think so—but in any case I haven't got the time, as I have to do so much for my health. This is apparently like the Sibylline books: the fewer there are left, the higher the value. I of course rely more and more on Anna's care, just as Mephistopheles once remarked:

> In the end we depend
> On the creatures we made.

In any case, it was very wise to have made her. I wish I could tell you in person how much I have your well-being at heart.

Your old
Freud

* mockery.—*Tr.*

280   *To* THOMAS MANN

June 1935

FOR THOMAS MANN'S SIXTIETH BIRTHDAY

Dear Thomas Mann

Please accept a heartfelt message of affection on your sixtieth birthday. I am one of your "oldest" readers and admirers; I
could wish you a very long and happy life, as is the custom on such
occasions. But I shall refrain from doing so; the bestowal of wishes
is trivial and seems to me a regression into the era when mankind
believed in the magic omnipotence of thought. My most personal
experience, moreover, tends to make me consider it a good thing
when merciful fate puts a timely end to our span of life.

Nor do I consider it worthy of imitation when, on a festive occasion such as this, affection overrides respect and obliges the
hero of the day to listen to speeches which overwhelm him with
praise as a human being and analyze as well as criticize him as
an artist. I don't want to be guilty of such arrogance. But I will
permit myself something else: in the name of countless numbers
of your contemporaries I wish to express the confidence that you
will never do or say anything—an author's words, after all, are
deeds—that is cowardly or base, and that even at a time which
blurs judgment you will choose the right way and show it to
others.

Yours very sincerely
Freud

281   *To* VICTOR WITTKOWSKI

Vienna IX, Berggasse 19
January 6, 1936

Dear Sir

A few months after our revered friend Romain Rolland has
reached his seventieth birthday, I shall be eighty. I am afraid this
must be my answer to your suggestion. I would like to give, but

have nothing to give, and the production of something new in response to need and occasion is not forthcoming at this time of life, at least not from me. As recently as a year ago I succeeded in writing something[1] that would have been of special interest to R. R., but it suffered from one defect which prevented it from being published, and since then my ability to produce has dried up. It is probably too late for it to revive again.

If there is something that makes this refusal easier for me, it is that "all reference to politics" has to be excluded. Under this paralyzing restriction—not being allowed to follow my urge to praise his courage of conviction, his love of truth and his tolerance—I couldn't do anything, even if I were in my prime.

On January 29 I shall tell him in a few lines that I am thinking of him with affection. Thanks to your letter I now know the date.

With best wishes
Yours sincerely
Freud

[1] See note 2 to letter of Sept. 30, 1934.

282 *To* BARBARA LOW*

Vienna XIX, Strassergasse 47
April 19, 1936

Dear Barbara Low

I know that you have not thought that the death of your brother-in-law David[1] had left me untroubled, because I had not written at once and directly to you and your sister. On the contrary, it concerned me in a quite special way, which I cannot explain to anyone who is not eighty likewise. My age gives me a right to a special relation* to death. Eder belonged to the people one loves without having to trouble about them. One's heart warmed at the thought of him; of course one did not think of him often enough and for that one feels sorry today. This is the danger with people one knows to be so entirely faithful, true and reliable. One thinks one can pause whenever one likes; when one meets

* Written in English.
[1] The psychoanalyst Dr. David Eder (1868-1936).
* *Verhältnis.*

again one will enjoy intimacy again. It was so indeed, but why, one asks now, did it happen so rarely? The world is becoming so sad that it is destined to speedy destruction—this is the only palliative for me. I can easily imagine how he, too, must have suffered under the bitterness of these times. We were both Jews and knew of each other that we carried that miraculous thing in common, which —inaccessible to any analysis so far—makes the Jew. But I am certainly wrong to throw the shadows of my dark mood on the younger and stronger ones. For you and so many others the world goes on and should bring you better things.

With heartfelt greetings to your mourning sister

Your faithful

Freud

283 *To* ALBERT EINSTEIN

Vienna IX, Berggasse 19
May 3, 1936

Dear Professor Einstein

You struggle in vain against my answering your charming letter.[1] I really must tell you how delighted I am to learn of the change in your judgment, or at least a move in that direction. Of course I always knew that you admired me only "out of politeness," and that you are convinced by very few of my assertions. But I have often asked myself what indeed there is to admire about them if they are not true—i.e., if they do not contain a high degree of truth. Incidentally, don't you think I should have been far better treated if my doctrines had incorporated a greater percentage of error and folly?

You are so much younger than I; and I may hope that by the time you have reached my age you will have become a disciple of mine. Since I shall not be here to learn of this, I am now anticipating the satisfaction. (You realize what I have in mind: "In proud fore-feeling of such lofty bliss I now enjoy,"[2] etc.)

With warm devotion and unswerving admiration

Your

Sigm. Freud

[1] For Freud's eightieth birthday.
[2] Quotation from Goethe's *Faust*, II.

284 *To* STEFAN ZWEIG

Vienna XIX, Strassergasse 47
May 18, 1936

Dear Stefan Zweig

I hope you will forgive me for not answering you before. The demanding and exhausting days are over at last.

I reread your letter before writing this. It rings so simple and true that I almost forgot it was written by a master of style. It nearly convinced me of my importance. Not that I doubt the content of truth in my doctrines, but I find it difficult to believe that they could exert a demonstrable influence on the development of the immediate future. As a result I appear to myself much less important than you represent me, and prefer to dwell on what I recognize with far greater certainty, namely your extremely kind sentiments displayed by the trouble you took for the celebration of my birthday. The beautiful address[1] which you and Thomas Mann composed together and Mann's speech[2] in Vienna were the two events which almost reconcile me to having grown so old. For, although I have been exceptionally happy in my home, with my wife and children and in particular with one daughter who to a rare extent satisfies all the expectations of a father, I nevertheless cannot reconcile myself to the wretchedness and helplessness of old age, and look forward with a kind of longing to the transition into nonexistence. Whatever happens, I am unable to spare my loved ones the pain of separation.

The exceptional position I occupy with you will then also come to an end. For I believe that in the gallery of remarkable human beings which you have set up in your panopticon—as I sometimes jokingly call it—I am certainly not the most interesting, but nevertheless the only *living* person. Perhaps it is to this fact that I owe so much of the warmth of your sympathy. For with the biographer

---

[1] An address in honor of Freud's eightieth birthday, signed by distinguished people from all over the world.

[2] "Freud und die Zukunft" ("Freud and the Future"), lecture given by Thomas Mann in honor of Freud's eightieth birthday in Vienna, 1936.

as with the psychoanalyst we find phenomena which come under the heading of "Transference."

In warm gratitude

Your

Sigm. Freud

285 *To* ARNOLD ZWEIG

Vienna XIX, Strassergasse 47

May 31, 1936

Dear Arnold Zweig

"The affection of this world is indeed mixed with cruelty."[1] Every half hour of the past two weeks has been taken up with completing printed cards of thanks like the enclosed: a few words or sentences under the signature, most of them stilted and forced, and only today, the first of "the pleasant feast of Whitsuntide,"[2] am I finding the time to write you a letter, aroused by the threat that you wish to become my biographer. You who have so many more attractive and important things to do, who can appoint kings and survey the brutal folly of mankind from the height of a watchtower! No, I am far too fond of you to allow such a thing to happen. Anyone turning biographer commits himself to lies, to concealment, to hypocrisy, to flattery, and even to hiding his own lack of understanding, for biographical truth is not to be had, and even if it were it couldn't be used.

Truth is unobtainable; humanity does not deserve it, and incidentally, wasn't our Prince Hamlet right when he asked whether anyone would escape a whipping if he got what he deserved?

Thomas Mann's visit, the address he presented to me, and the public lecture he delivered for the celebration, were gratifying and impressive events. Even the Viennese colleagues celebrated me and betrayed by all manner of signs how difficult they found it. The Minister of Education sent a formal message of polite congratulation, whereupon the newspapers were forbidden on pain of confiscation to publish this act of sympathy within the country. In addition foreign and domestic newspapers printed numerous articles

---

[1] Author unidentified.
[2] Quotation from Goethe's "Reineke Fuchs."

expressing hatred and repudiation. Thus one could declare with satisfaction that sincerity has as yet not disappeared completely from the world.

For me this date naturally marks no epoch; I am the same as I was before. Among the not very numerous gifts of antiques your very strange signet ring gives me pleasure. Thinking of you warmly, I await your further news.

Your

Freud

286 *To* LUDWIG BINSWANGER

Vienna IX, Berggasse 19

October 8, 1936

Dear Friend

Your lecture[1] was a pleasant surprise. Those who listened to it and reported to me were apparently unaffected; it must have been too difficult for them. Reading it I enjoyed your beautiful style, your erudition, the breadth of your horizon, your tact in contradicting me. When it comes to praise one can take unlimited quantities, as everyone knows.

Of course I don't believe you. I have always lived on the ground floor and in the basement of the building—you maintain that on changing one's viewpoint one can also see an upper floor housing such distinguished guests as religion, art, and others. You are not the only one; most cultivated specimens of *homo natura* think likewise. In this respect you are the conservative, I the revolutionary. If I had another life of work ahead of me, I would dare to offer even those high-born people a home in my lowly hut. I already found one for religion when I stumbled on the category "neurosis of mankind." But we are probably talking at cross purposes and it will be centuries before our dispute is settled.

In cordial friendship and with greetings to your wife

Your

Freud

[1] "Freuds Auffassung des Menschen im Lichte der Anthropologie" ("Freud's Conception of Man in the Light of Anthropology"), lecture given in honor of Freud's eightieth birthday in Vienna.

287 *To* THOMAS MANN

Vienna IX, Berggasse 19
November 29, 1936

Dear Friend

The beneficent personal impressions of your last visit to Vienna keep coming back to my mind. Not long ago I laid aside your new volume of the Joseph legend[1] with the melancholy thought that this beautiful experience is now over and that I shall probably not be able to read the sequel.

The effect of this story combined with the idea of the "lived vita" in your lecture and the mythological prototype has started within me a trend of thought which I am making the pretext of a talk with you as though you were sitting opposite me here in my study, but without wishing to provoke a polite reply, let alone a detailed appreciation. I myself do not take the experiment very seriously, but it does have for me a certain attraction, something like the cracking of a whip for an ex-coachman.

I keep wondering if there isn't a figure in history for whom the life of Joseph was a mythical prototype, allowing us to detect the phantasy of Joseph as the secret daemonic motor behind the scenes of his complex life.

I am thinking of Napoleon I.

(a) He was a Corsican, the second son of a large family of brothers and sisters. His eldest brother was called Joseph, and this fact, as chance and necessity are wont to combine in human life, was fateful for him. In a Corsican family the privilege of the eldest is guarded with particularly sacred awe. (I think Alphonse Daudet once described this in a novel. In *Le Nabob?* Or am I mistaken? Was it in some other book? In Balzac?) By this Corsican tradition a normal human relationship becomes exaggerated. The elder brother is the natural rival; the younger one feels for him an elemental, unfathomably deep hostility for which in later life the expressions "death wish" and "murderous intent" may be found appropriate. To eliminate Joseph, to take his place, to become Joseph himself, must have been Napoleon's strongest emotion

[1] *Josef in Ägypten* (*Joseph in Egypt*).

as a small child. It is strange, no doubt, and yet it has been correctly observed, that just these very excessive, infantile impulses tend to turn into their opposite. The hated rival becomes the loved one. This was the case with Napoleon. We assume that he started out with an ardent hatred of Joseph, but we learn that later on he loved him more than any other human being and could hardly find a fault with this worthless, unreliable man. Thus the original hatred had been overcompensated, but the early aggression released was only waiting to be transferred to other objects. Hundreds of thousands of unknown individuals had to atone for the fact that this little tyrant had spared his first enemy.

(b) On another level the young Napoleon was tenderly tied to his mother and concerned to replace his prematurely deceased father by caring for his brothers and sisters. As soon as he became a general it was suggested that he marry a young widow, older than himself, who possessed both influence and rank. There were a number of things to be said against her, but what probably decided him was that her name was Josephine. Owing to this name he could transfer to her part of the tender attachment he felt for his elder brother. She did not love him, treated him badly, betrayed him, but he the despot, as a rule cynically cold toward women, clung to her passionately and forgave her everything; nothing she did could arouse his anger.

(c) The infatuation for Josephine B. was undoubtedly brought about by the name, but of course it was not an identification with Joseph. This emerges most clearly in the famous expedition to Egypt. Where else could one go but Egypt if one were Joseph and wanted to loom large in the brothers' eyes? If we were to examine the political reasons for this enterprise of the young general, we would probably find that they were nothing but the willful rationalization of a fantastic idea. It was this campaign, by the way, that marked the beginning of Egypt's rediscovery.

(d) The intention which drove Napoleon to Egypt was to be realized in his later life in Europe. He took care of his brothers by making them kings and princes. The good-for-nothing Jerome was perhaps his Benjamin. And then he forsook his myth; he allowed himself to be swayed by practical considerations, to repudiate the beloved Josephine. With this act his decline began. The great destroyer now worked on his self-destruction. The rash, poorly prepared campaign against Russia led to his downfall. It was like a self-punishment for his disloyalty to Josephine, for the regression from his love to his original hostility toward Joseph. And here too, although contrary to Napoleon's intention, fate repeated another chapter in the Joseph legend. Joseph's dream—that the sun, the moon, and the stars should bow down to him—led to his being cast into the pit. . . .

My daughter reminds me that I have already divulged to you this interpretation of the daemonic man after you read your essay here. She is right, of course. I had forgotten, and the idea revived after reading your book. And now I hesitate whether to hold onto these lines or send them to you after all with many apologies.

Cordially

Your

Freud

288 *To* MARIE BONAPARTE

Vienna

December 6, 1936

My dear Marie

Just received your card from Athens and your manuscript of the Topsy book.[1] I love it; it is so movingly genuine and true. It is not an analytical work, of course, but the analyst's thirst for truth and knowledge can be perceived behind this production, too. It really explains why one can love an animal like Topsy (or Jo-fi)[2] with such extraordinary intensity: affection without ambivalence, the simplicity of a life free from the almost unbearable conflicts of civilization, the beauty of an existence complete in itself; and yet, despite all divergence in the organic development, that feeling of an intimate affinity, of an undisputed solidarity. Often when stroking Jo-fi I have caught myself humming a melody which, unmusical as I am, I can't help recognizing as the aria from *Don Giovanni:*

A bond of friendship

Unites us both. . . .

---

[1] *Topsy, Chow-Chow au Poil d'Or* (*Topsy, the Chow with the Golden Hair*), Paris, 1937.

[2] Freud's chow at this time.

And if you, at the youthful age of fifty-four, can't help thinking so often of death, are you surprised that at 80½ I keep brooding on whether I shall reach the age of my father and brother, or even that of my mother, tortured as I am by the conflict between the desire for rest, the dread of renewed suffering (which a prolonged life would mean), and by the anticipation of sorrow at being separated from everything to which I am still attached?

Warm greetings to you (and Topsy) from

Your

Freud

289 *To* MAX EITINGON

Vienna IX, Berggasse 19

February 5, 1937

Thirty years, dear Friend,
is a long time, even for an old man like me. I thank you for drawing my attention to this our anniversary. You will have to be content with the recognition that during this time you have played a most laudable role in our movement and that on the most diverse occasions and in various concerns you have been personally as close to me as few others. There have been no opportunities for expressing gratitude in tangible form except by transferring to you the ring I have been wearing for years on my finger. But many a thing can be understood without words, implicitly so to speak.

Having recovered from the most recent damage and able once more to smoke to a certain extent, I have even started writing again. Just minor things: a fragment that could be detached from the work on Moses (known to you and Arnold Zweig) has been completed. The more important things connected with it must of course remain unsaid. A brief technical essay[1] which is slowly taking shape has the function of helping me fill the many free hours with which my dwindling analytical practice has presented me. It is understandable that patients don't surge toward an analyst of such an unreliable age. I would willingly abandon professional work altogether, but accustomed as I am to looking after so many

[1] Probably "Konstruktionen in der Analyse," 1937. ("Constructions in Analysis," Standard Ed., 23.)

people I find it hard to bear the uncertainty of not knowing how long one's savings will last. Let us hope for a timely hand from fate.

The most gratifying thing in my suroundings is Anna's capacity for work and her consistent achievement.

In the hope that your domestic solitude will not be prolonged,
Cordially
Your
Freud

290   *To* MARIE BONAPARTE

Grinzing
August 13, 1937

My dear Marie

I can answer you without delay, for I have little to do. Two days ago "Moses II" was finished and laid aside, and the best way to forget one's minor ailments is by exchanging thoughts with friends.

To the writer immortality evidently means being loved by any number of anonymous people. Well, I know I won't mourn your death, for you will survive me by years, and over mine I hope you will quickly console yourself and let me live in your friendly memory—the only form of limited immortality I recognize.

The moment a man questions the meaning and value of life, he is sick, since objectively neither has any existence; by asking this question one is merely admitting to a store of unsatisfied libido to which something else must have happened, a kind of fermentation leading to sadness and depression. I am afraid these explanations of mine are not very wonderful. Perhaps because I am too pessimistic. I have an advertisement floating about in my head which I consider the boldest and most successful piece of American publicity: "Why live, if you can be buried for ten dollars?"

Lün[1] has taken refuge with me after having been given a bath.

[1] Another of Freud's chows.

If I understand her right, she wants me to thank you warmly for the greeting. Does Topsy realize she is being translated?[2]

Write again soon.

Affectionately
Your
Freud

[2] Marie Bonaparte's *Topsy* (see note 1 to letter of Dec. 6, 1936), translated by Sigmund and Anna Freud, was published by Allert de Lange, Amsterdam, 1939, under the title *Topsy, der goldhaarige Chow.*

291 *To* MARTIN FREUD

Grinzing
August 16, 1937

Dear Martin

I am enjoying your description of Capri and your lovely life there. Yesterday, Sunday, I was reviving memories of our stay in Capri with your uncle.[1] It was very hot then, and we were the only guests in September. The boatman who rowed us into the Blue Grotto told us horror stories of a certain Timperio, I think, who once haunted the island. Vesuvius also was active, producing a smoke cloud by day and a fire cloud by night, just like the God of Exodus in the Bible. For Jehovah (Jahve) was a volcanic god, as you will learn from the second essay on Moses, which is now finished, awaiting your return.

Quite a good summer here, occasionally interrupted by lapses into April or premonitions of November. Also by ill-health. . . .

You will be confronted with a minor business matter concerning the *Verlag* when you return. Nothing important.

On September 1 my friend Emanuel Löwy,[2] even he, will be eighty. He presented me with a beautiful etching by Dürer on the identical occasion. Even if he hadn't I would not like the day to pass unnoticed. But it is difficult to find him a present. I have

[1] Alexander Freud.
[2] Emanuel Löwy, Professor of Archaeology at the Universities of Rome and Vienna, a lifelong friend of Freud's.

nothing but the collected works, but with his poor eyesight he can hardly read any more. . . .

Anna is enjoying her so-called vacation—i.e., she is playing with the little babies[3] instead of the big ones. . . .

I hope the days will continue to be most beautiful until you return to our truly moderate climate.

Affectionately
Papa

[3] Anna was working in a nursery.

292 *To* STEFAN ZWEIG

Vienna IX, Berggasse 19
October 17, 1937

Dear Stefan Zweig

It is hard to say whether your welcome letter gave me more pain or pleasure. Like you I suffer from the times we live in, and like you I find comfort only in the feeling of solidarity with a few others in the assurance that the same things have remained precious to us, and that the same values appear to us incontestable.

But in all friendship I feel entitled to envy you for being able to offer resistance through your excellent work. May you succeed more and more! I am looking forward with pleasure to your *Magellan*.[1]

My work lies behind me, as you say yourself. No one can predict how posterity will assess it. I myself am not so certain; scientific research and doubt are inseparable, and I have surely not discovered more than a small fragment of truth. The immediate future looks grim, for psychoanalysis as well. In any case I am not likely to experience anything enjoyable during the weeks and months I may still have to live.

I have started complaining quite against my will. What I wanted was to come closer to you in a human way rather than be admired like a rock against which the waves break in vain. But even if my defiance remains silent, it remains defiance nevertheless and— *impavidum ferient ruinae*.[2]

[1] *Der Erdumsegler Magellan.* (*Conqueror of the Seas: The Story of Magellan*, Viking Press, New York, 1938.)

[2] Horace, *Odes*, III, 3, 7: "The falling ruins leave him undismayed."

I hope you won't keep me waiting too long for your next courageous and excellent books.

With cordial greetings
Your old Freud

293 ANONYMOUS

Vienna IX, Berggasse 19
December 14, 1937

Dear *Herr Doktor*

I must add a few words to my thanks for sending me your
valuable little book. Not only because you evidently appreciate
analysis and also mention my name several times in a friendly
manner, but above all because your book contains so many remarks
which cannot help appearing important and pertinent to a Jewish
reader.

Several years ago I started asking myself how the Jews acquired
their particular character, and following my usual custom I went
back to the earliest beginnings. I did not get far. I was astounded
to find that already the first so to speak embryonic experience of
the race, the influence of the man Moses and the exodus from
Egypt, conditioned the entire further development up to the present day—like a regular trauma of early childhood in the case history
of a neurotic individual. To begin with, there is the temporal conception of life and the conquest of magic thought, the rejection of
mysticism, both of which can be traced back to Moses himself and—
although not with all the historical certainty that could be desired—
perhaps a little further. Two essays[1] in this year's volume of
Imago contain part of my findings; the most important section I
had to withold. I would be pleased if you would have a look at
these essays.

With kind regards
Yours sincerely
Freud

---

[1] "Moses ein Ägypter" und "Wenn Moses ein Ägypter war . . . ," *Imago*, 1937.
("Moses an Egyptian" and "If Moses was an Egyptian . . . ," Parts I and II of
Moses and Monotheism, Standard Ed., 23.)

294 *To* ERNST FREUD

Vienna IX, Berggasse 19
January 17, 1938

Dear Ernst

Your letter pleased me no less than the lovely Persian glass dish you sent me. In return I have had mailed to you an essay on Moses, one of my few recent works which may arouse general interest. I am afraid this interest may exceed justified proportions and be blown up into something sensational. But I may be mistaken. It is my first appearance as a historian; late enough! I don't anticipate a friendly reception from the scientific critics—Jewry will be very offended.

I don't know what you read about me in the *Evening Standard*. Probably nothing but lies. From me you can learn that I am not at all well and that I am beginning to find life pretty inconvenient. This is not surprising; what else can one expect?

My best wishes for the opening of Hidden House![1] It is typically Jewish not to renounce anything and to replace what has been lost. Moses, who in my opinion left a lasting imprint on the Jewish character, was the first to set an example. In our present difficult times your existence in England stands out boldly against all the misery around us. Whenever I think of your success I feel pleased and full of hope for the chances of the next generation.

You know about the rest of us. Mama is bearing up magnificently, your aunt is about to undergo an operation for cataract which we hope will turn out well. Anna is splendid, in spirits, achievement, and in all human relationships. It is amazing how clear and independent her scientific work has become. If she had more ambition . . . but perhaps it is better like this for her later life.

My warmest greetings to your good Lux and the three big boys about whom their grandfather knows alas too little. How I would like to say: See you in Grinzing!

Papa

---

[1] A house on the east coast of England, substitute for one Ernst Freud's family had owned on the German island of Hiddensee.

295 *To* MAX EITINGON

Vienna IX, Berggasse 19
February 6, 1938

Dear Friend

I have often asked myself in bewilderment if it is absolutely necessary for newspapers to tell lies so regularly and without restraint. In any case it is good to know that you didn't believe the news this time, either. Our brave and in its way decent government is now more energetic than hitherto in keeping the Nazis at bay, although in view of the latest events in Germany no one can be sure what is going to happen.

My last operation performed by Pichler[1] two weeks ago has produced the usual reactions, but for a week now I have been able to work—and chew—again. Since the microscopic findings on the removed tissue were suspicious this time, Pichler has threatened me with another operation in the near future, but hasn't yet come to a decision. This isn't very pleasant of course, but the operation itself, thanks to Evipan, has become ideally painless and undangerous, and there is no need to worry about more than the immediate future.

We are following the reports about all that is happening in the "Holy Land" with great concern. I am glad to hear that Mirra is nevertheless more cheerful. Sometimes one can't help thinking of Master Anton's concluding words in one of Hebbel's dramas: "I no longer understand this world."[2] Have you read that Jews in Germany are to be forbidden to give their children German names? They can only retaliate by demanding that the Nazis refrain from using the popular names of John, Joseph, and Mary.

Cordially
Your
Freud

[1] Professor Dr. Hans Pichler, oral surgeon, who treated Freud for many years.
[2] Quotation from the drama *Maria Magdalene*, by the German playwright Friedrich Hebbel (1813-1863).

296 *To* ALEXANDER FREUD

Vienna IX, Berggasse 19
April 19, 1938

Dear Brother

Your seventy-second birthday finds us on the verge of separating[1] after long years of living together. I hope it is not going to be a separation forever, but the future—always uncertain—is at the moment especially difficult to foresee.

I would like you to take over the good cigars which have been accumulating with me over the years, as you can still indulge in such pleasure, I no longer.

The rest—you will know what I mean—is silence.

Affectionately as ever
Your
Sigm.

[1] Freud was awaiting his emigration.

297 *To* ERNST FREUD

Vienna IX, Berggasse 19
May 12, 1938

Dear Ernst

I am writing to you for no particular reason because here I am sitting inactive and helpless while Anna runs here and there, coping with all the authorities, attending to all the business details. One can "already see the journey."[1] All we are waiting for is the Inland Revenue's "all clear," which is supposed to arrive within a week.

Two prospects keep me going in these grim times: to rejoin you all and—to die in freedom.[*] I sometimes compare myself with the old Jacob who, when a very old man, was taken by his children to Egypt, as Thomas Mann is to describe in his next novel.[2] Let us

[1] Oft-quoted statement of Freud's daughter Sophie, when a child.
[*] These four words were written in English.
[2] *Joseph der Ernährer*, Berman Fischer Verlag, Stockholm.

hope that it won't also be followed by an exodus from Egypt. It is high time that Ahasuerus came to rest somewhere.

To what extent we old people will succeed in coping with the difficulties of a new country remains to be seen. You will help us in this. Compared to being liberated nothing is of any importance. Anna will certainly find it easy to manage, and this is the main thing, because for us old people between seventy-three and eighty-two the whole undertaking would have made no sense.

If I were to arrive as a rich man, I would start a new collection with the assistance of your brother-in-law.[3] As it is I will have to be content with the two small figures which the Princess rescued on her first visit, and those which she bought during her last stay in Athens and is keeping for me in Paris. How much of my own collection I can have sent on is very uncertain.[4] The whole thing reminds me of the man trying to rescue a bird cage from the burning house.

I could go on writing like this for hours, but you will be too busy to read it. So here are fond greetings to you, Lux, and all the boys, from

Papa

[3] Hans M. Calmann, London art dealer.
[4] The entire collection actually reached England.

298 *To* ERNEST JONES

Vienna IX, Berggasse 19
May 13, 1938

Dear Jones

Anna tells me she concludes from your last letter that you expect an answer to yours for my birthday, which is why I am writing—but also because I am sitting here in my study with nothing whatever to do and generally useless. We had decided not to take any notice of this birthday, to postpone it until the sixth of June, July, August and so on; in short to a date after our liberation, and in fact I haven't answered any of the letters, telegrams, etc. Now it looks as though we shall land in England in May after all. I say it "looks," for in spite of all promises, uncertainty rules the day. With her touching devotion Princess Marie telephoned the day before yesterday to say she is planning to come to Vienna on Monday (that is, May 16) to escort us across the frontier as far as Paris; yesterday we had to ask her not to be so optimistic since we still cannot fix the day of our departure.

Another reason why I didn't answer before lies in today's general inhibition to write. Perhaps you will remember that I once traced back the so-called "physiological feeble-mindedness of women" (Möbius)[1] to the fact that women were forbidden to think about sex. As a result they acquired a dislike for thinking altogether. Imagine how such a censorship must affect me who have always been in the habit of expressing freely what I believe. But the first sentence you would have heard from me at Victoria Station would have told you how much pleasure your friendly letter has given me.

I wish I could arrive in England in a better condition. It is true that I am coming with my own doctor, but I am in need of several doctors and soon after my arrival I will have to find an aurist and consult the jaw specialist whose name Pichler has given me. Now and again one thinks that "Le jeu ne vaut pas la chandelle," yet although one is right one must not agree with oneself. The advantage the emigration promises Anna justifies all our little sacrifices. For us old people (73–77–82) emigrating wouldn't have been worth while.

Anna is indefatigably active, not only for us but for countless others. I hope that in England she will also be able to do much for analysis, but she will not intrude.

Perhaps we will meet soon after all!

Affectionately
Your
Freud

[1] Dr. Paul Möbius (1853-1907), German neurologist.

299   *To* MAX EITINGON

39 Elsworthy Road
London N. W. 3
June 6, 1938

Dear Friend

I haven't been giving you much news these past weeks. To make up for this, I am writing you the first letter from the

new house, even before the new notepaper has arrived. Everything
is still unreal, as in a dream. This could be a wonderful fulfilment
of a wish dream if we hadn't found Minna seriously ill with a high
temperature on our arrival. The outcome is still uncertain. You
probably know that we didn't all leave at the same time. Dorothy
was the first, Minna on May 5, Martin on the fourteenth, Mathilde
and Robert on the twenty-fourth, ourselves incidentally not until
the Saturday before Whitsunday, June 3. With us came Paula[1]
and Lün, the latter as far as Dover, where she was taken into quar-
antine by a friendly vet. Dr. Schur,[2] our family doctor, was to have
accompanied us with his family, but at the eleventh hour he was
unfortunate enough to require an appendix operation, with the
result that we had to content ourselves with the protection of the
lady pediatrician,[3] whom Anna brought along. She looked after me
very well, and in fact the difficulties of the journey did manifest
themselves in my case with painful heart fatigue, for which I was
given liberal doses of nitroglycerin and strychnine. The tedious customs investigation in Kehl we were spared by a miracle. Over the
Rhine bridge and we were free! The reception in Paris at the Gare
de l'Est was friendly, somewhat noisy with journalists and photog-
raphers. We spent from 10 A.M to 10 P.M in Marie's[4] house. She
surpassed herself in tender care and attention, returned to us part
of our money and refused to allow me to continue the journey
without some new Greek terracotta figures. We crossed the Channel by the ferryboat and caught sight of the sea for the first time
at Dover. And soon we were in Victoria station, where the im-
migration officers gave us priority. Our reception in London was a
very cordial one. The more serious papers have been printing brief
and friendly lines of welcome. All kinds of fuss is bound to be in
store for us.

To go back, Ernst and my nephew Harry[5] were in Paris to receive
us. Jones met us at Victoria and drove us back through the beautiful city of London to our new house, 39 Elsworthy Road. If you
know London: it is quite far north, beyond Regents Park at the

---

[1] Paula Fichtl, for over thirty years in the service of the Freud family and friend
of four generations.

[2] Dr. Max Schur—see List of Addressees and letters of June 28, 1930 and June
26, 1938.

[3] Dr. Josefine Stross.

[4] Princess Marie Bonaparte.

[5] Alexander Freud's son.

foot of Primrose Hill. From my window I see nothing but greenery, which begins with a charming little garden surrounded by trees. So it is as though we were living in Grinzing where Gauleiter Bürckel has just moved into the house opposite. The house itself is elegantly furnished. The rooms upstairs, which I cannot reach without a sedan chair, are said to be particularly beautiful; on the ground floor a bedroom, a study and dining room have been arranged for us—Martha and me—quite beautiful and comfortable enough. Ernst of course is responsible for the choice and the furnishing of the house, but we can't stay here for more than a few months and will have to rent another house to be ready for the arrival of our furniture.

It is hardly an accident that [in this letter] I have remained so matter-of-fact up to now. The emotional climate of these days is hard to grasp, almost indescribable. The feeling of triumph on being liberated is too strongly mixed with sorrow, for in spite of everything I still greatly loved the prison from which I have been released. The enchantment of the new surroundings (which make one want to shout "Heil Hitler!") is blended with discontent caused by little peculiarities of the strange environment; the happy anticipations of a new life are dampened by the question: how long will a fatigued heart be able to accomplish any work?—Under the impact of the illness on the floor above me (I haven't been allowed to see her [Minna] yet) the pain in the heart turns into an unmistakable depression. But all the children, our own as well as the adopted ones, are charming. Mathilde is as efficient here as Anna was in Vienna; Ernst really is what he has been called, "a tower of strength," Lux and the children are worthy of him; the two men, Martin and Robert, are holding their heads high again. Am I to be the only one who doesn't cooperate and lets his family down? My wife, moreover, has remained healthy and undaunted.

We have become popular in London overnight. "We know all about you," says the bank manager; and the chauffeur who drives Anna remarks: "Oh, it's Dr. Freud's place." We are inundated with flowers. Now you may write again—and whatever you like. Letters are not opened.

Affectionate regards to you and Mirra.

Your
Freud

300 ANONYMOUS

39 Elsworthy Road
London N.W. 3
June 18, 1938

Dear Sir

It is only natural that we should try to help one another in these times. But I believe that if in doing so we don't keep to the truth we shall soon lose all credit.

You ask me for confirmation that Dr. E. W. was in sole charge of his department at the Viennese Clinic. Now, it so happens that I hardly know Dr. W. by name, certainly not personally, and I have no idea where or in what capacity he worked in Vienna; as a result I cannot give you any such confirmation.

Yours sincerely
Freud

301 *To* ALEXANDER FREUD

39 Elsworthy Road
London N.W. 3
June 22, 1938

Dear Alex

I really feel a strong reluctance to answer your letter. You will soon hear why. The fact is, things are going very well for us, too well I would say if it weren't for an injured heart and an irritated bladder reminding one of the impermanence of human happiness. This England—you will soon see for yourself—is in spite of everything that strikes one as foreign, peculiar, and difficult, and of this there is quite enough—a blessed, a happy country inhabited by well-meaning, hospitable people. At least this is the impression of the first weeks. Our reception was cordial beyond words. We were wafted up on the wings of a mass psychosis. (I feel compelled to express myself poetically.) After the third day the mail delivered letters correctly to "Dr. Freud, London" or "Overlooking

Regents Park"; a taxi driver bringing Anna home exclaimed on
seeing the number of the house: "Oh, it's Dr. Freud's place." The
newspapers have made us popular. We have been inundated with
flowers and could easily have suffered serious indigestion from
fruit and sweets. As for the letters—I have been working like a
writing-coolie for two whole weeks trying to sift the chaff from
the wheat and (forgive the lame comparison) answer the latter.
There were letters from friends, a surprising number from complete strangers who simply wanted to express their delight at our
having escaped to safety and who expect nothing in return. In addition of course hordes of autograph hunters, cranks, lunatics, and
pious men who send tracts and texts from the Gospels which promise salvation, attempt to convert the unbeliever and shed light
on the future of Israel. Not to mention the learned societies of
which I am already a member, and the endless number of Jewish
associations wanting to make me an honorary member. In short,
for the first time and late in life I have experienced what it is to be
famous.

There only remains to report that Minna, who has evidently
been suffering from the same bronchial pneumonia which you recently had, is beginning to recover; that Robert and Mathilde are
running the house very well, that Martha is really enjoying her
life, and that Anna is as usual working for herself and others.
Harry we often see. As for you two, I advise you to wait patiently
until you can leave your Swiss exile; the prospects the Nazis hold
out to you seem surprisingly favorable. Can one really imagine
those scoundrels capable of it? Is there still a remnant of decency
or justice left in them? Let me know the moment you have reason
to think so.

Affectionate greetings to you both.
Your
Sigm.

302 *To* STEFAN ZWEIG

39 Elsworthy Road
London N.W. 3
July 20, 1938

Dear Stefan Zweig
    I really have reason to thank you for the introduction which
brought me yesterday's visitors. For until then I was inclined to

look upon surrealists, who have apparently chosen me for their patron saint, as absolute (let us say 95 per cent, like alcohol) cranks. The young Spaniard,[1] however, with his candid fanatical eyes and his undeniable technical mastery, has made me reconsider my opinion. It would in fact be very interesting to investigate analytically how a picture like this came to be painted. From the critical point of view it could still be maintained that the notion of art defies expansion as long as the quantitative proportion of unconscious material and preconscious treatment does not remain within definite limits. In any case these are serious psychological problems.

As for the other visitor, I like putting obstacles in the path of a candidate in order to test the seriousness of his intention and increase his spirit of sacrifice. Psychoanalysis is like a woman who wants to be seduced but knows she will be underrated unless she offers resistance. If your Mr. J. takes too long thinking it over, he can go to someone else later on, to Jones or to my daughter.

I am told that on going away you left something behind—gloves, etc. You realize that this means a promise to return.

Cordially
Your
Freud

[1] Salvador Dali, born 1904.

303 *To* ALFRED INDRA

39 Elsworthy Road
London N.W. 3
July 20, 1938

Dear Mr. Indra

Your news that the foreign exchange office demands the sale of the Dutch gulden on deposit in Zürich came to me as a painful surprise. On this subject I have the following remarks to make:

You yourself were present at the negotiations during which the Gestapo gave us the assurance that we could keep this account, and you will probably remember their reasoning, which sounded considerate. Officially it was granted to give us the means of building up a new existence in a foreign country, which as a matter of fact is made possible by the permission to practice psychoanalysis unhampered in England. Now I can imagine a situation in which the right to dispose of this foreign currency could become of vital necessity to us.

We emigrated on June 5, now more than seven weeks ago; since then we have been awaiting the arrival of our household effects, my books and the small collection of antiques which were released on the payment of a certain sum. So far nothing has arrived. If this transferable property is witheld, I shall be left with no other choice but to use Dutch gulden for the purchase of new furniture, linen, fixtures, and books, notwithstanding the consequences alluded to in your letter, for we cannot last long in furnished apartments which are beyond our means.

As soon as we are in possession of our Viennese effects I authorize you, sir, to offer the Dutch gulden to the foreign exchange office. Admittedly this would mean a great sacrifice for me. *One would have thought that an official promise, such as that given, would be honored.* I heard today from the Commissar Dr. Sauerwald that he will try to arrange for the Dutch gulden to be replaced by another offer. This meets with my full approval.

If the Dutch gulden really have to be sold, the sum will probably be deposited to our credit in a blocked account. Could you not arrange, sir, to have the money transferred to the account of Princess Marie of Greece, who has paid the emigration tax for us? In this event we would be enabled to settle one of our debts.

As regards the point touched upon in your letter—i.e., money owed to me as an author by foreign publishers—I must remark that this item has always been the most fluctuating and uncertain of my sources of income. At the present time I have no outstanding royalties, I am not owed anything by foreign publishers. I do not expect any payments of this nature before the end of the year. Nor do I know the names of these publishers by heart. Neither I nor my son,[1] the director of the *Verlag*, was permitted to bring out any publishers' statements, and I haven't with me any of my translated works from which I could find the publishers' names. In any case, owing to the decline in sales in all countries, no large sums will be involved before the end of the year.

With many thanks for your efforts on our behalf

Yours sincerely
Freud

[1] Martin Freud.

304 *To* MAX SCHUR

39 Elsworthy Road
London N.W. 3
July 26, 1938

Dear Doctor

You are right; everything has changed. I am even prepared for your half-yearly bill to be presented to me in another currency. But it is not only for me that things have changed, and I wouldn't dream of starting to economize with, of all people, my doctor.

Cordially
Your
Freud

305 *To* MARIE BONAPARTE

20 Maresfield Gardens
London N.W. 3
October 4, 1938

My dear Marie

It is only right and proper that my first letters from "home" should be addressed to you. It won't be a long one, for I can hardly write, no better than I can speak or smoke.

This operation was the worst since 1923 and has taken a great deal out of me. I am dreadfully tired, and feel weak when I move. I have actually started work with three patients, but it isn't easy. The after-effects are supposed to be over in six weeks, and I am only at the end of the fourth.

The house is very beautiful. It suits me that you have postponed your visit for a while; by that time it will be ready for you to see. There is only one great drawback, no spare bedroom; this was sacrificed to the elevator; or rather there is a chance of a room, but it is a sad one. Aunt Minna is inhabiting the room as a patient, and it is still uncertain how long she will need it. It strikes us as superfluous that at the moment Martin is also ill with a coli infection and a high temperature in the hospital which Minna has just left.

Everything here is rather strange, difficult, and often bewildering, but all the same it is the only country we can live in, France being impossible on account of the language. During the days when war seemed imminent the behavior on all sides was exemplary, and it is wonderful to see how, now that the intoxication of peace has subsided, people as well as Parliament are coming back to their senses and facing the painful truth. We too of course are thankful for the bit of peace, but we cannot take any pleasure in it.

The views of the beautiful castle in Brittany were the last sign I had from you. I trust that you are back in Paris again and that I shall hear from you soon.

With warm greetings to you, the Prince,[1] and the young couple[2]

Your rather old
Freud

[1] Marie Bonaparte's husband. Prince George of Greece and Denmark.
[2] Their daughter, Princess Eugénie, and her husband.

306 *To* YVETTE GUILBERT

20 Maresfield Gardens
London N.W. 3
October 24, 1938

My dear Friends

The warmth of your letter gave me great pleasure and the confidence with which you promise a visit in May 1939 has moved me greatly. At my age, however, every postponement has a sad connotation.[*] It was privation enough during these last years not to have been granted an hour's rejuvenation by the magic spell of Yvette.

Most affectionately
Your
Freud

[*] This word was written in English.

307 *To* CHARLES SINGER

20 Maresfield Gardens
London N.W. 3
October 31, 1938

Dear Sir

I am glad to have the opportunity of establishing at least
a written contact with you, and I am grateful for the permission to
do so in the German language. At the same time I wish to thank
you for the pamphlet and the report which my son Ernst has passed
on to me.

The reason for our correspondence is certainly strange enough.
My little book, at present at the printer's, bears the title *Moses and
Monotheism*,[1] as I trust you will be seeing for yourself next spring.
It contains an investigation based on analytical assumptions of the
origin of religion, specifically Jewish monotheism, and is essentially
a sequel to and an expansion of another work which I published
twenty-five years ago under the title *Totem and Taboo*. New ideas
do not come easily to an old man; there is nothing left for him to
do but repeat himself.

It can be called an attack on religion only in so far as any scientific
investigation of religious belief presupposes disbelief. Neither in my
private life nor in my writings have I ever made a secret of my being
an out-and-out unbeliever. Anyone considering the book from this
point of view will have to admit that it is only Jewry and not
Christianity which has reason to feel offended by its conclusions.
For only a few incidental remarks, which say nothing that hasn't
been said before, allude to Christianity. At most one can quote the
old adage: "Caught together, hanged together!"

Needless to say, I don't like offending my own people, either.
But what can I do about it? I have spent my whole life standing up
for what I have considered to be the scientific truth, even when it
was uncomfortable and unpleasant for my fellow men. I cannot end
up with an act of disavowal. Your letter contains the assurance
which testifies to your superior intelligence, that everything I write

[1] *Der Mann Moses und die monotheistische Religion* (*Moses and Monotheism*),
London and New York, 1939; Standard Ed., 23.

is bound to cause misunderstanding and—may I add—indignation. Well, we Jews have been reproached for growing cowardly in the course of the centuries. (Once upon a time we were a valiant nation.) In this transformation I had no share. So I must risk it.

Very sincerely yours
Freud

308 *To* DAVID BAUMGARDT

20 Maresfield Gardens
London N.W. 3
November 3, 1938

Dear Professor Baumgardt

I am confirming the receipt of your impressive book[1] immediately after its arrival. If I were to wait until I had read it, it might take too long. But read it I certainly will, and I hope to learn much from it and even to find in it the solution to the little problem for which I was going to ask you myself.

One of the teachers during my scientific childhood once assured me that he never read anything by people whom he did not know, and that after having made their personal acquaintance he could usually spare himself the reading of their publications. Applying this to our case, I say that after having made your acquaintance I can read your book without fear and compunction.

With warm greetings to you and your wife
Your
Freud

[1] *Der Kampf um den Lebenssinn unter den Vorläufern der modernen Ethik* (*The Search for the Meaning of Life among the Precursors of Modern Ethics*), Leipzg, 1933.

309 *To* MARIE BONAPARTE

20 Maresfield Gardens
London N.W. 3
November 12, 1938

My dear Marie

I am always prepared to acknowledge, in addition to your indefatigable diligence, the self-effacement with which you give your energy to the introductions and popular expositions of psycho-analysis.

And yet you claim to be so very ambitious and to long for immortality at any price! Well, your actions testify to a nobler character.

Your comments on "time and space" have come off better than mine would have—although so far as time is concerned I hadn't fully informed you of my ideas. Nor anyone else. A certain dislike of my subjective tendency to grant the imagination too free a rein has always held me back. If you still want to know, I will tell you next time you come.

My state of health still hasn't taken a decisive turn. The sequestrum of bone will not come away and the discomfort remains the same. Surgeons are certainly cruel fellows. . . .

The latest horrifying events in Germany aggravate the problem of what to do about the four old women[1] between seventy-five and eighty. To maintain them in England is beyond our powers. The assets we left behind for them on our departure, some 160,000 Austrian schillings, may have been confiscated already, and are certain to be lost if they leave. We have been considering a domicile [for them] on the French Riviera, at Nice or somewhere in the neighborhood. But would this be possible?

I am still quite unproductive. I can write letters, but nothing more. Anna has given three public lectures which have been highly praised, even for their language.

With warmest greetings

Your

Freud

[1] Four of Freud's five sisters, unable to leave Austria in time, were deported and killed by the Nazis.

310 *To* TIME AND TIDE*

20 Maresfield Gardens
London N.W. 3
November 16, 1938

To the Editor of
Time and Tide

I came to Vienna as a child of four years from a small town in Moravia. After seventy-eight years of assiduous work I had to leave my home, saw the scientific society I had founded, dissolved, our institutions destroyed, our printing Press ("Verlag") taken over by the invaders, the books I had published confiscated or reduced to pulp, my children expelled from their professions. Don't you think you ought to reserve the columns of your special number for the utterances of non-Jewish people, less personally involved than myself?

In this connection my mind gets hold of an old French saying:

> Le bruit est pour le fat,
> La plainte est pour le sot;
> L'honnête homme trompé
> S'en va et ne dit mot.[1]

I feel deeply affected by the passage in your letter acknowledging "a certain growth of anti-Semitism even in this country." Ought this present persecution not rather give rise to a wave of sympathy in this country?

Respectfully yours
Sigm. Freud

* Written in English.
[1] Quotation from "La Coquette Corrigée," by the French poet La Noue (1701-1761):

> A fuss becomes the fop,
> Complaining fits the fool;
> An honest man deceived
> Will turn his back in silence.

311 *To* RACHEL BERDACH (BARDI)

London N.W. 3
December 27, 1938

Dear Madam (or Miss?)

Your mysterious and beautiful book[1] has pleased me to an extent that makes me unsure of my judgment. I wonder whether it is the transfiguration of Jewish suffering or surprise that so much psychoanalytical insight should have existed at the court of the brilliant and despotic Staufer [Friedrich II, Von Hohenstaufen] which makes me say that I haven't read anything so substantial and poetically accomplished for a long time!

And with it such a diffident letter! Can it be that your modesty causes you to underrate your own value? Who are you? Where did you acquire all the knowledge expressed in your book? Judging by the priority you grant to death, one is led to conclude that you are very young.

Won't you give me the pleasure of paying me a visit one day? I have time in the mornings.

Yours very sincerely
Freud

[1] *Der Kaiser, die Weisen und der Tod* (*The Kaiser, the Wise Men and Death*), 1937.

312 *To* ERNEST JONES

20 Maresfield Gardens
London N.W. 3
March 7, 1939

Dear Jones

It still remains surprising to me how little we human beings can foresee the future. When, shortly before the war, you told me about founding a Psychoanalytical Society in London, I could not foresee that a quarter of a century later I would be living so near to it and to you; even less would I have thought it possible that I, in spite of living so near, would not be taking part in your celebration.

But in our helplessness we have to accept everything as fate brings it to us. Thus I have to content myself with sending your celebrating Society a cordial greeting and hearty good wishes from afar and yet so near. The events of recent years have made London the principal site and center of the psychoanalytical movement. May the Society carry out the functions thus falling to it in the most brilliant manner.

Your old
Sigm. Freud

313 *To* MARIE BONAPARTE

20 Maresfield Gardens
London N.W. 3
April 28, 1939

My dear Marie

I haven't written to you for a long time, while you have been bathing in the blue sea. I assume you know why, can even detect it in my handwriting. (Even the pen is no longer the same; like my doctor and other external organs, it has left me.) I am not well; my illness and the aftermath of the treatment are responsible for this condition, but in what proportion I don't know. People are trying to lull me into an atmosphere of optimism by saying that the carcinoma is shrinking and the symptoms of reacting to the treatment are temporary. I don't believe it and don't like being deceived.

You know that Anna won't be coming to the meeting in Paris because she cannot leave me; I am growing increasingly incapable of looking after myself and more dependent on her. Some kind of intervention that would cut short this cruel process would be very welcome. Should I still look forward to seeing you again at the beginning of May? Your going to St. Tropez has reassured me that Eugénie is well; otherwise you wouldn't have left Paris.

And herewith I greet you warmly; my thoughts are often with you.

Your
Freud

314 *To* HERBERT GEORGE WELLS*

20 Maresfield Gardens
London N.W. 3
July 16, 1939

Dear Mr. Wells

Your letter starts with the question how I am. My answer is I am not too well, but I am glad of the chance to see you and the Baroness[1] again and happy to learn that you are intending a great satisfaction[2] for me. Indeed, you cannot have known that since I first came over to England as a boy of eighteen years, it became an intense wish phantasy of mine to settle in this country and become an Englishman. Two of my half brothers had done so fifteen years before.

But an infantile phantasy needs a bit of examination before it can be admitted to reality. Now my condition is the following: There are two judgments on my case: one of them represented by my physicians maintains the hope that the combined radium and X-ray treatment I am now undergoing will cure me of the last attack of my malignant growth and leave me free to meet other adventures in life. Perhaps they only say so officially. There is another party much less hopeful to which I myself adhere in view of my actual pains and troubles. Now let us suppose the fact that you know the affair of the Act of Parliament cannot go through before half a year or more. In such a case, I expect you would prefer to drop your intention. So I have an interest not only in seeing you but also in your seeing me.

As regards the time available for your visit, I see that you are ready to call on me any afternoon except the eighteenth. Sunday afternoon after four o'clock would be best for me. If I should not be in a condition to see you, I would let you know Sunday morning by telephone.

* Written in English.
[1] Baroness Moura von Budberg.
[2] Wells intended asking Commander Locker-Lampson, M.P., to introduce an Act of Parliament conferring immediate British citizenship on Freud.

With the expression of my heartiest thanks, and my compliments to the Baroness,

Yours sincerely
Sigmund Freud

315 *To* ALBRECHT SCHAEFFER

20 Maresfield Gardens
London N.W. 3
September 19, 1939

Dear Mr. Schaeffer

What an unexpected and welcome letter! How often have I thought of my poet during these in some respects empty times, and wondered to what corner of the wild upheaval current events in his fatherland have cast him! It gave me deep pleasure to learn that what I had feared has not occurred and that you have found such an invaluable partner in your wife.

Not everything I could tell you about myself would coincide with your wishes. But I am more than eighty-three years old, thus actually overdue, and there is really nothing left for me but to follow your poem's advice: Wait, wait.[1]

Yours very sincerely
Freud

[1] Freud died on the night of September 22-23 at 3 o'clock in the morning, three days after this letter was written.

Bibliography
and Acknowledgments

The following publications have previously printed letters included in this volume:

The Standard Edition of the Complete Psychoanalytical Works of Sigmund Freud, edited and translated by James Strachey (Hogarth Press, London): the letters to
> Hinterberger (1907)
> B'nai B'rith (May 6, 1926)
> The Mayor of Příbor-Freiberg (October 25, 1931)

Origins of Psychoanalysis, edited by Marie Bonaparte, Anna Freud, and Ernst Kris (Basic Books, New York, 1950):
> The seven letters to Wilhelm Fliess, which appear in that volume under the numbers 6, 44, 50, 88, 93, 137, 152.

The Life and Work of Sigmund Freud, by Ernest Jones (Hogarth Press, London, and Basic Books, New York): the letters to
> Lou Andreas-Salomé (July 28, 1929; May 16, 1935)
> Mathilde Breuer (May 13, 1926)
> The Family (September 22, 1907)
> Richard Flatter (March 30, 1930)
> Stefan Zweig (May 18, 1936)
> and extracts from various other letters.

Die Neue Rundschau (S. Fischer Verlag, 1955/1):
> The two letters to Arthur Schnitzler.

Erinnerungen an Sigmund Freud, by Ludwig Binswanger (Berne, 1956):
> Extracts from the letters to Ludwig Binswanger.

Some ten letters to various addressees have appeared in books and journals without special authorization.

BIBLIOGRAPHY AND ACKNOWLEDGMENTS

The letters to Professor C. G. Jung are included in this collection by arrangement with the Jung Archive.

I have to thank the Sigmund Freud Archives, New York, for their assistance in getting in touch with a number of addressees.

The Library of the Hebrew University in Jerusalem for the photocopies of the letters to Judah Leon Magnes, Enrico Morselli, Israel Spanier Wechsler, and Stefan Zweig. Mr. Ernst Pfeiffer (Helmstedt) for the photocopies of the letters to Lou Andreas-Salomé. Professor Philipp Merlan (Scripps College, Claremont, Calif.) for the letters to Elise and Heinrich Gomperz.

I am also indebted to the Institute of Germanic Languages and Literatures (University of London) and their secretary-librarian Robert Pick, Ph.D., for their help in tracing the origin of many quotations; to my sister Anna, Dorothy Burlingham, and my wife, who advised me in the selection of the letters, and to my wife and Lena Neumann for their cooperation in collecting and arranging the footnotes and in correcting the manuscript.

List of Addressees

6   BONAPARTE, Marie (H.R.H. Princess George of Greece)
221, 288, 290, 305, 309, 313
b. St. Cloud, nr. Paris, 1882. Founder of the Société Psycho-
analytique de Paris in 1925. Lives in St. Cloud.

1   BRAUN-VOGELSTEIN, Julie   232
b. Stettin, Germany, 1883. German writer; lives in New
York.

4   BREUER, Josef, M.D.   46, 102, 105, 112
b. Vienna, 1842, d. Vienna, 1925. Physiologist and physician;
Member of the Austrian Academy of Science; author with
Freud of *Studies on Hysteria*.

1   BREUER, Mathilde (née Altmann), wife of Dr. Josef Breuer
222
b. Vienna, 1846, d. Vienna, 1931.

1   CARRINGTON, Hereward Hubert Lavington   192
b. Jersey, Channel Islands, 1880. American psychologist;
lives in New York.

1   DYER-BENNET, Richard S.   237
Major in British army; lives in Barnehurst, England.

1   EINSTEIN, Albert, Professor of Physics   283
b. Ulm, Germany, 1879, d. Princeton, N.J., 1955. Propounder
of the theory of relativity; Member of the Academy of
Science, Berlin; awarded the Nobel Prize in 1921.

8   EITINGON, Max, M.D.   162, 184, 195, 245, 266, 289, 295,
299
b. Galicia, Poland, 1881, d. Jerusalem, 1943. German psychoanalyst; founder of the Berliner Psychoanalytic Poli-
clinic in 1920, and of the Palestinian Psychoanalytical
Association in 1933.

2   ELLIS, Henry Havelock   224, 233
b. Croydon, England, 1859, d. London, 1939. English scientist and author; specialist in the psychology of sex.

9   FERENCZI, Sandor, M.D.   141, 143, 148, 163, 183, 187,
235, 254, 256
b. Miskolcz, Hungary, 1873, d. Budapest, 1933. Founder of
the Hungarian Psychoanalytical Association in 1913.

1   FLATTER, Richard, LL.D.   249
b. 1891. Austrian author and translator of Shakespeare into German; lives in Vienna.

7   FLIESS, Wilhelm, M.D.   106, 110, 111, 113, 114, 116, 119
b. Arnswalde, Germany, 1858, d. Berlin, 1928. Propounder of the theory of periodicity in human life.

1   FLUSS, Emil   1
Friend of Freud in his youth.

1   FREUD, Adolfine (Freud's sister)   252
b. Vienna, 1862, d. Theresienstadt, 1942.

4   FREUD, Alexander (Freud's brother)   121, 179, 296, 301
b. Vienna, 1866, d. Toronto, 1943. Professor at the Academy of Exports, Vienna.

2   FREUD, Amalie (Freud's mother; née Nathansohn)   185, 204
b. Odessa, 1835, d. Vienna, 1930.

3   FREUD, Anna, LL.D., *h.c.* (Freud's youngest daughter) 140, 158, 159
b. Vienna, 1895. Director of the Hampstead Child Therapy Clinic, London.

4   FREUD, Ernst (Freud's youngest son)   194, 196, 273, 294, 297
b. Vienna, 1892. London architect.

1   FREUD, Lucie (Lux) (née Brasch; Freud's daughter-in-law; married to Ernst Freud)   194
b. Berlin, 1896; lives in London.

1   FREUD, Margit (Freud's niece; daughter of Freud's sister, Marie; married to a cousin, Moritz Freud)   229
b. Vienna, 1889. Journalist in Copenhagen.

8   FREUD, Martha (née Bernays; Freud's wife); *see also* BERNAYS, Martha.   108, 109, 129, 133, 138, 147, 156, 157

1   FREUD, Martin (Jean Martin), LL.D. (Freud's eldest son) 291
b. Vienna, 1889; lives in London.

1   FREUD, Mathilde (Freud's eldest daughter; married Robert Hollitscher)   137
b. Vienna, 1887; lives in London.

1   FREUD, Sophie (Freud's second daughter; married Max Halberstadt) 153
b. Vienna, 1893, d. Hamburg, 1920.

6   FREUD family 120, 128, 130, 131, 132, 142

2   FREUND, Anton von, Ph.D. 181, 182
b. Budapest, 1880, d. Vienna, 1920. Industrialist.

2   FREUND, Rozsi von (née Brody; married to Anton von Freund) 181, 189, 227
b. Gödöllö, Hungary, 1887; lives in London.

3   GOMPERZ, Elise (married to *Hofrat* Professor Theodor Gomperz) 117, 118, 165

1   GOMPERZ, Heinrich, Ph.D. 115
b. Vienna, 1873, d. Los Angeles, 1942. Professor of Philosophy, University of Vienna.

4   GRODDECK, Georg, M.D. 176, 188, 201, 212
b. Baden-Baden, 1866, d. Baden-Baden, 1934. German author and psychoanalyst.

2   GUILBERT, Yvette 259, 306
b. Paris, 1866, d. Aix-en-Provence, 1943. French *diseuse*.

3   HALBERSTADT, Max (Freud's son-in-law) 152, 154, 155
b. Hamburg, 1882, d. Johannesburg, 1940. Photographer.

1   HALBERSTADT, Max and Sophie (née Freud) 175

1   HALL, Granville Stanley, M.D. 166
b. Ashfield, Mass., 1844, d. Worcester, Mass., 1924. Professor of Psychology; President of Clark University, Worcester, Mass.

1   HINTERBERGER 135
Bookseller in Vienna.

1   HITSCHMANN, Eduard, M.D. 171
b. Vienna, 1871, d. Boston, 1957. Psychoanalyst.

1   HOOPER, Franklin 211
b. Worcester, Mass., 1862, d. nr. Saranac Lake, N.Y., 1940. Editor of the *Encyclopaedia Britannica*.

1   INDRA, Alfred, LL.D. 303
Viennese lawyer.

5   JONES, Ernest, M.D.   216, 225, 238, 298, 312
      b. Gowerton (Wales), 1879, d. London, 1958. Founder of
      the British Psychoanalytical Society in 1913. Author of *The
      Life and Work of Sigmund Freud.*

1   JONES, Herbert and Loe (née Kann)   167

7   JUNG, Carl Gustav, M.D.   124, 125, 126, 127, 136, 145, 160
      b. Basel, 1875. Professor of Psychology, lives in Zürich.

1   KEYSERLING, Count Hermann   268
      b. Kovno, Lithuania, 1880, d. Innsbruck, 1946. German
      philosopher and writer.

1   KNÖPFMACHER, Wilhelm, M.D.   2
      Friend of Freud in his youth.

1   KRAUS, Karl   122
      b. Gitschin, Czechoslovakia, 1874, d. Vienna, 1936. Austrian
      satirist and critic.

1   LAMPL, Hans, M.D.   218
      b. Mauer, nr. Vienna, 1889, d. in Holland, 1958. Psycho-
      analyst.

      LAMPL de GROOT, Jeanne   218
      b. 1895 in Holland. Psychoanalyst in Amsterdam.

1   LEVY, Katá (née Freund)   203
      b. Budapest, 1883. Psychoanalyst in London.

      LEVY, Lajos, M.D.   203
      b. Budapest, 1875. Director of the Jewish Hospital in Buda-
      pest; lives in London.

1   LIPSCHÜTZ, Alexander, M.D.   262
      b. Riga, Latvia, 1883. Director of the Institute of Experi-
      mental Medicine and the National Health Service in
      Santiago, Chile.

1   LÖWY, Heinrich, M.Sc.   250
      b. Austria, 1884. Professor of Physics and Geophysics.

1   LOW, Barbara   282
      b. 1877, d. London, 1955. Psychoanalyst.

1   MAGNES, Judah Leon, Rabbi   272
      b. 1877, d. Jerusalem, 1948. Chancellor, later President of
      the Hebrew University in Jerusalem.

1   SCHILLER, Max, M.D. (married to Yvette Guilbert)   260
       b. Vienna, 1860, d. Paris, 1952.

2   SCHNITZLER, Arthur, M.D.   123, 197
       b. Vienna, 1862, d. Vienna, 1931. Austrian dramatist and
       writer.

2   SCHUR, Max, M.D., 251, 304
       b. Stanislawow, Poland, 1897. Physician, now psychoanalyst
       in New York.

1   SIMMEL, Ernst, M.D.   236
       b. Breslau, Germany, 1882, d. Los Angeles, 1947. Director
       of the Tegel Psychoanalytical Sanatorium, nr. Berlin.

1   SINGER, Charles   307
       b. 1876, d. 1960. Professor of History of Science at the
       University of London.

1   STEINIG, Leon   267
       b. Trembowla, Austria-Hungary, 1898. Is now with the In-
       ternational Atomic Energy Agency in Vienna.

1   STEKEL, Wilhelm, M.D.   206
       b. 1868, d. London, 1940. Viennese psychoanalyst.

1   STRUCK, Hermann   168
       b. Hamburg, 1876, d. Tel Aviv, 1944. German artist.

1   TANDLER, Julius, M.D.   213
       b. Iglau, Czechoslovakia, 1869, d. Vienna, 1937. Professor
       of Anatomy at the University of Vienna, head of the Health
       Department.

1   TIME AND TIDE   310
       English periodical, edited by Lady Rhondda.

2   VIERECK, George Sylvester   234, 270
       b. Munich, 1884. Journalist and writer.

1   VOIGTLÄNDER, Else, Ph.D.   149
       For some years a member of the Viennese Psychoanalytical
       Association.

1   WECHSLER, Israel Spanier, M.D.   240
       b. Lespedi, Rumania, 1886. Professor of Neurology at
       Columbia University, New York.

1   WEISS, Edoardo, M.D.   269
     b. Italy. Founder of the Italian Psychoanalytical Association
     in 1936; lives in Chicago.

1   WELLS, Herbert George   314
     b. Bromley, 1866, d. London, 1946. English author.

2   WITTELS, Fritz, Ph.D.   205, 209
     b. Vienna, 1880, d. New York, 1950. Austrian writer, psycho-
     analyst.

1   WITTKOWSKI, Victor   281
     b. Güstrow, Germany, 1909. Lives in Cologne.

7   ZWEIG, Arnold, Ph.D.   230, 255, 257, 264, 276, 278, 285
     b. Glogau, Germany, 1887. German author; lives in East
     Berlin.

8   ZWEIG, Stefan, Ph.D.   139, 191, 207, 258, 265, 284, 292, 302
     b. Vienna, 1881, d. Brazil, 1942. Austrian author.

6   ANON.   180, 247, 275, 277, 293, 300